Managing Technological Innovation and Entrepreneurship

Michael J. C. Martin

Reston Publishing Company, Inc.
Reston, Virginia
A Prentice-Hall Company

Library of Congress Cataloging in Publication Data

Martin, Michael J. C.
   Managing technological innovation and entrepreneurship.

   Includes bibliographical references.
   1. Technological innovations—Management. 2. High
technology industries—Management. 3. Industrial
management. 4. Entrepreneur. I. Title.
HD45.M35   1984   658.4'06   83-21220
ISBN 0-8359-4201-5

Copyright © 1984 by
Reston Publishing Company, Inc.
Reston, Virginia 22090
A Prentice-Hall Company

10 9 8 7 6 5 4 3 2 1

PRINTED IN THE UNITED STATES OF AMERICA

**To Paula,
David, and Kay**

Contents

Preface viii

Part 1: The Technological Innovation Process

1 **Prologue: The Technological Innovation Process: Three Illustrative Examples   2**

1.1 Invention and Innovation, 2   **1.2** Marconi and the Development of Radio Telegraphy, 3   **1.3** The EMI CAT Scanner, 9   **1.4** The Fiber Tip Porous Pen, 13

2 **Some Frameworks for Viewing the Process   17**

2.1 Introduction, 17   **2.2** Bright's Treatment of the Innovation Process and the Innovation Chain Equation, 17   **2.3** Time and Cost Characteristics of the Innovation Process, 20   **2.4** Kuhn's and Popper's Treatments of the Growth of Scientific Knowledge, 22 **2.5** The Abernathy-Utterback Treatment of the Growth of High-Technology Industries, 25   **2.6** Sahal's Treatment of Technological Evolution, 28   **2.7** Revolutionary and Normal Innovations, 29 **2.8** Innovations as "Technological Mutations," 31   **2.9** Criteria for Evaluating Innovations, 34   **2.10** Three Innovations Briefly Reexamined, 36

Part 2: The Corporate Setting

3 **The Corporate Technological Base   40**

3.1 Introduction, 40   **3.2** Innovation Process and Technological Base, 41   **3.3** Nondirected or Fundamental Research, 41 **3.4** Contract Research and Development, 42   **3.5** Directed or Applied Research and Primary Development, 43   **3.6** Further Development and Design Engineering, 43   **3.7** Pilot or Prototype Production, 43 **3.8** Full Production, 44   **3.9** Market Development, 44 **3.10** Patents and Licensing Services, 45   **3.11** Quality, Reliability, and Technical "After Sales" Services, 45   **3.12** Education, Training, and Advisory Information Services, 45   **3.13** Managing the Technological Innovation Base

4 **Technological Strategies   47**

4.1 Introduction, 47   **4.2** General Considerations—The Innovation Management Function, 48   **4.3** The Technology-Market or

iv

Technological Opportunities Matrix, 50   **4.4** Technological
Strategies, 52   **4.5** Offensive Strategy—Leadership or "First-to-
Market" Policy, 52   **4.6** Defensive Strategy—"Follow-the-Leader"
Policy, 59   **4.7** Imitative Strategy—"Me-Too" Policy, 61
4.8 Applications Engineering—Interstitial Strategy, 62
4.9 Dependent Strategies—"Branch Plant" Policy, 63
4.10 Absorbent Strategies, 64   **4.11** Traditional Strategies, 65
4.12 Other (Nontechnologically) Innovative Strategies, 66

5   **Forecasting Future Technology Markets   70**

5.1 Introduction, 70   **5.2** The Emergence of Technological
Forecasting: The Concept of Future History, 71   **5.3** Technology
Monitoring, 74   **5.4** The "S"-Shaped Logistic Curve, 75
5.5 Envelope Curves and Trend Extrapolation, 77   **5.6** Precursor
Trends, 79   **5.7** Fisher-Pry Substitution Model, 80   **5.8** Some
Pitfalls of Extrapolative Techniques, 81   **5.9** Systems Dynamics
Models, 82   **5.10** Delphi Method, 84   **5.11** Creativity Stimulating
Techniques, 87   **5.12** Morphological Analysis and Mapping, 87
5.13 Relevance Trees, 89   **5.14** Cross-Impact Matrices, 91
5.15 Composite Forecasts—Scenarios, 93   **5.16** Generating
Scenarios, 95   **5.17** Scenarios and Inventing the Future, 98
5.18 The Validity of Technological Forecasts, 99
5.19 Organizational Location of the TF Function, 101

6   **Environmental Concerns and Technology Assessment   106**

6.1 Introduction, 106   **6.2** The Technology Assessment Process and
Environmental Impact Statements, 108   **6.3** Organizational
Framework for Conducting Technology Assessments, 113

Part 3: The R & D Setting

7   **R & D Management: Some General Considerations   119**

7.1 Introduction, 119   **7.2** Organizational and Geographical
Locations of R & D, 119   **7.3** R & D Budget Setting—Pragmatic
Aspects, 121   **7.4** R & D Planning and Budget Setting—Normative
Aspects, 126

8   **Creative Thinking   134**

8.1 The Need for Creative Thinking, 134   **8.2** The Creative Process—
The Socratic Daemon, 135   **8.3** Identifying Creative People, 144
8.4 Creating a Creative Climate, 147   **8.5** Creativity Stimulating and
Enhancing Techniques, 149   **8.6** Techniques for Stimulating
Convergent Thinking, 150   **8.7** Techniques for Stimulating Divergent
Thinking, 151

9   **Project Evaluation**   **159**

9.1 Introduction, 159   **9.2** Revolutionary Innovations—White's Approach, 160   **9.3** Radical Normal Innovations, 166 **9.4** Incremental Normal Innovations, 166   **9.5** Evaluation as a Concomitant Process, 172

10   **Project Selection Approaches**   **180**

10.1 Introduction, 180   **10.2** Choice of Decision Takers, 180 **10.3** Project Selection Techniques, 181   **10.4** Selection Committee Procedure—The Q Sort Method, 191

11   **Project Management and Control**   **197**

11.1 Introduction, 197   **11.2** Budgetary and Accounting Aspects, 197 **11.3** Project Progress as an Uncertainty Reduction Process, 198 **11.4** Estimation Procedures in Project Evaluations, 199   **11.5** Project Control—Behavioral Aspects, 204   **11.6** Project Control—Analytical Aspects, 206

12   **Project Needs and the Personal Needs of R & D Staff**   **212**

12.1 Introduction, 212   **12.2** Conflict Between Individual and Organizational Values in R & D, 212   **12.3** Development Alternatives for R & D Staff, 214   **12.4** The Dual-Ladder Structure, 216 **12.5** Staffing Needs of the R & D Function, 222   **12.6** Matching Individual and Organizational Needs in R & D, 226   **12.7** Matrix Structure, 229

Part 4: The Operations Setting

13   **Transferring the Project from R & D to Production**   **234**

13.1 Introduction, 234   **13.2** Specific Problems of Technology Transfer, 235   **13.3** Management Framework for Transfer, 240 **13.4** Matrix Structures, 241   **13.5** Learning Curves and the Technological Progress Function, 245   **13.6** New Venture Operations, 249

Part 5: The Entrepreneurial Setting

14   **Small Is Beautiful: Spin-Offs, Innovation, and Entrepreneurship in Small Businesses**   **252**

14.1 Introduction, 252   **14.2** The Economic Role of Small High-Technology Business, 253   **14.3** The Spin-Off Phenomenon, 255 **14.4** Characteristics of Technological Entrepreneurs, 261

14.5 Illustrations of Technological Entrepreneurs, 265
14.6 Entrepreneurial Venture Initiation, 267

15 **Creating the New Technological Venture** **273**

15.1 Introduction, 273   **15.2** Stages in the New Venture Creation Process, 274   **15.3** The Formation of the Venture Team, 276 **15.4** The Venture Proposal and Business Plan, 280   **15.5** Sources of Financial Support, 285   **15.6** Common Pitfalls of Initial Operations, 292   **15.7** Consolidation and Growth to Maturity, 298 **15.8** Illustrative Examples, 299   **15.9** Final Comments, 301

16 **Small in Large Is Also Beautiful: Stimulating Intracorporate Entrepreneurship** **304**

16.1 Introduction, 304   **16.2** Dynamic Conservatism Versus Entrepreneurship in Large Organizations, 305   **16.3** Texas Instruments—Objectives, Strategies, and Tactics—OST, 306 **16.4** Separate Venture Operations, 314   **16.5** R & D Venture Organizations, 315   **16.6** Internal Individually or Jointly Sponsored Ventures, 316   **16.7** External Joint Ventures, 319   **16.8** External Supported Ventures or Severed Ventures, 319   **16.9** Problems of Venture Development and Intracorporate Entrepreneurship, 320

17 **Epilogue: Institutionalizing Technological Innovation and Entrepreneurship** **327**

17.1 In Search of the "Secret Weapon" of Great Companies, 327 **17.2** Final Comments, 332

Index **335**

Preface

Man is limited not so much by his tools as by his vision. Historians tell us that the notion of the earth as round had been discussed for five hundred years before Columbus' time. What Columbus did was to translate an abstract concept into its practical implications.

Richard Tanner Pascale and Anthony G. Athos, **The Art of Japanese Management.**

The declining economic growth rate, coupled with high inflation, and the growing unemployment of the last decade, as well as the impact of Le Defi Japonais, has aroused both public and business concern for technological innovation. As a result of much research performed since the 1950s, the complex technological and socioeconomic process whereby a technological invention is converted into a socially useful and commercially successful new product is now less obscure than it was a generation ago. This improved understanding is of little practical value unless it is disseminated to present and future innovation managers through the education system. This requirement raises another issue. Managers have displayed a historic "technology-aversion"[1] so that the management of technology has been a neglected subject in both the management and engineering schools of the universities. Consequently, the management and engineering educational systems have evolved with largely separate languages and cultures. If technological innovation is to sustain future economic growth and employment in competitive international markets, the gap between these two cultures must be bridged by individuals capable of effectively managing the process. The issues and problems involved have been discussed in two specially convened conferences on the subject in the United States[2] and Canada.[3] It is sometimes argued that individuals without prior engineering or science backgrounds cannot become effective technological managers. However, as Skinner points out, individuals without such technological backgrounds can manage technology, provided they

> *can learn to understand and deal effectively with the technology of their industry....What is usually sufficient is a framework consisting of a few basic concepts and a set of questions that lead to the acquisition of that knowledge and those insights needed by a manager.[4]*

The purpose of this book is to help bridge the gap between management and engineering educational systems by providing such a framework to future and present technological innovation managers with or without engineering/science backgrounds. The book has grown out of my experiences developing and teaching courses on Managing Technological Innovation and Entrepreneurship to mixed classes of business (mainly MBA), engineering, and science students from the senior undergraduate to the postdoctoral level.[5] The book can therefore be used as a required text for similar or related one-semester courses in management/business, engineering, and science schools of higher education. It could also be used on related extension/postexperience courses. Alternative course frameworks, and so forth, are discussed in teaching notes provided by the publisher. The book is also written for professional readers, that is, technological managers and entrepreneurs in high-technology firms, research and development (R & D) professionals and managers in both company and government laboratories, and administrators in government agencies responsible for stimulating and regulating high technology.

Chapter 1 begins with three examples of the innovation-entrepreneurship process to illustrate the well-founded contention that it is much more than scientific or engineering invention. Chapter 2 offers some conceptual frameworks for viewing the process congruent with that contention. Chapters 3 through 6 view the innovation-entrepreneurship process from the corporate-general management perspective. By definition, a high-technology business must be technology based. Traditionally, this technological base has been viewed rather narrowly as "engineering" or "R & D" versus "manufacturing" and "sales." In chapter 3 a description of the technological base of a company is provided which is broad enough to encompass the requirements of the process, to provide a factual background to later chapters. Chapter 4 begins by arguing that corporate management's responsibility to stimulate and manage innovation creates the need for an innovation management function and for technological strategies that permeate the organization. It discusses the alternative technological strategies which may be pursued by a company. Technological and social forecasting techniques now used to aid the long-term planning of innovation are reviewed in chapter 5. These days, everyone accepts that innovations must be evaluated to identify potentially harmful impacts on individuals and the ecological and social environment. Chapter 6 briefly discusses procedures for conducting technology assessments and for developing environmental impact statements for prospective innovations.

Chapters 7 through 12 deal with the R & D setting of the process. Chapter 7 briefly discusses the overall organization, budgeting, and planning of the R & D activities in larger companies. R & D management has both behavioral and technicoeconomic aspects and each are dealt with in turn. Chapter 12 also outlines the alternative career development paths available to scientists and engineers and so may be of specific interest to readers who are contemplating a career change from R & D to management. Chapter 13 moves to the operations setting and discusses the problems of transferring technological know-how from an R & D to a production environment.

Chapters 14 through 16 focus on the entrepreneurial aspects of the process. Chapter 14 discusses the important role that successful, growing high-technology companies play in generating economic wealth and jobs. It discusses the "spin-off" phenomenon, notably associated with Silicon Valley in California and Route 128 in Massachusetts. The success of a "spin-off" is markedly dependent upon the entrepreneurial aptitudes and skills of its founders. Therefore, the personality and biographical characteristics of technological entrepreneurs are examined.

Many scientists, engineers, and technological managers employed in larger companies, government, or universities aspire to set up their own high technology businesses, possibly to exploit their own inventive ideas commercially. Chapter 15 examines, in some detail, the problems, pitfalls, and rewards of setting up a new high-technology venture. Clearly larger high-technology companies do not wish to lose all their inventions to spin-off ventures. Furthermore, innovation requires entrepreneurship to succeed in large as well as small companies. The approaches that established companies have taken to sustain entrepreneurship are discussed in chapter 16. The book then concludes with brief comments on the overall management of the technological innovation-entrepreneurship process.

I have included fairly extensive literature citations to support the observations in most chapters, for readers who wish to explore individual topics in more depth. I suggest that they be ignored in a "first reading" so that the gists of the discussions can be followed; then, if required, they can be consulted later according to the reader's specialist interests. I also apologize to those writers whose works have not been cited. The volume of literature available precludes the citation of all the important contributions that have been made in the field. Since the book focuses on the innovation-entrepreneurship process at the individual firm level, I have excluded discussion of public policy aspects of the subject and the relationship between innovation and long waves in world economies. These important (but contentious) topics are discussed in more specialized texts[6,7] and I do not think I could give them sufficient space to discuss them adequately here. For the same reasons, I have excluded discussion of the impact of government regulations on patterns of industrial evolution. Such impacts (notably within the automobile and pharmaceutical industries) have also been discussed extensively elsewhere.[8,9,10]

No book is written without help from one's students and colleagues. I should like to express my thanks to the 200 to 300 bright, mature, and industrious students who have made teaching my course such a pleasure (for me anyway!). I am sure some of their ideas have "spun off" here! I also thank Tom Clarke for stimulating my interest and providing an excellent bibliography in the field,[11] Dr. David Othen, and two unnamed reviewers for their many helpful comments. Needless to say, all the faults are my own. Finally, I thank Marilyn Taylor of Dalhousie University, Caroline Morse, and others for their heroic efforts in typing successive manuscripts, and Ann Mohan and her colleagues at Reston for converting the final manuscript into a finished product.

Alvin Toffler has described education as a vision of the future. It is often claimed that the age of the heroic inventor-entrepreneur is past, but the evidence of the recent explosive developments in the solid-state electronics and computer industries suggests otherwise. If this book helps just one individual to become a successful technological entrepreneur who creates satisfying employment and products for others, it will have been worth it.

References

1. Wickham Skinner, "Technology and the Manager," in *Manufacturing in the Corporate Strategy* (New York: John Wiley & Sons, 1978).

2. Dwight M. Baumann, ed., *National Conference for Deans of Engineering and Business* (Washington, D.C.: National Science Foundation, 1981).

3. Thomas E. Clarke et al., eds., *T.I.M.E. for Canada Workshop* (Ottawa, Ont.: Department of Industry, Trade and Commerce, Technological Innovation Studies Program Report No. 70, 1979).

4. Skinner, "Technology and the Manager," *op. cit.*

5. Michael Martin, "Teaching Business Students Technological Innovation Management," *Business Graduate* 13, No. 2 (May 1983): 62-63.

6. Roy Rothwell and Walter Zegreld, *Industrial Innovation and Public Policy: Preparing for the 1980's and the 1990's* (Westport, Ct.: Greenwood Press, 1981).

7. Christopher Freeman, John Clark and Luc Soete, *Unemployment and Technical Innovation: A Study of Long Waves and Economic Development* (Westport, Ct.: Greenwood Press, 1982).

8. William J. Abernathy, *The Productivity Dilemma: Roadblock to Innovation in the Automobile Industry* (Baltimore: Johns Hopkins University Press, 1978).

9. Jerome E. Schnee, "International Shifts in Innovative Activity: The Case of Pharmaceuticals," *Columbia Journal of World Business* 13, no. 1 (Spring 1978): 112-22.

10. Roy Rothwell, "The Impact of Regulation of Innovation: Some Data," *Technological Forecasting and Social Change* 17, no. 1 (May 1980): 7-34.

11. Thomas E. Clarke, *R & D Management Bibliography* (Vancouver, B.C.: Stargate Consultants, 1981).

Part 1

The Technological Innovation Process

Part 1 describes the technological innovation process. It begins with three illustrative examples. It then examines Bright's treatment of the process. Kuhn's and Popper's treatments of the evolution of scientific knowledge are discussed and linked to more recent treatments of technological evolution by Abernathy–Utterback and Sahal. These treatments provide a framework for viewing the process in the corporate setting.

1

Prologue: The Technological Innovation Process: Three Illustrative Examples

It is a little like playing poker when it is the only game you want to play and all you know about the game is that the deck is stacked. You don't know how it is stacked but it is stacked and you must play.
Monte C. Throdahl, senior vice-president, Monsanto Co., on the innovation process. Untitled presentation in Dwight M. Baumann (Ed.), **National Conference for Deans of Business and Engineering**

1.1
Invention and Innovation

In the Preface, I implied that general managers of high-technology businesses have a responsiblity to nurture a continuous succession of new products, based upon a continued succession of technological innovations, and therefore need to be familiar with the innovation process. I also suggested that students of science, engineering, and management who plan (or have already started) to pursue careers in high-technology industries should recognize the main features of this process, particularly if they aspire to reach senior/general management positions or set up their own independent high-technology business later. Therefore to begin, we will examine the process in a little detail to identify the combination of elements required to generate a commercially successful technological product.

Most high-technology companies are "science based" in that their innovations develop out of new ideas and inventions generated by scientific activity, either within their own R & D function or elsewhere. However, at the outset, it is important to recognize the distinction between scientific *invention* and technological *innovation*. A scientific invention may be viewed as a new idea or concept generated by R & D, but this *invention* only becomes an *innovation* when it is transformed into a socially usable product. Laypersons, probably because of the mystique that surrounds science, generally view invention as a relatively rare event and assume that once it has occurred, the process of innovation can be completed in a straightforward manner. In actuality, the converse situation pertains here. All who have worked in R & D will agree that although it is intellectually and emotionally demanding and frustrating, even so the R & D community is quite prolific in generating inventions, and companies can rarely afford to fund all promising R & D projects (see chapter 2, section 2.3). It is the subsequent path to technological innovation that is typically fraught with numerous obstacles to be overcome, if the R & D invention is to be commercially successful.

Managing the innovation process may perhaps be compared with breeding and training a horse to finish in the first three places in the English Grand National or other such steeplechases. Although many horses may be bred and trained to come under "starters-orders," in any year only three horses perform well enough over the fences to win the first three places in the race and to win prize money for their owners and bets for their backers. Managing the innovation process in larger companies essentially requires breeding a stable of promising R & D inventions, some of which are developed into product innovations in the expectation that they will prove to be commercially successful "winners."

To illustrate some of the features and obstacles of this innovation process, three examples will be given. There are numerous other cases and research studies cited in the literature, two of the most useful of which are Layton et al. and Langrish et al.[1] Pilkington describes his personal experiences in pioneering the development of the float glass process to commercial success and so presents innovation through the eyes of the innovator.[2] Several examples are also described in a special issue of *Research Management,* one of which is included below.[3]

1.2
Marconi and the Development of Radio Telegraphy

The main sources of historical material in this section are Baker and Donaldson.[4]

Radio grew out of what is now called experimental and theoretical physics. Its origins can be traced back to the "science" of classical Greece as well as China, but the foundation was laid by Dr. Humphrey Gilbert, physician to Elizabeth I, who really began the systemic scientific study of the phenomena of electricity and magnetism. Human knowledge of the properties of these phenomena grew steadily between the sixteenth and nineteenth centuries, but it was Michael Faraday who first discovered an interrelationship be-

4

Prologue:
The
Technological
Innovation
Process:
Three
Illustrative
Examples

tween the two of them. Like many brilliant experimental physicists, Faraday was no mathematician, and it was James Clerk Maxwell who provided the mathematical formulation of this interrelationship with his electromagnetic field theory, including the proof of Maxwell's equations and the existence of electromagnetic waves. Heinrich Hertz provided the experimental verification of the existence of these waves, Sir William Crookes suggested their potential utility as communication tools, and Sir Oliver Lodge provided a practical demonstration of their use in a lecture at the Royal Institution in 1894. By this time, numerous inventive scientists and engineers were exploring and experimenting with the properties of this promising new tool.

Technological innovation typically requires the collaborative effects of numerous workers, but often its ultimate success can be attributed to the sustained efforts of one or more inventor-entrepreneurs who pioneer the development of a new high-technology industry. The inventor-entrepreneur is twice-blessed by Providence with both technological and commercial creative skills and has the vision, capability, and energy to improve upon and commercially exploit an invention. Examples of such inventor-entrepreneurs are Alfred Nobel (dynamite), W.H. Perkin (aniline dyes), Herbert Dow (industrial electrolysis), Alexander Graham Bell (the telephone), and Thomas Edison (electric light and power supply). Such a role can be credited to Marconi in the development of radio-telegraphy.

Guglielmo Marconi was born in 1874 in Bologna, Italy, of Irish and Italian parents. His mother was Irish, born Annie Jameson and member of the famous Irish whiskey family, and his father was a well-to-do Italian landowner and businessman. It seems reasonable to conjecture that Guglielmo's family's business background influenced his entrepreneurial outlook (see chapter 14 for a discussion of this factor). He was educated in England and Italy, studied physics at the university, and conducted experimental work under Professor Righi of the University of Bologna. He displayed inventive skills from an early age and, having read the work of Hertz, produced a crude form of wireless communication when he was only twenty-two years old. He quickly recognized the potential utility of wireless telegraphy in ship-to-ship or ship-to-shore communication where wired telegraphy was obviously impossible. He therefore offered to demonstrate his invention to the Italian government, but it was disinterested. Acting upon the advice of the Jameson family, he decided to pursue his ideas in Britain—the home of the world's largest mercantile fleet and Queen Victoria's Royal Navy, and where his mother's family connections might prove useful.

Once residing in England, he made contact with other workers in the field and began a program of practical demonstrations of wireless communication to various British government departments in

1896. In 1897 he filed the first patents for wireless telegraphy which were held by the Wireless Telegraph and Signal Co. Ltd, which he founded and capitalized in the City of London in that year. It had a nominal capital of 100,000 £1 shares of which Marconi held 60,000. The remaining 40,000 were sold in a public issue yielding £40,000 which after £15,000 start-up expenses, provided the first £25,000 of working capital. Thus began a decade and a half of unremitting and frustrating efforts to establish the embryonic innovation. Marconi quickly allied himself with other talented scientists, engineers, and businessmen and filed or purchased a succession of patents that appeared crucial to the technological development of radio telegraphy, including Dr. (later Sir) J. A. Fleming's* development of the thermionic diode, based upon earlier work by Thomas Edison.

Throughout this period, Marconi was punctilious in maintaining the technological integrity of his company, never publicizing new inventions until they had been fully proven, never making promises that he could not "deliver." In his choice of public demonstrations of wireless communication, he did display a considerable flair for "publicity," however. By 1898, he had established wireless intercommunications between Bournemouth (in southern England) and the Isle of Wight, 14.5 miles away. In the winter of that year, the eminent Victorian statesman, William Ewart Gladstone, lay dying in Bournemouth surrounded by the world's press corps. This event could be compared to the week prior to the death of Sir Winston Churchill sixty-seven years later. Unfortunately, a winter snowstorm hit southern England, severing wired telegraphic and telephonic communication between Bournemouth and London, so the reporters were unable to file their latest reports of the stateman's sinking health to their newspaper offices in London. Marconi quickly determined that wired telegraphic or telephone services (via undersea cable) were still maintained between the Isle of Wight and London. He therefore allowed the press corps to file their reports by *wireless* communication from Bournemouth to the Isle of Wight and thence to London. This action doubtlessly won him many friends amongst the media people of the day.

The Isle of Wight was to provide him with the site for invaluable publicity again in the summer of that year. The Prince of Wales was convalescing from an injury aboard a yacht moored a few miles off that island. Queen Victoria who was resident at Osborne House on the island at the time naturally wished to be kept informed of her

---

*There is an interesting parallel between the invention of the diode tube and the discovery of penicillin. Both were invented or discovered fortuitously when the researchers were trying to solve other problems and initially remained unexploited. The author is unaware whether the developer of the diode tube and the discoverer of penicillin were related.

6

Prologue:
The
Technological
Innovation
Process:
Three
Illustrative
Examples

heir's progress. Intervening hills made visual signal communication impossible, so crude wireless stations were quickly erected on both the yacht and at Osborne House to report the Prince's progress, and over sixteen days 150 messages were exchanged. Doubtlessly, on this occasion, Queen Victoria was more than "amused."

Although these two incidents could be labeled as fortuitous, but useful, "publicity stunts," they helped to establish the usefulness of wireless telegraphy and, by the early 1900s, it had proved to be of limited marine use and Marconi was engaged in a succession of demonstration projects of increasing import. By 1901, Marconi was ready to demonstrate trans-Atlantic radio communication. Despite the fact that on 17 September 1901 a severe gale destroyed the antennae masts (costing £50,000) at the British transmission station at Poldhu, Cornwall, a temporary replacement was quickly installed and on 12 December of that year a signal was successfully transmitted from that station to a receiving antenna on Signal Hill, St John's, Newfoundland. The first trans-Atlantic wireless communication had been achieved.

Unfortunately, any euphoria arising from this success was short-lived. The first reaction to the successful transmission was that solicitors for the Anglo-American Telegraph Co. claimed that Marconi had infringed that company's monopoly over communications in Newfoundland. Although this claim was, to say the least, disputable, Marconi decided to accept the offer of Alexander Graham Bell (the inventor of the telephone) of facilities in Cape Breton, Nova Scotia. Both the Canadian federal government and the government of the Province of Nova Scotia supported his work, and on 5 December 1902 signals were successfully transmitted across the Atlantic in an easterly direction from Glace Bay, Nova Scotia, to Cornwall. On 16 December the following messages were transmitted:

> *His Majesty the King. May I be permitted by means of first wireless message to congratulate Your Majesty on success of Marconi's great invention connecting Canada and England. Minto [Lord Minto, governor-general of Canada].*

> *Lord Knollys, Buckingham Palace, London. Upon occasion of first wireless telegraphic communication across Atlantic Ocean, may I be permitted to present by means of this wireless telegram transmitted from Canada to England my respectful homage to His Majesty the King.*
> *G. Marconi, Glace Bay.*

At the same time, Marconi was establishing another station at Cape Cod, Massachusetts, with the support of the U.S. government. On 18 January 1903 the following message was transmitted from Cape Cod to Cornwall:

His Majesty King Edward the Seventh (by Marconi's trans-Atlantic wireless telegraph) in taking advantage of the wonderful triumph of scientific research and ingenuity which has been achieved in perfecting a system of wireless telegraphy I send on behalf of the American people most cordial greetings and good wishes to you and all the people of the British Empire.
Theodore Roosevelt. White House, Washington.

The age of the trans-Atlantic wireless communication had arrived!

It would be gratifying to report that the fortunes of Guglielmo Marconi and the company (now called the Marconi-Wireless Telegraph Co. Ltd.) were now made, but this was not so. The path of technological innovation, like that of true love, seldom runs smoothly. Wireless telegraphy was certainly attracting the support of governments and the business community. Several governments were by now providing limited support for demonstrations of its marine and military utility, and Marconi was able to launch a U.S. company that later became the Radio Corporation of America (RCA). The path to ultimate technological and commercial success was still strewn with many obstacles, however. First, Marconi and his colleagues recognized that many technological problems had yet to be solved, so most of the earnings of his companies were channeled back into what is now called R&D. Second, as in Newfoundland, the established wired telegraphy services were reluctant to surrender their monopolies and markets to the fledgling innovation, so continued efforts were required to change national legislations to allow competition from wireless communication. Third, Marconi faced competition from other wireless telegraphy companies (notably Telefunken in Germany) who were pursuing their own technological and commercial developments. In 1906, Dr. Lee de Forest in the United States added a third electrode to the diode tube to produce the triode tube—a device capable of signal amplification as well as rectification. Dr. de Forest, who was not affiliated with the Marconi companies, patented his invention and Marconi recognized that it could be of crucial importance in the development of radio telegraphy. He therefore sued de Forest, claiming that the triode did not incorporate a new inventive principle, but was a variation of the existing diode tube and therefore an infringement of Fleming's patent. Several years of protracted litigation followed, which imposed a further drain on Marconi's time, energy, and financial resources. In 1908 the ordinary shares of the company—for which investors paid up to four pounds—dropped to six shillings and three pence (about a dollar and a quarter), so it was becoming increasingly difficult to raise further capital from the stock market. Fortunately in 1910 and 1912 the affairs of the company were transformed by two chance blessings that came in tragic guises.

8

Prologue:
The
Technological
Innovation
Process:
Three
Illustrative
Examples

In 1910 the wife murderer Dr. Albert Crippen evaded arrest by the English police by boarding the SS *Montrose* bound for Canada, traveling under a false name and with his mistress disguised as a boy. The murder inquiry and subsequent search for Dr. Crippen, after the police dug up the dismembered remains of Mrs. Crippen in the couple's London home, received widespread publicity. While at sea headed for Canada, the skipper of the *Montrose* recognized his infamous passenger and radioed this information back to London. Scotland Yard sent a detective on a faster ship to arrest Dr. Crippen on his arrival in North America. The details of the gory murder, compounded with lust and adultery, dramatic circumstances of Crippen's arrest, and his subsequent trial and execution whetted public interest in radio-telegraphy. This interest was intensified with the occurrence of a major maritime tragedy a short time later.

In April 1912, the "unsinkable" RMS *Titanic* set sail on its maiden voyage from Southampton to New York. On Sunday night, 14 April the ship struck an iceberg and sank with the loss of 1,517 lives—constituting the greatest peacetime maritime disaster in history. Ironically, the SS *Californian* was close by, but because the radio operator had gone off duty after working sixteen continuous hours, the ship did not pick up the *Titanic*'s radioed distress signals. Had the *Californian* had a radio operator "on watch" most of the lives lost would have been saved. Prior to that tragedy, wireless telegraphy was being adopted only cautiously by society in general and maritime interests in particular. Because of the publicity surrounding the *Titanic*'s maiden voyage and social prestige of some of the passengers drowned, political and social pressures forced the shipping industry to install wireless telegraphy and require a twenty-four-hour radio watch on all ocean-going vessels.* From then on, despite further setbacks (including the "Marconi scandal" of 1912-13, surrounding the terms of a proposed British government contract to be awarded to the company), the company's fortunes steadily prospered. World War I witnessed the widespread adoption of radio communication in many applications and, shortly after that war the birth of the radio broadcast industry. The Marconi companies played important roles in most of these developments and grew to a leading and respected presence in the radio and later, the electronics industries.

---

*Marconi may have been indirectly associated with another maritime tragedy. Some historians believe that the reason the German government authorized the sinking of the *Lusitania* in 1915 was because it wrongly thought Marconi to be aboard. In fairness to that government, it should be stated that it warned embarking passengers in New York that the *Lusitania* would be torpedoed once it entered British territorial waters.

The EMI CAT Scanner

Most of the material for this case is drawn mainly from Powell's works.[5]

By the early 1970s EMI or Electrical and Musical Industries Ltd. was a British-based multinational corporation with sales in excess of 1 billion dollars and diverse product bases. These ranged from the Golden Egg Group Ltd. (a leading British restaurant and hotel chain) through serious and popular phonograph records (including music by the Beatles), movies (including *Murder on the Orient Express*) to high-technology electronics products. EMI had initiated half a century of stereophonic records going back to the 1920s. In 1936 EMI developed the first high-definition television system upon which current television technology is still based. EMI also pioneered airborne radar during World War II and developed the first all solid-state computer in the United Kingdom in 1952.

Ninety percent of EMI's R&D expenditure was marked for specific projects, but the balance was used to fund promising "ideas" generated by R&D scientists and engineers. In 1967, Godfrey Houndsfield, one of EMI's senior research engineers who had played a key role in developing the above solid-state computer, was working on the problem of programming the computer to recognize the written word and display it on a screen. During his pattern recognition research he discovered how inefficient the current methods for collecting and storing information were. He realized, for example, that only 1 percent of the information carried on an X-ray beam transmitted through a patient was put to clinical use, the "shadow" picture produced by the differential opacity of X-rays to body tissues being extremely inefficient. He surmised that there should be a way of utilizing this "wasted" data by exploiting computer technology and therefore obtained financial support from the "ideas" fund to explore a different approach. Dr. (now Sir) Godfrey Houndsfield subsequently shared a Nobel prize in medicine for this work.

Houndsfield passed a narrow pencil beam of rays through a pig's head (the "patient") detecting the radiation transmitted (and thereby that absorbed) and storing the information in computer memory. The radiation source and detectors were mounted opposite each other on a common gantry, and all three moved in translation. The pencil beam of rays passed through a thin "slice" or plane of the subject, and a total scan of the whole slice was completed, by the repetition of the irradiation-detection process 160 times as the gantry linearly traversed the subject (see figure 1.1). The gantry was then successively rotated one degree at a time and the scan repeated,

10

Prologue:
The
Technological
Innovation
Process:
Three
Illustrative
Examples

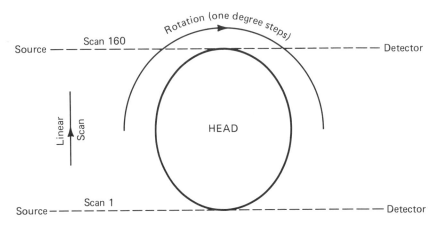

Figure 1.1 CAT Scanner Principle.

until a half-rotation had been completed, by which time 160 by 180 or 28,000 observations had been made. Houndsfield developed computer software to process this data and display an integrated "picture" of the cross-sectional slice of the pig's brain, from these 180 multi-angled scans. In this first laboratory "lash-up" it took nine days of irradiation and two and a half hours of computer time to produce one picture. Pictures of further slices could be obtained by changing the traversing plane.

The results obtained were sufficiently promising for the British Ministry of Health to support further work with a small development grant, and in October 1971 a program of clinical trials was launched at an internationally recognized center in neurosurgery and medicine, the Atkinson Morley Hospital in Wimbledon. Work was confined to brain scans, because in those early days of development, it was the only bodily organ that could be kept still long enough for the scanning procedure to be completed. Within six months, Dr. James Ambrose, the consultant radiologist who supervised these trials, was absolutely convinced that the technique of computerized axial tomography (CAT) constituted the greatest advance in radiology since the discovery of X-rays in 1895.

The advantages of the new technique over the existing methods were the following:

1. The CAT scans provided diagnostic detail and accuracy far superior to that offered by conventional techniques and so offered the potential for diagnostic refinement and improved therapy.

2. The technique was noninvasive (it did not require foreign substances to be injected into the brain to produce a diagnostic picture), was safe, and was not unpleasant for the patient.

3. The technique was rapid (by then, scan times had been reduced to a few minutes) and straightforward and could be frequently performed as an outpatient procedure. Thus it was likely to be cost-effective.

By April 1972, EMI recognized that it had developed an "invention" of major humanitarian, medical, and commercial promise, but also recognized that the continuance of the innovation process presented them with a dilemma. Clearly the inventive promise shown warranted continuation on humanitarian grounds, but was fraught with commercial risks. EMI estimated that a further £6 million of investment would be required to create a viable business operation to manufacture and market CAT scanners. The domestic British "market" centrally directed through the National Health Service would be too small to justify this investment, so that major international markets (notably the United States) would have to be created. Even then, it was expected that it would take some time to recover the initial investment. Moreover, the marketing of CAT scanners internationally created its own problems. Despite being a large corporation in the electronics industry, EMI lacked a major market presence and track record in medical electronics in the big markets in the United States as well as in West Germany and Japan, where EMI would face some of its major traditional competitors such as Siemans, Philips, CGR, and General Electric. Given the complexity of the scanner technology, EMI would have to establish a sales and servicing network to persuade users (hospitals) to purchase an item of equipment at a price of about one quarter of a million dollars which could be expected to have a radical impact upon radiological practice when, at this point in time, its efficiency had been demonstrated only to one doctor who had performed a clinical trial on seventy patients! Discussions with the medical profession in Britain and Europe indicated that the North American market would be the critical one, so further extensive investigations were conducted following the sound military (and commercial) maxim that time and effort spent on reconnaissance is seldom wasted. Given the novelty of the technology, it was virtually impossible to make an informed estimate of potential sales, and the estimate made after these discussions was for sales of 6,000 over a decade beginning in 1973.

Although by 1972 the corporation had built up a substantial patent application portfolio, it had yet to be granted one patent and so had no real idea of the strength and depth of the patent protection. Once the impact of the innovation was generally recognized, EMI could expect strong competition from a number of potential sources—the traditional competitors in medical electronics cited earlier, plus small entrepreneurially oriented companies with appropriate skills and possibly major corporations in the pharmaceutical

12

Prologue:
The
Technological
Innovation
Process:
Three
Illustrative
Examples

industry. The latter were used in incurring high R&D expenditures to support pharmaceutical innovations, but were cutting back on their R&D investment in the United States because of changes in government regulations for the testing of new drugs, so they too might seek to invest "spare" R&D dollars in promoting non-pharmaceutical medical innovation. EMI did consider licensing the invention to one of the established medical electronic suppliers (and was "courted" to do so by most of them), but decided to "go it alone." EMI reasoned that it had established a three-year lead in the technology, which could be exploited to establish its credibility and market presence. EMI therefore established a manufacturing capability in the London area and concentrated its major marketing effort in the United States. It identified the top thirty institutions in the radiological field there and concentrated on installing EMI scanners in half of these. EMI reasoned that the firsthand experience that the doctors and students would gain from the successful operation of EMI scanners in these institutions would generate a supportive momentum throughout the whole country.

An important step in the implementation of this market strategy was taken in November 1972. The corporation set up an exhibit at the Radiological Society of America Congress in Chicago, and Dr. Ambrose presented a paper on the clinical results he had obtained. The positive impact on the radiological profession was immediate and demand for CAT scanners took off. By 1975, the scanner market was estimated at £40 million per year and EMI had an order backlog of £55 million. It was anticipated that the corporation would secure a worldwide market share of at least £100 million per year and possibly several times this figure. During 1972–76 EMI's turnover in electronics was quadrupled to £207 million, and pretax profits increased from £1 million to £26 million. The London stock market began to mutter "another Xerox." Initially, because of lengthy scan times, only brain scanners could be manufactured and marketed, but in the mid-seventies body scanners were produced. By 1978 the scan time had been reduced to three seconds.

During the mid-seventies, as had been anticipated, other companies entered the scanner market. Ohio Nuclear was a small entrepreneurial company that quickly launched machines, and GE established itself in the market by developing its own capabilities. Thus EMI faced toughening competition. More important, however, was the changing attitude of the U.S. government. The enthusiastic acceptance of CAT scanning by radiologists, coupled with the competitive characteristics of the U.S. health care market, meant that many hospitals sought to purchase their "own" scanners rather than share facilities with a competitor, regardless of the underlying health care needs. For instance, two hospitals in the same U.S. city *both* stated that they would buy an EMI scanner provided they were the first hospital in the city to receive one. EMI displayed the wisdom of Solomon and sent two teams to that particular city to install a scanner simultaneously in each hospital!

Faced with a continued inflation in health care costs, the U.S. government felt that there was an unnecessary proliferation of these expensive units and in 1976, the newly elected President, Jimmy Carter, introduced a "certificate of need" requirement before a hospital could purchase a scanner. This effectively reduced the U.S. scanner market by one-half, just at the time that EMI was expanding its production to maintain its strong market penetration. During the late 1970s the corporation ran into other problems too. EMI's American phonograph record sales fell off because it had failed to maintain a stable of talented young recording artists and was reluctant to record and market the profanities of the "punk rock" fashion. EMI was also losing money in its Australian TV set manufacturing operation. In December 1977 EMI's ordinary share dropped by 17 percent in one week and this slide continued over the next year or so, as the financial community speculated about the corporation's future. Given the underlying strength and resources of the corporation, its declining financial status made it a potentially attractive "takeover" prospect. Speculation ended in late 1979 when it was announced that the Thorn Group Ltd. had merged with EMI to form Thorn-EMI Ltd. The new corporation immediately began to rationalize its operations and indicated its willingness to withdraw from the CAT scanner business. Both GE and Toshiba were keen to purchase EMI's scanner operations and in April 1980 the sale of EMI Medical to GE was announced. In the press release announcing its withdrawal from medical electronics, Mr. Peter Laister, managing director of Thorn-EMI, said:

> "It is obviously a great disappointment to all of those who have associated with this business to have to withdraw from this field. EMI's highly innovative work—especially in CT scanning—has helped to transform important areas of medical diagnosis and treatment and has contributed greatly to improved health care around the world. Although the net cost of this withdrawal will be substantial, it should be borne in mind that in the two years up to 30 June 1979 losses on these activities had totalled £26 million since which time losses continue to be incurred."[6]

1.4
The Fiber Tip Porous Pen

This case description is based on Peper.[7]

In 1947 Sidney Rosenthal was granted a seventeen-year U.S. patent for his felt-tipped pen which he exploited using the name

14

Prologue:
The
Technological
Innovation
Process:
Three
Illustrative
Examples

Magic Marker.™ The felt tip was too broad and coarse to be readily used as an everyday writing instrument, so that the latter market was dominated by the traditional metal-nibbed fountain and ball-point pens. In 1964 a new pen with a bonded, porous, fiber tip was imported from Japan and launched on the U.S. market. Despite little advertising and promotion, the vivid bold line produced by this tip attracted attention, and it was clear that fiber-tipped pens would capture a significant portion of the writing instrument market. The Paper Mate Division of the Gillette Company manufactured and marketed only ball-point pens at that time, and recognized the new product as a major threat to its markets that must be met with an effective response.

The key elements of the fiber-tip pen follow:

1. An ink reservoir absorbed into a loosely packed cylinder of fibers.

2. A writing point that consists of relatively densely packed and firmly bonded fibers in contact with the reservoir.

3. A liquid junction that forms and leaves a permanent trace of the path movement when the tip contacts and moves across the paper.

4. A cylindrical housing that can be gripped by the writer, can support the lip, and can provide a sealed enclosure for the reservoir to prevent the ink from soiling the writer's hand and keep its solvent (mainly water) from evaporating.

The first fiber-tipped pens launched on the market suffered from three clear disadvantages.

1. As the fibers wore down in use, the tips became glazed over, no longer freely depositing ink, and the tips lost the appealing writing line that they produced when fresh.

2. If the tip were inadvertently left uncovered for an hour, the ink dried and clogged the tip, making it unusable.

3. Finally, the polystyrene copolymer housing was relatively highly permeable to water vapor, so that the solvent evaporated during storage and the pens were unusable within six to twelve months of manufacture.

Paper Mate management reasoned that if the company could quickly produce a fiber-tipped pen without these defects, they could convert a competitive threat into a business opportunity. The process and product technologies for developing fiber-tipped pens are quite different from those for ball-points, and the company possessed expertise only in the latter, so it faced a considerable technological challenge. It could, however, draw upon the technological resources of its parent Gillette Company. The Gillette R & D function had expertise in the areas of textile fibers and capillary phenomena, and it was quickly decided that nylon was better than the acrylic material used in the Japanese fiber tips. Having chosen the fiber material, the researchers' next problem was to devise a manufacturing process to bond it into a porous rod with a tip ground at its end. This was the most difficult technological problem to solve, and two approaches were adopted simultaneously: bonding the fiber using an adhesive, and sintering.

Both approaches were pursued unsuccessfully for six months with Dr. Peper following the first one. He was then transferred to the Paper Mate factory in Santa Monica, California, and he continued there with the sintering approach because this appeared to offer the better prospects for ultimate success. Unfortunately, it proved impossible to achieve the right degree of bonding by sintering alone so Dr. Peper tried strengthening weakly sintered rods by saturating them with an epoxy adhesive in a volatile solvent. This yielded promising results, and then the sintering step was eliminated and compressed bundles of nylon fibers were heated to a temperature at which they take a set and were then saturated with adhesive solution. This approach was found to be more controllable than sintering and an acceptable fiber rod could be produced by finding the optimum combination of fiber packing density and concentration of adhesive in solution.

Simultaneously with the development of a suitable fiber material, suitable ink formulations were explored. The inks in the existing pens were found to have too low humectant levels, so these were increased to concentrations until the most efficient levels were found. With this ink, a tip could be left exposed overnight without the ink drying and clogging it, and the pen would write well throughout the life of ink supply. Improvements were also made in the pen body. Polypropylene has a much lower water vapor permeability than the styrene copolymers then in use, so the pen body parts were molded from the former material, although it was more difficult to maintain consistency with it. The vapor tightness and reliability of the seal between cap and pen barrel were improved with a new design, so the shelf life of the finished product was further extended. The snapping noise made by the cap-barrel latching arrangement indicated to the consumer that a satisfactory seal had been made.

The complete new product was launched on the U.S. market as the Flair™ pen in the summer of 1966 in time for the back-to-school promotional cycle. Its recognizably superior quality made it an immediate commercial success, and demand consistently outpaced production capacity. Its consumer acceptance is reflected by the fact

16

Prologue:
The
Technological
Innovation
Process:
Three
Illustrative
Examples

that the brand name Flair is now generically identified with porous-tipped pens in the United States.

We shall return to these three examples briefly later in chapter 2 after we have identified some of the major features of the innovation process.

References

1. Christopher Layton et al., *Ten Innovations* (London: George Allen & Unwin, 1972); J. Langish et al., *Wealth from Knowledge: A Study of Innovation in Industry* (New York: John Wiley & Sons, 1972).

2. L.A.B. Pilkington, "The Float Glass Process," *Proceedings of the Royal Society* series A, 314, No. 1516 (December 1969), 1-25.

3. "Living Case Histories in Innovation," *Research Management, XXIII,* no. 6 (November 1980).

4. W. J. Baker, *A History of the Marconi Company* (London: Methuen & Co, 1970); Frances Donaldson, *The Marconi Scandal* (New York: Harcourt Brace Jovanovich, 1962).

5. J. A. Powell, Lubbock lecture, Oxford University (1977); J. A. Powell, "Exploiting a Technological Breakthrough" Talk presented to the 1978 Top Management Forum, Management Centre Europe, Copenhagen and Brussels (September 20–22, 1978).

6. Thorn-EMI Press Release.

7. Henry Peper, "Fiber Tip Porous Pens—A Two Prong Attack Produced a Reliable Process and Product," *Research Management, XXIII,* no. 6 (November 1980), 19–21.

2

Some Frameworks
for Viewing the Process

A technological innovation is like a river—its growth and development depending on its tributaries and on the conditions it encounters on the way. The tributaries to an innovation are inventions, technologies and scientific discoveries; the conditions are the vagaries of the market place.
Ernst Braun and Stuart MacDonald, **Revolution in Miniature**

2.1
Introduction

In chapter 1, I illustrated, with examples, the contention that technological innovation is a complex socioeconomic and technological process that often extends over several decades or longer, requiring substantial "front-end" financial investment and not a little luck. In this chapter we will consider some conceptual frameworks for viewing this process. We begin with Bright's treatment of the process, which we then link first with Kuhn's and Popper's contrasting views of the evolution of scientific knowledge and then with more recent work by Abernathy and Utterback and Sahal on the evolution of high-technology industries. These considerations provide us with an overall framework for discussing the innovation process at the individual firm level.

2.2
Bright's Treatment of the Innovation
Process and the Innovation
Chain Equation

Bright[1] provides an eight-stage conceptual treatment of the innovation process beginning with the following definition:

17

The process of technological innovation embraces that sequence of activities by which technical knowledge is translated into a physical reality and becomes used on a scale having substantial societal impact. This definition includes more than the act of invention; it includes initiation of the technical idea, acquisition of necessary knowledge, its transformation into usable hardware or procedure, its introduction into society, and its diffusion and adoption to the point where its impact is "significant."

Bright's stages (slightly modified here) are as follows:

Stage 1. The innovation begins in one or both of two ways. One is by suggestion and/or discovery; that is, from the speculations and/or discoveries of scientists, or possibly craftsmen* in pursuing their activities. Another way is by the perception of an environmental or market need or opportunity. Many commercially successful innovations arise, at least partially, from such perceptions and this important factor will be discussed in greater detail later.

Stage 2. This is the proposed theory or design concept; that is, the synthesizing of existing knowledge and techniques to provide the theoretical basis for the technical concept. This synthesis usually occurs after considerable trial and error.

Stage 3. This is verification of the theory or design concept followed by stage 4.

Stage 4. This is the laboratory demonstration of the applicability of the concept; that is, the development of the "breadboard" model.

Stage 5. Alternative versions of the concept are evaluated and developed to be defined as the full-scale approach. At this stage, a prototype is developed and subjected to field trials. Alternatively, a pilot production plant produces small quantities of the new product which may be submitted to test markets or clinical trials. It is followed by stage 6.

Stage 6. This is the commercial introduction or initial operational use of the innovation and is followed by stage 7.

Stage 7. This is the widespread adoption of the innovation when its scale and scope of usage are sufficient to generate substantial cash flows in the producing firms and significant societal impacts. This stage is succeeded by stage 8.

Stage 8. This is proliferation, when either the generic product (such as radar equipment to detect speeding motorists) or the

---

*Throughout this text, the masculine gender will be used to indicate both the feminine and masculine gender, unless the context indicates otherwise.

generic technology (such as radio microwave technology in cooking ovens) is adapted for use in newly defined markets.

Bright stresses that this generalized treatment simplifies a complex socioeconomic and technical process. Many innovations never survive to the eighth stage, while the progress of others is delayed by pursuing developments down technical or market "blind alleys." Thus, it is not a simple linear process because multiple feedback loops may be present as developments are recycled to earlier stages when unexpected difficulties arise. It may also incorporate feedforward loops since, in well-managed institutions, the potential for proliferation of the generic technology is evaluated at as early a stage as possible and certainly by stage 4. The distinctions between the stages are ill defined because several stages may be occurring simultaneously and are better viewed as overlapping "phases" rather than distinct "stages," particularly if the innovation incorporates two or more scientific inventions. I shall describe them as phases from now on. The phases may perhaps be viewed as analogous to Shakespeare's "Seven Ages of Man" insofar as they represent the achievement of a defined state of growth. Bright's application of his treatment to wireless telegraphy is shown in table 2.1.

Stage/ Process	Date	Individual	Activity
1	1846	Michael Faraday	Observation leading to scientific suggestion.
2	1864	James Clerk Maxwell	Electromagnetic wave theory.
3	1886	Heinrich Hertz	Experimental detection of electromagnetic waves.
	1892	Sir William Crookes	Suggests their use in wireless communication.
4	1894	Oliver Lodge	Laboratory demonstration of use.
5	1896	Guglielmo Marconi	First patent and field trial.
6	1897		Commercial introduction.
7	1910-12		Increasing adoption— Crippen-Titanic effect.
8	Later	Many	Proliferation—radio industry, radar, TV industry, etc.

TABLE 2.1 The Innovation of Wireless Telegraphy

The innovation process may also be expressed succinctly using a chemical analogy, as a chain process expressed in the innovation chain equation in figure 2.1. Commercially successful innovations require the synthesis of scientific, engineering, entrepreneurial, and management skills, combined with a social need and a supportive sociopolitical environment, if a sustained chain reaction is to be achieved. To preserve the logic of this "life cycle" approach to the innovation process, we must add a ninth stage—namely, the "death" of the innovation, which typically occurs when it is superseded by a new one. This point will be discussed further in chapters 4 and 5.

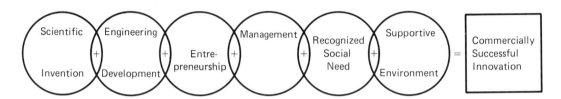

Figure 2.1 The Innovation Chain Equation.

2.3
Time and Cost Characteristics of the Innovation Process

Table 2.1. shows that it took sixty to seventy years for wireless telegraphy to complete the first seven phases of the innovation process. Bright cites other examples in which the total times to complete the innovation process range from sixteen and a half years for integrated circuits (ICs) to eighty-seven years for the electron beam welder. These figures suggest, and this is confirmed by a historical analysis of other examples, that the process takes usually at least a decade and frequently much longer for major innovations. More minor innovations (we shall distinguish between major and minor innovations later, in section 2.8) may take shorter lengths of time, but an organization must recognize that if it plans to participate in a major innovation at an early phase in the process, it may expect to invest corporate resources in this effort for at least a decade, before expecting significant fiscal rewards for its efforts.

Some observers believe that the time scale of the process is shortening, but others disagree with this view and examples of both decreases and increases can be cited.[2] Many believe that this time is shortening in the computer industry, but this apparent shortening may be explained by the fact that the time is measured from a point when the inventions have been completed and only applications

21
Time and
Cost
Characteristics
of the
Innovation
Process

remain. Whatever applies in the computer industry, there is evidence that changes in U.S. federal regulations for the testing and introduction of new drugs has both lengthened the time scales and decreased the rate of innovation in the U.S. ethical pharmaceutical industry.[3]

Although the rate of innovation has fallen in the U.S. ethical pharmaceutical industry, it has unquestionably increased in the computer industry and this creates a further problem. Successive new systems are developed and marketed by the computer industry too rapidly for the users, so that the purchaser of an up-to-date computer installation finds that it is obsolete and needs replacing, before its high capital cost can be fully amortized. Thus the substantial "front-end" financial investment that must be made by the innovator must be matched by the corresponding investments of the users, and this is a real concern for both manufacturers and users in that industry and others.

The cash flow profile associated with the innovation process is similar to that for any new product or process development except that the time scale is longer and both the technical and commercial uncertainties are greater. This profile is illustrated in figure 2.2. The technological innovation process clearly acquires a substantial "front-end" financial investment before any positive cash flow accrues. Also R & D costs typically only represent 5 to 10 percent of this "front-end" investment. The balance is spent on pilot/prototype design and trials, production tooling and engineering and production and marketing start-up operations. Naturally, an innovation will not be pursued unless it is anticipated that the revenue generated during its life will ensure an acceptable return on investment (ROI) before it becomes obsolete. Since R & D requires no more than

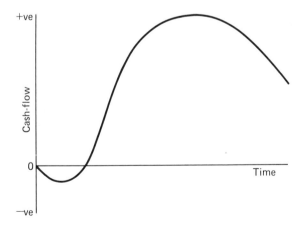

Figure 2.2 Cash-Flow Profile.

10 percent of the total cost of innovation and a good R & D laboratory may generate numerous promising inventions, some may be abandoned because the company lacks the financial resources to exploit them. This fact has an important impact on the behavior of individuals in high-technology companies as will become apparent in later chapters.

Most technological innovations offer a technically improved method of providing a given product or service and hence supersede an earlier innovation. Therefore, to preserve logical consistency, we must expect any innovation to be superseded by a successor sooner or later. In fact, the innovation process in the evolution of technology may be viewed as analogous to the emergence of a new species in biological evolution. Both innovations and a new species proliferate if they identify social and ecological niches in which they display superior performance or survival characteristics. Both face the ultimate prospect of displacement through socioeconomic conditions or natural selection. In fact, an individual innovation must be viewed in the context of general technological evolution as discussed in the remainder of this chapter, beginning with Kuhn's and Popper's treatments of the social-psychology and methodology of science.

2.4
Kuhn's and Popper's Treatments of the Growth of Scientific Knowledge

Kuhn's Paradigm

Kuhn provides a seminal descriptive treatment of the history and social-psychology of science, based upon his notion of *paradigm*.[4] Although he does not attempt to define the term rigorously, or indeed employ it in a consistent manner, he does invest it with an operationally useful meaning. Traditionally, scientists have used the terms *theory, law* or *model* to describe their conceptual treatments of a given body of knowledge. Despite their value and acceptability, these terms tend to be limited operationally to the labeling of specific novel development (such as relativity or quantum theory) usually attributed to an individual scientist (such as Albert Einstein or Max Planck), rather than the generalized framework which develops from them (such as relativistic or quantum physics). Kuhn uses the term *paradigm* to describe both levels of scientific development. That is, it is used to describe both the novel achievement that can usually be attributed to an individual scientist, and also the consequent novel framework of attitudes, assumptions, and approaches to scientific research that derive from it. He suggests that there are two stages in the evolutionary development of a given branch of science.

The first is the *preparadigm stage* when there is no single generally accepted conceptual treatment of the phenomena in a field of
study, but a number of competing schools of thought. Kuhn describes the field of optics prior to Newton as lacking in a common
body of belief, so each worker in physical optics felt forced to build
his field anew from its foundations. Each felt relatively free to
choose his own sets of supporting observations and experiments,
since there was no standard set of methods and phenomena that
every optical worker felt forced to employ and explain.

The second stage is the *paradigm acquisition stage* which marks
the achievement of a level of maturity of the science. A paradigm is
proposed and generally accepted which provides the conceptual
framework for further investigations in the field. Thus in optics, the
paradigm was provided by Newton's Opticks which claimed that
light was material corpuscles. From that time, workers in the field
sought for evidence of the pressure of light corpuscles impinging on
solid bodies—evidence that had not been sought by earlier wave
theorists.

The acquisition of a paradigm characterizes scientific maturity
and the acceptance of an agreed conceptual framework or set of rules
by which scientists in the field can plan and conduct their research
activities. These rules remain in force unless or until the paradigm is
replaced by a new one. This evolution leads Kuhn to define two types
of scientific activity in a mature science.

The first is *normal or puzzle-solving science,* which constitutes
the majority of research activity. Scientists conduct research based
on the accepted paradigm as indicated in the previous paragraph.
The universally accepted paradigm facilitates the conducting of
specialist or esoteric "puzzle-solving" research (where the "puzzle"
is defined within the framework of the paradigm), with research
results communicated within the field through the specialist journals. In fact, the "state of the art" or "body of knowledge" is defined
within the framework of the paradigm, so that expansion of this
body of knowledge is achieved incrementally by problem- or puzzle-
solving within this framework.

The second activity is *revolutionary science,* which occurs much
more rarely within a given field. The activity occurs when a paradigm is first acquired (as in the paradigm acquisition stage cited
earlier) or more frequently, when one paradigm is replaced with a
new one (and, in this sense, a revolution occurs). The history of
science is replete with examples of such revolutions, including the
Copernican revolution in astronomy in the seventeenth century and
the transition from classical to modern physics around the beginning of the twentieth century. Others may spring readily to the
minds of scientifically educated readers.

Although Kuhn provides a quite realistic description of the
social-psychology of scientific activity, the distinction between

23
Kuhn's and
Popper's
Treatments
of the
Growth of
Scientific
Knowledge

normal and revolutionary science is by no means clear-cut, and detailed study of scientific innovations suggests there is a continuous spectrum of puzzle-solving/paradigm shift activities between these polar extremes. Kuhn's treatment is reflected in more recent treatments of technological evolution which we shall examine shortly, after we have considered Popper's evolutionary treatment of the methodology of science.

Popper's Methodology

Whereas Kuhn focuses on the social-psychology of scientific evolution, Popper focuses on its methodology. He describes his concept of scientific method elsewhere, and it is succinctly summarized in "The Rationality of Scientific Revolutions."[5] A scientist always begins with a problem (P_1) which requires solution (figure 2.3.). This may be a practical problem or an inconsistency between current theory and observation. The scientist then seeks to solve the problem or remove the inconsistency by postulating a new tentative theory (TT) or *conjecture*. The layperson's view of scientific method is that the scientist then seeks to confirm this tentative theory (hypothesis) by suitable experimentation or observation, but Popper argues that a converse methodology obtains. Rather than seeking to confirm the tentative theory, the professionally competent scientist devises experiments or observations that seek to disprove or *refute* it by an error-elimination process (EE). Only after the tentative theory or conjecture has withstood the severest error elimination or refutation that can be devised, does the scientist accept it. Furthermore, a conjecture is only likely to be generally accepted after it has withstood the severest attempts to refute it by other scientists in the field. Thus a theory is accepted by a process of *conjecture and refutation* rather than "confirmation" of hypothesis by experiment or observation. Popper's methodology is preferable to the traditional view because, in the last analysis, it is logically impossible rigorously to confirm a theory by experiment or observation. Even if it is confirmed a million times there always remains the possibility of a

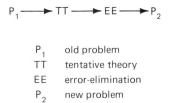

$$P_1 \longrightarrow TT \longrightarrow EE \longrightarrow P_2$$

P_1	old problem
TT	tentative theory
EE	error-elimination
P_2	new problem

Figure 2.3 Popperian Scientific Evolution.

25

The
Abernathy-
Utterback
Treatment
of the
Growth of
High-
Technology
Industries

negative outcome of a new experiment or observation in the future and, as we stated earlier in the paragraph, new tentative theories are often developed in response to the inconsistencies between current theory and observation. In Popper's methodology, a theory or conjecture is never confirmed or proved; it is at best, as yet, unrefuted. Thus the highest acceptance a theory can enjoy is *"an-as-yet-unrefuted conjecture."* Once a theory is accepted it generates a new problem (P_2), so Popper represents the process by the sequence of transformations illustrated in figure 2.3.

P$_2$ is attacked in the same manner, so scientific knowledge continually evolves from old problems to new problems by means of the conjecture and refutation process. Popper reasons that this evolution is similar to Darwinian biological evolution. Problem solving through an error-elimination learning process is a primal activity of the biological organism as well as science, where, for the former, the problem to be solved is "survival." A new organism produced by genetic mutation may be compared to a tentative theory or conjecture, and Darwinian natural selection can be compared to the refutation process. A new organism survives and multiplies in an ecological niche through a successful error-elimination learning process, otherwise it dies out. Thus a successful conjecture that survives the refutation process is like a successful species that survives the natural selection process, and both may be represented by the above sequence of transformations. The difference is that whereas the organism itself dies if the mutation is unsuccessful, for the scientist the unsuccessful theory dies in his stead!

Popper and others have applied this methodology not only to science, but also to history and social change, art, and the "body-mind" problem in philosophy; that is, to both the biological and cultural evolution of Man. Later (section 2.9) I shall suggest that it may be applied to the evaluation of prospective innovations, but first we will examine the process of technological (as against scientific) evolution.

2.5
The Abernathy-Utterback Treatment of the Growth of High-Technology Industries

Abernathy and Utterback provide a descriptive treatment of the evolution of high-technology industries generated by major technological innovations (such as the automobile or computer), which appears implicitly to reflect the Kuhnian treatment.[6] They suggest that such industries evolve through the following three stages or states:

Fluid State

In the initial stages of development, product designs (such as early automobiles and computers) are fluid, manufacturing processes are loosely and adaptively organized, and both product and process may be subject to major changes relatively frequently. In this state, alternative product designs will proliferate in the absence of any consensus on the design ideas. Thus it corresponds to the preparadigm stage in scientific evolution. Furthermore, the revolutionary technology itself will be in a fluid state of development generating uncertainties concerning technological as well as product performances. The potential markets for the generic technology will still remain to be fully identified and the acceptability of the technology in these markets is, as yet, unknown. The extent to which developments are spurred by technology push, as against market pull (see section 2.8), may vary with each innovation. However, the successful companies are likely to be ones that display both entrepreneurial and engineering flair to exploit the synergy between the market and the technology. Because this is a period of "trial and error," both technically and commercially, relatively high expenditures on R & D are unlikely to offer commensurate economic benefits. The successful developments of market niches in which the innovation will be accepted are likely to promote rapid progress.

Dominant Design and Transition State

After a period of fluidity marked by developments through "trial and error," there emerges a *dominant design*. The Model T Ford and IBM 360 are examples of the dominant designs in the automobile and computer industries, respectively. These represent superior product designs that provide the "standards" for the industry to follow and correspond to the paradigm acquisition stage in scientific evolution. The establishment of a dominant design tends to stimulate, delineate, and coalesce the markets for the revolutionary technology so that market uncertainty is reduced. The company or companies that launch the dominant design on the market enjoy competitive advantages, but do not necessarily secure monopolistic or oligopolistic positions. Emphasis is placed upon performance maximizing so another company with a strong enough technological base can reap good profits by developing incremental technological improvements in the dominant design in terms of performance, reliability, and after sales servicing. Japanese companies have been outstandingly successful in this respect in the electronics industry. Alternatively, a company may seek a market niche for the, by now, more generally acceptable technology, which the leading companies are disinterested in serving. These two approaches are discussed more fully in chapter 4 as "me-too" and "interstitial" strategies.

Specific or Mature State **27**

The
Abernathy-
Utterback
Treatment
of the
Growth of
High-
Technology
Industries

Manufacturing methods and product designs become standardized based on the dominant design. Outputs increase and hence unit costs decrease both through the increases in manufacturing efficiency facilitated by standardized procedures and processes and through increased sales in the larger markets created by the dominant design. Innovations are now incremental and the industry becomes more rigidly organized or in a *specific state*. Once the industry develops into this specific or "mature" state, the scope and sources of innovation change. For composite products, such as automobiles, product assembly is performed by an oligopolistic group of manufacturers (GM, Ford, and so forth), using components supplied by vertically integrated subsidiaries or independent suppliers. At the product assembly level, further innovations are cost reducing and focus on production process rather than product improvements. It should be recognized, however, that such incremental cost-reducing innovations can offer substantive profits, because they apply over a large sales volume.

Once an industry has reached a specific state, there is also a strong internal resistance to further substantive innovation. The industry is enjoying the benefits of its "mature" position on the technological learning curve (see chapter 13, section 13.5), with little technological uncertainty, relatively low costs, and a large established market. The majority of the work force (from senior manager to craft level) will have invested their personal career skills in the industry, and will probably feel "threatened" by change. Capital investment in plant and equipment that could be made obsolete by change may be high. Thus, the investments commitment in technological, fiscal, and human capital presents barriers to change. If it does not enter through a component supplier, any substantial innovation is likely to be stimulated by external factors. Changes in the political and social environments could occur which would impose changes on the product technology. For example, the North American automobile industry is now being stimulated into adopting innovation because of pollution and gasoline consumption standards that have been legally imposed on the automobile. Invasion of an industry by "outsiders" based on generic technologies that may themselves be in the fluid state could also occur. This happened in the period after World War II with the invasion of the chemical industry into the textile industry, through the substitution of synthetic for vegetable and animal fibers. It also recently occurred in the watch industry, with the electronic watch (based on IC technology) replacing the mechanical watch. This innovation strategy is discussed further in chapter 4.

Abernathy and Utterback do not claim this treatment to be rigorously applicable to all industries, and Sahal has proposed a more extensive treatment based upon a larger sample of industries.

2.6
Sahal's Treatment of Technological Evolution

The Abernathy-Utterback treatment implies that industries become technologically moribund on reaching a mature state, and de Bresson and Townsend[7] argue that, while providing a useful building block, that treatment fails to provide an integrative framework for understanding total technological evolution. They cite chemicals, pharmaceuticals, food products and others, as industries that have evolved further beyond a mature state. Sahal provides a more extensive treatment based upon a larger sample of industries.[8] His treatment appears implicitly to reflect the Kuhnian and Popperian treatments of scientific evolution we reviewed earlier.

Sahal views a given technology as a self-organizing system that evolves by trial-and-error learning. The important features of his treatment relevant to our present discussions are discussed in the following sections:

Technological Guideposts

Through the interplay of the self-organizing system formed by a given technology and its sustained application and development there emerges a pattern of artifact design or *technological guidepost* which charts the course of further innovative activities. The concepts of dominant design and technological guidepost appear similar, both corresponding to a paradigm acquisition in Kuhnian scientific evolution. However, thereafter, the Abernathy-Utterback and Sahal treatments differ, because Sahal postulates a continued growth through step-wise improvements in capabilities.

Step-wise Improvements in Capabilities

Rather than progressing to a specific or mature state, Sahal argues that a technology continues to evolve through a succession of step-wise improvements in its capabilities. This step-wise growth is well documented in the literature (see chapter 5). Whereas the Abernathy-Utterback treatment implies that a technology becomes moribund on reaching maturity, Sahal implies that this "maturity" is merely a plateau or interlude in a continuing evolution, which is a product of past and stimulus for future innovations. The durations of such interludes can vary considerably. The successive computer generations reflect step-wise growths in computing capabilities, and we have witnessed four such generations in about twenty-five years, but other technologies have remained on a plateau for several decades. Sahal cites the Fordson farm tractor as a design which lasted for twenty-five years (circa 1920–40). A step-wise improvement is typically consolidated with a new technological guidepost. Sahal also makes the important point that the "timing" of the introduction of a step-wise scale growth is dependent upon the scale of operations in the user environment *as well as* internal developments in the technology, and advances occur through simultaneous coevolutions at both levels. Haustein, Maie, and Uhlmann cite the steamboat *Great Eastern* as an example to illustrate this point.[9] This ship, introduced in the mid-nineteenth century, was 100 times more powerful and was seven times heavier than existing ships at that time. Unfortunately, it proved to be a commercial failure which forced its owners into bankruptcy because the port and service facilities available at the time were insufficient to handle it. These authors also point out that notable electrical inventions (water heaters, hearths, and motors) were displayed at a Vienna exhibition in 1883. However, their exploitation as innovations was delayed for decades awaiting the development of comprehensive electrical generation and distribution systems. Thus, "timing" in innovation management is most important, as an innovation may sometimes fail through being introduced "too soon" rather than "too late."

Sahal emphasizes that there is one crucial difference between biological and technological evolution. Whereas distinct biological species cannot interbreed, step-wise technological growth is quite frequently achieved by the *creative symbiosis* of two (or more) previously unrelated technologies (such as the use of nuclear power in marine propulsion and solid-state electronics in numerical control systems). Further progress is then based on a new coalition of technologies, possibly pioneered by invaders from other industries. We cited such an example at the end of the previous section with the invasion of the textile industry by the chemical industry, since synthetic fibers are a product of the creative symbiosis of textile and chemical engineering technologies.

2.7
Revolutionary and Normal Innovations

Kuhn describes a revolutionary paradigm in science as an achievement which possesses two characteristics: (a) It is sufficiently unprecedented to attract an enduring group of adherents from competing modes of scientific activities and (b) It is sufficiently open-ended to leave all sorts of problems to solve.

If we substitute "scientific, engineering, entrepreneurial, and management activities" for "scientific activities" in the above, these words are equally applicable to technological (as against scientific) innovation. Thus we may integrate the Kuhn, Abernathy-Utterback and Sahal treatments and define two types of innovations.

Revolutionary Innovations

These may be based upon major inventions that create a new industry (e.g., the transistor) or alternatively, a step-wise improvement in capability (e.g., the change from discrete transistors to ICs) in a mature industry. They may also be associated with a creative symbiosis of previously unrelated technologies. They invoke new paradigmic frameworks for technological (as against scientific) puzzle-solving expressed in the dominant design or technological guidepost. Revolutionary technological innovations are comparatively rare, but do occur more frequently than their scientific counterparts. Like their scientific counterparts they readily attract an enduring group of adherents from competing modes of technological activity. This may take the form of invasions of mature industries by "outsiders."

Furthermore, the revolutionary innovations create opportunities for small entrepreneurial firms to establish themselves as major corporate entities. This is dramatically illustrated in the growths of Texas Instruments and Intel, based upon the revolutionary innovations of the transistor and ICs respectively. Thus top managements of companies in mature industries need to be continuously sensitive to the "threats" of revolutionary innovations. This makes it desirable for mature companies to engage in "technology monitoring" to detect the "signals of technological change." This topic is discussed later in chapter 5 (section 5.3).

Paradoxically, revolutionary innovations must also be approached circumspectly. They may require a substantial front-end commitment of corporate resources and, as the Great Eastern example cited in the previous section illustrates, can cause commercial disasters if introduced prematurely. Also, because of the unpredictable markets created by revolutionary innovations, a company introducing one may experience failure for reasons largely outside its control. This was EMI's experience with the CAT scanner. These are two reasons why some companies prefer to follow defensive technological strategies (see chapter 4).

Normal Innovations

These constitute the large majority of innovations and correspond to puzzle-solving activities in Kuhnian normal science. They are typically incremental performance-improving and/or cost-reducing innovations conducted within the framework of the established technological paradigm and facilitate cumulative trial-and-error learning in technological evolution.

As with Kuhnian science, the distinction between revolutionary and normal innovations in technology is by no means clear-cut. We may view normal innovations as embracing a continuum from minor incremental innovations to occasionally major and radical innovations which, though substantive, fall short of inducing Kuhnian paradigm changes. Such radical innovations may present considerable challenges and opportunities to the industries involved.

Having examined the Kuhnian aspects of technological evolution which has been independently explored more extensively by Dosi,[10] we continue by suggesting a Popperian framework for evaluating individual prospective innovations within a firm.

2.8
Innovations as "Technological Mutations"

In stage 1 of his treatment of the innovation process (section 2.2), Bright suggests that an innovation begins with either discovery or the perception of an environmental or market need or opportunity. This underlying distinction is reflected in the innovation literature in the distinction between *technology push* and *market pull*. The former implies that a new invention is "pushed" through the R & D, production, and sales functions onto the market without proper consideration of whether it satisfies a user need—as shown in figure 2.4 (*a*). In contrast, an innovation based upon market pull has been developed by the R & D function in response to an identified market need as shown in figure 2.4 (*b*). We have just distinguished between revolutionary and normal innovations. The former are major inventions and innovations (such as radio and the computer) for which there is no manifest need and which were created by "technology push" or the visions and achievement drives of inventor-entrepreneurs such as Marconi. Once this essentially Kuhnian technological revolution occurs and the latent need becomes manifest through

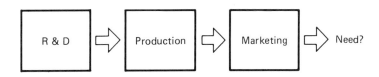

Figure 2.4 (a) Technology Push.

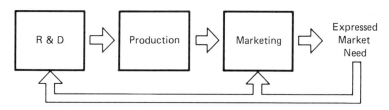

Figure 2.4 (b) Market Pull.

social recognition, "market pull" stimulates the proliferation of normal incremental innovations to satisfy evolving specialist user needs and the emergent evolution of a new high-technology industry. Because revolutionary innovations are relatively rare and most innovations are normal and incremental, it is hardly surprising that there is some evidence that innovations based upon technology push are less likely to be successful than those based upon market pull. Other writers however, have suggested that such a distinction is too simplistic.

We suggest here that this distinction between technology push and market pull can be best viewed in the context of our extension of the Popperian evolutionary treatment. The technology push sequence illustrated in figure 2.4. (*a*) is similar to that for a biological organism which has experienced a genetic mutation (except that the latter occurs randomly). The success or failure of each is determined by a trial-and-error selection process—the first in the marketplace and the second in the ecosphere. Clearly this sequence may be viewed as inefficient and fatal to both the "invention" and "organism" if the trials yield unsuccessful outcomes. Biological organisms have no choice but to evolve in this Darwinian manner, because they lack a higher intelligence which can perform a preliminary evaluation of genetic mutations (although genetic engineers can now do so). Baruch makes a crucial methodological distinction between biological and technological evolution.[11] The former is a *linear* Darwinian error-elimination process since the adaptive characteristics acquired by an organism in its life cannot be genetically transmitted to its offspring. In contrast, the development of an innovative new product is a conscious attempt by a company to adapt to or learn from its environment and past experience, and is thus a Lamarckian trial-and-error-elimination process, so it can institute "higher-intelligence" procedures for evaluating the innovation potential of inventions.

Effective high-technology companies will typically possess R & D functions which monitor and extend the state of the art in science and engineering to generate inventions with, as yet, undefined market needs. The companies will also possess the marketing capabilities and insights to identify both manifest and latent market opportunities for the evolving technological artifacts in the latter's changing economic, social-cultural, and political contexts. This observation that potential new technology must be judged in its contemporary socioeconomic and cultural context was also expressed by Graubar in his preface to the special issue of *Daedelus* on the problems and opportunities inherent in modern technology.[12]

> *Too many of us still tend...to think of technology in 19th Century terms. ...We should think of technology (and technological innovation) not just as a collection of artifacts, how-*

ever sophisticated and complex, but as a system whose social, cultural, intellectual, managerial and political components are seen as integral to it.

If we view "technology" and technological evolution in the manner Graubar suggests, we may conventionally label such a system as the "technosphere," by analogy with the "ecosphere" in biological evolution. Now, a biological organism can only test the selective advantage of a genetic mutation in the ecosphere by a Darwinian process of successful trial or death. However, just as a scientist can allow his unsuccessful theories to die in his stead in the conjectural-refutation process, so a company can conduct a Lamarckian evaluation of an invention using its prior "learned" experience and knowledge, prior to substantive resource commitment, to determine its selective advantage in the technosphere and its potential contribution to overall corporate objectives and goals. Companies will achieve success by possessing the judgment and resources to identify synergy between their technological, entrepreneurial, and managerial capabilities and this technosphere. This can alternatively be expressed by Schmookler's comparison with the blades of a pair of scissors.[13] Scissors will only cut with matching blades, and Schmookler suggests that a technological innovation and its market must match in a similar complementary manner.

This concept of synergy may be illustrated by an extension of figure 2.4 in figure 2.5. which views the company as an adaptive open system (consistent with Sahal's treatment) in the technosphere. The marketing function monitors this technosphere to identify new needs and opportunities and feeds information back to R & D and production. The R & D function monitors and interacts with the state-of-the-art in the relevant science and engineering fields to identify new technological trends and opportunities. Information from these two sources may be combined to generate new product ideas or inventions, or *technological mutations* which will meet market needs or opportunities. Li invokes the same analogy.[14] He compares the innovation process with the fundamental building block of biological evolution—the DNA double helix (see figure 2.6). One spiral of the helix constitutes the "technology push" or ensemble of scientific and engineering skills, while the other constitutes the "market pull" or ensemble of entrepreneurial, managerial, and marketing skills that are required. The notion of "technology-market synergy" introduced above is synonomous with the judicious synthesis of Li's helices. Thus, once we view technological as well as scientific growth as a Larmarckian evolution process we may view an invention which leads to a technological innovation in a company as corresponding to a genetic mutation in an organism. Indeed, Lorenz, the eminent ethologist, makes the same analogy when describing the genetic mutation process.

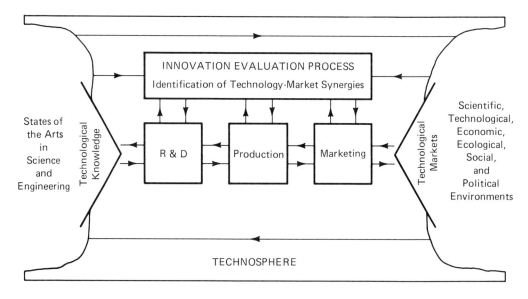

Figure 2.5 The Innovation Evaluation Process.

> *The process whereby a large modern industrial company,*
> *such as a chemical firm, invests a considerable part of its*
> *profits in its laboratories in order to promote new discoveries*
> *and thus new sources of profit is not so much a model as a*
> *specific case, of the [genetic mutation]* process that is going*
> *on in all living systems.*[15]

2.9
Criteria for Evaluating Innovations

As well as the notion of a "technological mutation," the innovation
chain equation (figure 2.1.) provides a fruitful analogy of the innovation process. All chains, whether physical, chemical, or abstract, are
only as strong as their weakest links, and in chapter 9, we shall cite
evidence of innovations that failed through the weaknesses of one or
more links in the chain. This analogy, coupled with the Popperian
conjecture-refutation evolutionary methodology, provide a useful
conceptual foundation for evaluating potential innovations, since
they may be invoked to postulate a series of challenging refutations,
or hurdles which must be overcome, if an innovation is to achieve
commercial success. They are also consistent with Graubar's comments, cited earlier.

---

*This author's addition.

34

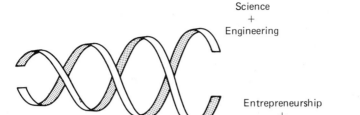

Science
+
Engineering

Entrepreneurship
+
Management and Marketing

Figure 2.6 The Innovation Double Helix.

1. As a technological conjecture, the invention must be able to withstand any scientific or engineering refutations with which it may be attacked. That is, it must be *demonstrably feasible technologically*. It also requires that the organization has the scientific and engineering resources to develop the invention.

2. It must be possible to produce and sell it profitably. This means it must satisfy a manifest or latent social or market need. If the need is latent rather than manifest, this need must be felt and accepted by potential users once the innovation is offered to society. That is, it must be *demonstrably feasible commercially*. It also requires that the organization has the entrepreneurial, managerial, and financial resources both to exploit the invention and to withstand the competition from other organizations which may also be seeking to exploit it.

3. Whatever hazards or disadvantages might be expected to occur from its continued and widespread use must be acceptable to individual users and society as a whole. That is, *any health, safety, and environmental impacts must be socially acceptable,* so these and other government regulative requirements must be satisfied.

4. Then, if its commercial success may be contingent upon regional, national, or international government policies, it should be ensured that these policies obtain. That is, *relevant government policies should be supportive.*

Even this broadened view is incomplete, however, and a further challenging refutation needs to be overcome, based upon considerations introduced in chapter 1 where I indicated that good R & D functions typically generate numerous inventions offering commercial promise. Innovation requires a substantial "front-end" fiscal investment which rises steeply as a project passes through the successive phases of the process, and a company can realistically spend only a limited proportion of its budget on innovative products. It can rarely afford to support all promising projects through to fruition when they may start to generate earnings. This means that the number of projects that can be financially supported often has to be reduced at each phase of the process despite their technological and commercial merit.

5. Therefore, a proposed innovation must be evaluated competitively with others to identify its *congruency with corporate objectives and goals,* and determine whether it offers a prudent investment of scarce corporate funds. Some companies hold an inventory of patents of their inventions which they are unable to exploit for this reason which others may purchase or license. As stated earlier, this "rationing" of corporate support for innovative opportunities has a significant impact on the behavior of individuals in technology-based companies, as will be demonstrated in later chapters.

2.10
Three Innovations Briefly Reexamined

To conclude this chapter, I will briefly reexamine the three examples described in chapter 1, in the light of the above evaluation scheme. It took Marconi about twenty years of sustained effort to overcome the hurdles (or refutations) to the exploitation of wireless telegraphy. It took until the 1910s and the "Crippen-*Titanic* effect," to convert that latent need for wireless telegraphy into a need recognized by both society in general and governments in particular. After that date, hurdles 2, 3, and 4 collapsed, and the Marconi companies were able to enjoy the rewards of their technological and entrepreneurial efforts.

In contrast, the latent need for computerized axial tomography was quickly converted into a manifest need by the presentation at the Radiological Congress in Chicago in 1972, and EMI appears to have quickly overcome the first two hurdles only to be defeated by hurdles 3 and 4. In the light of subsequent events, it might also be argued that investment in scanner technology was not congruent with corporate objectives and goals, so insufficient consideration was given to hurdle 5. EMI's experience with its CAT scanner illustrates the perils of being an offensive innovator (See chapter 4). It is of interest to read Abell's description of the overall development of the scanner market and Li's account of the development of scanner technology of GE who, as events turned out, purchased the EMI operation.[16] GE appears implicitly to have adopted an effective defensive strategy in the face of a revolutionary technological innovation in its area of technological expertise. Furthermore, by the judicious acquisition of the EMI scanner operation, it was able to secure a powerful position in the new market in the face of competition from others.

Paper Mate's development of the Flair pen also illustrates the advantage of defensive innovation. Faced with a new technological threat in its market, Paper Mate responded by developing its own technological innovation. By prompt and effective technological entrepreneurship, it provided competition to the Japanese import at hurdle 2 of the innovation process, and so converted a competitive threat into a technological and marketing opportunity.

Having reviewed some of the characteristic features of the technological innovation process, we will next study it in a corporate setting.

References

1. James R. Bright, "Some Management Lessons from Technological Innovation Research," *Long Range Planning* 2, no. 1 (September 1969), 36-41.

2. J. E. S. Parker, *The Economics of Innovation,* 2nd ed. (London: Longman Group, 1978), pp. 55-6.

3. Jerome E. Schnee, "International Shift in Innovative Activity: The Case of Pharmaceuticals," *Columbia Journal of World Business* 13, no. 1 (Spring 1973): 112-22; Roy Rothwell, "The Impact of Regulation on Innovation: Some U.S. Data," *Technological Forecasting and Social Change* 17, no. 1 (May 1980): 7-34.

4. Thomas S. Kuhn, *The Structure of Scientific Revolutions,* 2nd ed. (Chicago: University of Chicago Press, 1970).

5. Sir Karl R. Popper, *The Logic of Scientific Discovery* (London: Hutchinson, 1959); Sir Karl R. Popper, *Conjectures and Refutations* (London: Routledge & Kegan Paul, 1963). Sir Karl Popper, "The Rationality of Scientific Revolutions," in *Scientific Revolutions,* ed. Ian Hacking (Oxford: Oxford University Press, 1981), pp. 80-106.

6. W. J. Abernathy and J. M. Utterback, "Patterns of Industrial Innovation," *Technology Review* 80, no. 7 (June/July, 1978); 40-47; J. M. Utterback and W. J. Abernathy, "A Dynamic Model of Product and Process Innovation," *Omega* 3, no. 6 (1975): 639-56.

7. C. De Bresson and J. Townsend, "Multi-Variate Models for Innovation—Looking at the Abernathy-Utterback Model with Other Data," *Omega* 9, no. 4 (1981): 429-36.

8. Devendra Sahal, *Patterns of Technological Innovation* (Reading, Mass.: Addison-Wesley Publishing Co. 1981).

9. H. D. Haustein, H. Maie, and L. Uhlmann, *Innovation and Efficiency* (Laxenburg, Austria: International Institute for Applied Systems Analysis, Report no. RR-81-7, May 1981).

10. Giovanni Dosi, "Technological Paradigms and Technological Trajectories," *Research Policy* 11, no. 3 (June 1982); 147-62.

11. J. J. Baruch, foreword, in *Management of Research and Innovation,* ed. Burton V. Dean and Joel L. Goldbar, TIMS Studies in the Management Sciences 15 (1980), vii-ix.

12. S. R. Graubard, preface to "Modern Technology: Problem or Opportunity?" *Daedalus* 109, no. 1 (Winter 1980): v.

13. J. Schmookler, *Invention and Economic Growth* (Cambridge: Harvard University Press, 1966).

14. Y. T. Li, *Technological Innovation in Education and Industry* (New York: Van Nostrand Reinhold Co. 1980), pp. 81–83.

15. Konard Lorenz, *Behind the Mirror: A Search for a Natural History of Human Knowledge* (New York: Harcourt Brace Jovanovich, 1977), p. 27.

16. Derek F. Abell, *Defining the Business: The Starting Point of Strategic Planning* (Englewood Cliffs, N.J.: Prentice-Hall, 1980), chap. 5; Li, *Technological Innovation in Education and Industry,* chap. 3.

Part 2

The Corporate Setting

Part 2 begins with a description of the technological base of a high-technology company. It discusses the spectrum of alternative technological strategies available to such a company and techniques that may be used to define and "invent" its future. It concludes with a discussion of environmental concerns and technology assessment in innovation.

3

The Corporate
Technological Base

*But the first business of every theory is to clear up concep-
tions and ideas which have been jumbled together, and, we
may say, entangled and confused; and only when a right
understanding is established, as to names and conceptions,
can we hope to progress with clearness and facility, and be
certain that author and reader will always see things from
the same point of view.*
Carl von Clausewitz, **On War**, bk. 2, chap. 1, "Branches of
the Art of War"

3.1.
Introduction

The purpose of this text is to provide guidelines for the management
of the innovation process in high-technology organizations, and in
chapters 1 and 2 we examined the features and methodology of this
process. To reach commerical fruition, such innovations must typically occur within the framework of the extant technological base of
the organization. In this very short chapter we will define the components of this technological base insofar as it applies to the process
of technological innovation.

Traditionally, technological innovation appears to have been
largely bypassed in defining the management structures of high-
technology companies. Most companies build their structures around
the traditional functions of finance, marketing, personnel, R & D,
and production and many also define an engineering function, concerned with advanced design and development or replication of existing technological capabilities. At first sight, it might appear
appropriate to equate the technological base with the R & D and
engineering functions of the organization, but a consideration of the
material presented in chapters 1 and 2 indicates that such an equation is too simplistic. Although the R & D function provides the site
for its initial stages, the technological innovation process continues
and manifests itself in other functions of the organization. There-

fore, the technological base should be defined broadly enough to encompass both R & D and those other technological activities that have traditionally performed in the other functional areas, but which contribute to the commercially successful outcome of the innovation process. A corollary of this diffuse concept is that the components of the technological base should be reviewed holistically and should be designed to reflect the overall technological stance of the company. This notion will be made concrete when we examine technological strategies in the next chapter.

3.2
Innovation Process and Technological Base

We have just argued that the technological base manifests and defines itself through the enactment of the innovation process in the organization. Therefore, it is convenient to define such a base by sequentially examining the innovation process, but from a company viewpoint rather than the Bright-individual innovation or the industrywide approaches in chapter 2.

Table 3.1. repeats the first seven phases of Bright's treatment of innovation, but expands it to include the ancillary technology-related activities that may be required to promote it successfully. This sequence from research to full-scale production is elaborated further in chapter 9 in the context of the innovation evaluation process, and so is briefly outlined here.

3.3
Nondirected or Fundamental Research

As Bright indicates, the innovation process may begin with the invention of a new technological capability based upon a scientific and/or engineering discovery or suggestion. Such discoveries and suggestions occur, not infrequently, serendipitously, during the performance of fundamental research at the state of the art with the objective of expanding the body of scientific knowledge, rather than the development of new technological products. It can be argued that such fundamental research is the prerogative of the universities, government, and (possibly) industrywide research laboratories, rather than individual company R & D functions because it is not congruent with the latters' mission. However, some (particularly large) high-technology corporations that wish to maintain a posture of technological leadership within their industry believe it desirable to perform such research to keep up with the state of the art. This belief is discussed further in the next chapter.

Bright Process	R & D–Production Related Activities		Market Related Activities
1. Scientific suggestion. Perception of need.	Nondirected fundamental research		Market monitoring
2. Theory/design concept.	Directed applied research.	C o n t r a c t R & D	
3. Verification.	Primary development.		Initial market definition.
4. Laboratory demonstration.	Secondary development. Design engineering. Further development and design engineering.		Patenting. Development of market plan. Design of user education and advisory services.
5. Pilot/prototype to full scale production.	Pilot prototype to full scale production. Design of production system. Quality and reliability engineering. Design of technical "after-sales" services.		Test marketing. Refining product-market concept.
6. Commercial introduction.	Provision of "after-sales" services.		Market launch. Provision of user education and advisory services.
7. Widespread adoption.	Incremental product improvements.		Licensing.

TABLE 3.1 Innovation-Related Activities

3.4
Contract Research and Development

Many companies seek to subsidize their R & D efforts by performing contract R & D for outside agencies, notably government. Quite often it is possible to identify commonalities between the innovations which the company is striving to develop and the R & D needs of outside agencies. Thus, the performance of contract R & D for an outside agency may directly or indirectly solve some of the problems faced in developing the "in-house" innovation, and the cost to the company of the latter may thereby be reduced. This is seen most dramatically in the aerospace industry, where the know-how discovered in the development of successive generations of military aircraft has been exploited in the development of successive generations of their commercial counterparts. The commercial spin-offs are frequently

quoted by politicans and government officials to justify government
R & D expenditure in industry. Contract R & D may itself lead to
commercially successful innovations and is often included in a company's R & D project portfolio.

3.5
Directed or Applied
Research and Primary Development

A company's central R & D laboratory effort can be normally classified under this heading. Essentially, it represents phases 2 and 3 in
Bright's treatment whereby a scientific discovery or suggestion is
exploited to invent a novel technological capability which had the
potential for development as a new product. Research on the potential capability is advanced to the "breadboard model" or the primary
or experimental development stage when a laboratory-scale demonstration of the technical viability of innovation is achieved. If cost
and market analyses indicate that the potential innovation has at
least some minimum potential for commercial success, work can
proceed to the next phase in the process.

3.6
Further Development and Design
Engineering

The project now moves from a very small-scale laboratory realization toward increases in scale and/or quantities produced, dependent upon the technology. It may involve two or more development
steps in which increasing attention is given to cost and design considerations, both from the viewpoint of economy and ease of manu-
factoring/producing and market appeal. It may also be performed
in a divisional rather than central R & D facility (see chapter 7).

3.7
Pilot or Prototype Production

The project now moves from an R & D to a "production" environment. Production is scaled-up to the pilot run or initial prototype
levels. The move to a more "profit-oriented" production environment
provides an improved test of the viability of the project against more
demanding cost and time deadlines. As with development, it may
incorporate two or more scaling steps in which increasing outputs
are test marketed to selected customers, and customer responses are
evaluated.

3.8
Full Production

The final "scale-up" to full production is made. The technical feasibility of the project is proven. Then given that the product is accepted by the market, it becomes an addition to the product lines of the company.

3.9
Market Development

To ensure that the product satisfies a latent or manifest user need, parallel marketing activities will be undertaken. The market-product concept may be initially defined at phase 1 and certainly by phase 4 of the innovation process. Thereafter, a technical marketing function will design and develop a marketing strategy in conjunction with the technical product development, design, and production. These activities are indicated in the second column of table 3.1 and discussed later in chapter 9. The sequence "Nondirected Fundamental Research to Full Production" represents what might be called the central sequence in the innovation process, with each stage in the sequence requiring varying technological needs. These stages may

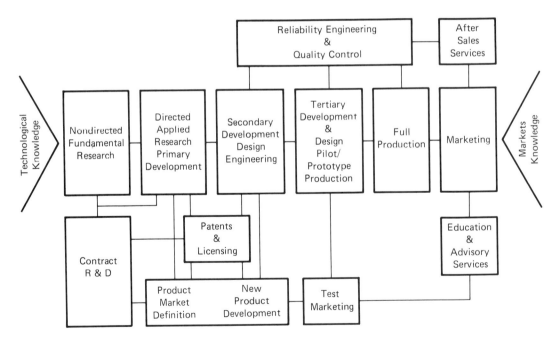

Figure 3.1 The Technological Base.

be viewed as components in the technological base as shown in figure 3.1, but for innovation to be successful, several other supporting components are required which will now be considered.

3.10
Patents and Licensing Services

The role of patents in the innovation process is debatable, but commercial outcomes are subject to the influences of the company's patent and licensing policies. It is more or less standard practice for a patent application to be filed at phase 3 of the innovation process if and when "inventive merit" can be demonstrated. Companies may use patents either to prevent competitors from marketing the same innovation (for example, Polaroid and Xerox) or as bargaining counters in negotiations with such competitors as potential partners. This suggests that the role of the patenting and licensing of innovations should be developed within the context of the overall technological strategy of the company.

3.11
Quality, Reliability, and Technical "After Sales" Services

Most high-technology products are manufactured to pre-determined quality and reliability standards and may require significant "after sales" servicing. These factors must be considered in relation to manufacturing and product design as well as marketing and distribution decisions. For example, Motorola designed the circuitry of their domestic color TV sets on a modular basis to enable the service repairmen to replace faulty modules easily and then promoted sales of the product by emphasizing ease of repairs and servicing using the "works-in-a-drawer" advertising slogan. Similarly, Maytag emphasized the reliability of their washing machines on TV commercials which humorously focused on the boredom of their repairmen through lack of repair work.

3.12
Education, Training, and Advisory Information Services

Ancillary to "after sales" services is the provision of "before and after" services which educate and advise the customer on the use of the product. The provision of such services can represent a substantial proportion of the product concept and its cost. This can be seen in the aerospace industry, where companies such as Boeing, Lockheed, and McDonnell-Douglas must expend considerable effort in training airline staffs in the use of new generations of aircraft. It is manifest similarly in the computer industry where the changing requirements of the technology dictate that the manufacturers are required to provide very extensive ranges of training and advisory services.

3.13
Managing the Technological Innovation Base

The above sections have outlined what may be viewed as the components of the technological base of the company. It can be seen that these are classified more or less independently of the conventional terminologies of finance, marketing, personnel, and production. Although these components may each lie within or impinge upon one or more of the above functions (as is illustrated in figure 3.1) it is argued that they provide a more meaningful framework for managing the technological innovation process in a company. They might be viewed as the components that primarily sustain the dynamic process of technological change within the company and its markets in contrast to the static manufacturing and marketing operation based upon unchanging technological capabilities.

Some (particularly small) companies, which may be either craft based or exploring and serving relatively undemanding specialized or local markets, may be able to survive based upon a fairly static capability. However, most high-technology companies in the developed world face a continuing and exacting demand to cope with technological change. Therefore, a holistic view of the dynamic technological innovation base should be taken if the company is to survive and compete in its changing technological environment. The manner of its competitiveness is reflected in the mode of management of the components defined above. In chapter 4, we define the spectrum of alternative technological strategies or stances available to the company and how they affect the structure and management of the components of its technological base.

4

Technological Strategies

A Prince or General who knows exactly how to organize his War according to his objective and means, who does neither too little nor too much, gives by that the greatest proof of his genius. But the effects of this talent are exhibited not so much by the invention of new modes of action, which might strike the eye immediately, as in the final result of the whole. It is the exact fulfillment of silent suppositions, it is the noiseless harmony of the whole action which we should admire, and which only makes itself known in the total result.

Carl von Clausewitz **On War,** *bk. 2, chap. 1, "Strategy"*

4.1
Introduction

It is platitudinous to observe that high-technology products have a finite life cycle as reflected in the innovation process discussed in chapters 1 and 2, particularly in the cash flow profile of figure 2.2. Whether they are ultimately profitable or otherwise, all such products sooner or later "die," so a high-technology company must maintain a continued succession of new products if it is not to "die" with its obsolete innovations. That is, quite simply a company

> *must "innovate or die." The process of innovation is fundamental to a healthy and viable organization. Those who do not innovate ultimately fail.*[1]

The process of technological innovation is central to organizational survival, but at the same time it is complex. As indicated in chapter 2, companies that view technological innovation as a simplistic linear process in which R & D inventions are *pushed* through to manufacturing and marketing (as shown in figure 2.4) are likely to fail commercially. Rather, innovation should be viewed as a closed-loop process (figure 2.5) in which needs or unoccupied niches in the rapidly changing technosphere, which match the technological capabilities and aspirations of the organization (that is, the technology-market synergies), are identified and exploited.

47

Ideally, this cybernetic open-systems process should be reflected in the company's technological policy or posture, the technological strategies it pursues, and the organizational climate it seeks to create. The purpose of this chapter is to take an overview of the considerations required to embed such a technological posture in the formulation and implementation of corporate strategies and plans. The balance of the book is essentially concerned with examining many of these considerations in further detail.

4.2
General Considerations—
The Innovation Management Function

Schematically, we may extend the cash flow profile of figure 2.2 in time, to incorporate a sequence or time series of innovations which maintain a steady positive cash flow for the company, as shown in figure 4.1. Obviously, this is an idealized and simplified treatment of how products are phased in and phased out of a company's manufacturing portfolio in real life, but it suffices for present purposes of exposition. Traditionally, we may view current operations and marketing management as being involved in phase 5 of the innovation process onwards, that is, the production, distribution, selling, and servicing of a technologically (though not necessarily commercially) proven and adopted product. Equally traditionally, we may view the earlier stages of the innovation process (insofar as they occur within the company) as being within the purview of R & D management. The open-systems view of innovations as exemplified in figure 2.5

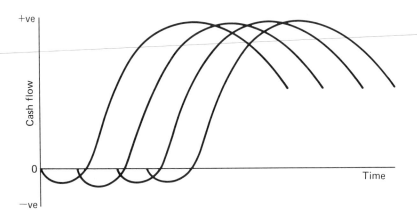

Figure 4.1 Sequence of Innovations.

clearly violates this linear managerial sequence. It suggests that we might benefit from open-systems management structures and "thinking" which reflect the nature of the innovation process, that is, that innovation should be a conscious, explicit, and *accountable* concern of all managers in the company, and a permanent item on the agendas of intermanagerial and interdepartmental meetings.

In practice, such a viewpoint tends towards utopianism. Most if not all of the time and energies of production and sales managers are likely to be consumed in the day-to-day pressures of making and selling the company's current product range, leaving them little or no time to meditate on and develop new product ideas. If open-systems (as against technology push) innovation is to occur, it is required to be the *explicit and major concern* of some managers in the company. It is therefore useful to define a nontraditional management function, called *innovation management,* in contrast to traditional production or manufacturing management and other operational functions such as marketing, finance, and personnel. The innovation management function should have a multifunctional management membership (from R & D, production, marketing and so forth) and report to or be part of general corporate management. Reflecting the ideas of Wills,[2] I suggest that a high-technology company requires a dichotomous structure reflecting its dual future-present or innovations-operations orientations conceptually illustrated in figure 4.2 (itself based upon Wills' Figure 1). This innovation management function should participate in the corporate planning process and the formulation and implementation of the firm's technological strategies. It should also ultimately be responsible for the evaluation of potential innovations or new products. The role of innovation management will be elaborated upon further in the next chapter.

Despite the need for innovation managers, it must be stressed that technological innovation should be a continued concern (as against a time-consuming preoccupation) of traditional line managers in the organization, because their long-term livelihoods depend upon it, and the corporate climate should communicate this concern. In fact, innovation management is perhaps best viewed as a visceral aspect of corporate planning since, in high-technology companies, *corporate plans* are really synonymous with *innovation plans.* It is now fairly well accepted that, if it is to succeed, planning must be a concern and include the involvement of line managers, and not just the "ivory-tower" activity of a corporate planning group at the corporate management level. The same imperative applies to innovation management, and how this imperative is achieved in successfully innovative companies, such as Texas Instruments and others, is discussed in chapters 16 and 17.

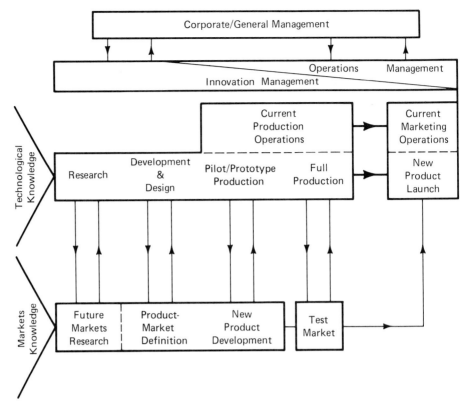

Figure 4.2 Innovation and Operations Management.

4.3
The Technology-Market or Technological Opportunities Matrix

From the time it started to become managerially fashionable in the 1960s, the formulation and implementation of corporate planning has been researched, taught, and applied extensively. We have just argued that innovation planning is a visceral aspect of corporate planning, but we have no desire to burden the informed reader with another treatise on corporate planning. Rather, we are concerned with the technological dimension of formulation and implementation of corporate plans. This formulation of a corporate technological policy or posture can be viewed metaphorically as the merging together of differing currents from both the external and internal environments of the company into a confluence or tide, which may be exploited through the company's technological navigation. These currents are illustrated in the left-hand side of figure 4.3. At any

50

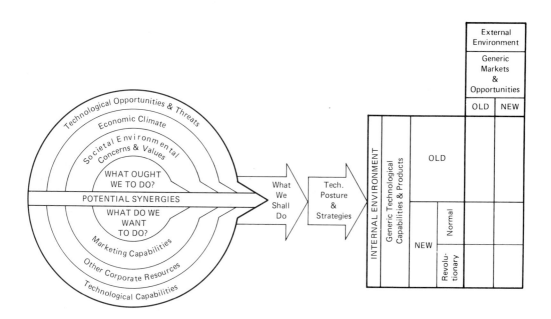

Figure 4.3 Strategy and the Technology-Market Matrix.

point in time, the company will be presented with a nexus of opportunities and threats implicit in its generic technological and industrial environment or technosphere. The extent to which it is able to meet and profitably exploit this nexus is a function of the stage reached in the industry life cycle, economic climate, current societal values, and concerns on the one hand and the company's technological, marketing, and other capabilities on the other. The notional matching of societal wants and needs to the company's aspirations (as expressed by stockholders, managers, and workers) generates the corporate policy or set of company goals. These goals may be expressed in societal and economic rather than technological terms, but in practice, in a high-technology company they will be achieved largely through the pursuit of specific technological strategies.

This exercise of technological entrepreneurship is essentially based upon matching technological capabilities to market opportunities, that is, identifying and exploiting technology-market synergies or occupying profitable niches in the continually evolving technosphere. This consideration may be expressed in the two dimensions of a technology-market or technology opportunity matrix as shown in the right in figure 4.3. The "technology" rows of the matrix distinguish between existing products or product lines or "old technology," and innovations or "new technology." The columns differentiate between "old" and "new" markets for products. Because this

51

chapter is primarily concerned with a review of technological
strategies, we are primarily interested in the two "new" technology
rows, but we shall comment upon the "old" row at the end of it.

4.4
Technological Strategies

The following classification of technological strategies is essentially
based upon ideas developed by Ansoff and Stewart and Freeman.[3]
At the outset, it should be stressed that the classification scheme
constitutes a typology which is a continuum rather than discrete.
Each strategy may shade into others, and a company may pursue
one or more strategies simultaneously, rather like "pure" and
"mixed" strategies in game theory. This consideration is particularly true if a company is a relatively large corporation with diversified
semiautonomous divisions, in which each pursues its own strategy.
As Freeman suggests, the typology may be viewed as analogous to
the famous psychologist Karl Gustave Jung's personality typology,
since a company, like an individual, is unlikely to conform to a single
type, but rather be a blend of several types. Pursuing the analogy
with human personality a little further, it could be argued that one
role of corporate management is to ensure that the company pursues
a blend of strategies that are "balanced" or complementary to each
other and to avoid "organizational schizophrenia." Just as there is a
"thin line" between human genius and madness, so the interpersonnel and interfunctional conflicts that are usually endemic to creative
and innovative organizations have the continued potential to destroy them. Indeed the "Route 128" and "Silicon Valley" phenomena
in Boston and California, respectively, are due in part to such interpersonal conflicts (see chapter 14). In outlining them, we will consider each strategy in turn, in roughly decreasing order of technological
and marketing challenge, relating it to the technology-market matrix, the third dimension of time, and the technological base functions discussed in chapter 3. I begin by describing the first of these,
the offensive strategy, in some detail.

4.5
Offensive Strategy—Leadership or
"First to Market" Policy

General Considerations

This is the most glamorous strategy and is notably identified with
well-known successful high-technology firms such as IBM in computers, RCA in television, Texas Instruments in semiconductors,

and DuPont in chemicals. All these were the first (in time) to market the revolutionary or radical innovations in their fields, and by definition they could claim to be exercising technological leadership in the industry, since they are initiating the start of an industry life cycle. Examples of unsuccessful attempts to achieve this leadership can also be quoted. The Comet airliner could be viewed as a revolutionary innovation that failed for technological reasons. In contrast, the Concorde appears to be technologically successful, but has failed for political, economic, and environmental reasons.

The introduction of a revolutionary or radical new technology is a comparatively infrequent event, and having been first to market it, a company will almost certainly wish to maintain its technological and market leadership indefinitely as the industry matures. It may thus seek to establish the dominant design in the industry and continue to be the first to market performance maximizing (and later cost-reducing) incremental innovations. We saw that in the fluid stage of the life cycle, the market is performance rather than price sensitive. It follows that companies can reap rich rewards by exploiting their technological virtuosity to produce a technologically superior product. The markets for the new technology are still not clearly defined and delineated, so these may be further explored. The ability to maintain such leadership and sustain a capability over some years demands a company technological base which exemplifies *all-round excellence.*

In chapter 3 we defined the technological base of the company, represented schematically in figure 3.1. This representation is recalled with the additional indication that the company seeks to apply revolutionary and normal new technology in its markets. We will now discuss the implications that the pursuit of an offensive strategy has upon the structure and climate of this base.

R & D Requirements

As was suggested in chapter 3, the maintenance of a nondirected research activity is deemed desirable if a company is to pursue this strategy. However, it has become fashionable to argue that the investment of corporate money and other scarce resources into nondirected research is commercially irresponsible and that research is best left to universities and government research establishments. It is argued that the results of such research, which could be viewed as covering the first two phases of the innovation process, can be readily monitored through continual literature surveys and the maintenance of continued social and professional relationships with these research establishments and the employment of staff from such institutions as consultants.

There can be little doubt that these monitoring activities *are and should* be undertaken by research staff in companies exercising technological leadership. However, the adage that invention occurs to the prepared mind is equally apropos, and many now believe that an individual can only achieve the level of mental "preparation" required, if he is active in fundamental research. This counterargument reasons that, given the esoteric nature of contemporary science and technology, its complex and tortuous interactions, the kaleidoscopic fluidity of the rapidly changing technosphere, and the competitive intensity in high-technology industry, the "spotting" of potential technology-market synergies can only be achieved through active participation in nondirected research. The validity of this counterproposition is difficult to prove or disprove, but it is unquestionable that some companies that have historically exercised a technological leadership role have also made outstanding contributions to fundamental knowledge.[4]

Whether or not it performs nonbasic research, it can be expected to perform applied research, and this leads us to the important consideration of climate and communication in the R & D functions of offensive innovators. These are discussed more fully in the article cited earlier by Ansoff and Stewart, who suggest that such innovators should operate R & D functions that work in close proximity to the state of the art and are research (R)-intensive.[5]

These authors, quite correctly, distinguish between the state of the art at different phases in the innovation process. They suggest that in research (corresponding to phases 1-3 in the innovation process) the state of the art denotes the frontier at which investigators seek to discover new phenomena or to devise new solutions to known problems. In contrast, they suggest that the state of the art in development (corresponding to phases 4 and 5 in the innovation process) denotes the situation where the validity of a theory or solution has already been proved, but has yet to yield a commercially successful application, thus focusing on both economics and technology. Thus we may perceive a number of overlapping successive phases in the state of the art, corresponding to the successive phases of the innovation process.

Research in offensive innovators is characterized by scarcity of precedent and low stability and predictability. Working in situations of ignorance and uncertainty at the frontiers of knowledge means that a precedence is scarce, and management cannot readily rely on historically based guidelines to assess the commercial viability of an innovation, or to control activities. Also such a company must anticipate rapid technical advances (either by itself, or its competitors), leading to major improvements in the performances and/or reductions in costs of products. Thus, it must perpetually face the possibility of unpredictable opportunities to exploit, or threats to jeopardize its position in the marketplace.

The characteristics of an R-intensive organization are described by the preceding authors as follows:

1. *Nondirective work assignments and indefinite objectives which are "broadcast" widely.* At the early stages of the innovation process, the solution (and possibly the problem) is unknown, so that alternative solutions must be searched for and evaluated. Given the characteristics of pioneering activities at the frontiers of knowledge, this implies nondirective work assignments and emphasis on individual contributions and scientific and technical insights, rather than highly structured tasks and roles. It also implies the broadcast of information on problems, market data, and possible solutions among technical staff to stimulate the generation of the widest range of possible solutions.

2. *Continuing evaluations of results and swift perception of significant outcomes.* Given the fluidity and unpredictability of the work situation, alternative solutions can be expected to be continually generated. At any time an approach or project may be superseded by a superior one arising either from within the group, or possibly from a competitor's efforts. Technical management should be swift to perceive significant results and maintain a continuous review of project activities to permit swift switches in approaches in the light of these results.

3. *Value innovation over efficiency.* Needless to say, the foregoing considerations hardly imply a crude economic approach to R & D activities. This latter view is unacceptable because the objective is to generate markedly better solutions to problems from those produced elsewhere in terms of clear market or profit advantages, rather than to perform R & D at minimum cost.

What is clear from the remarks of these authors is that the R & D functions of offensive innovators should be staffed by able individuals who should be given considerable freedom to produce results. Their recommendations were anticipated in the seminal text by Burns and Stalker.[6] They are also illustrated in Kidder's engrossing description of the development of a new minicomputer in Data General Corporation.[7] Put in other words, R & D management must practice *inspired adhocracy* rather than deadening bureaucracy!

Pilot/Prototype Production

For the offensive strategy to be successful, this inspiration and momentum must be carried right through the innovation process. If the company is to maintain its objective of being first to market an innovation, that innovation must be swept through the remaining stages as swiftly as effective problem solving allows. Although an offensive strategy implies an R-intensive company, this research emphasis is *not* at the expense of the development, design, manufacturing, and marketing functions of the firm. The company is R-intensive compared with companies following other strategies, as we shall see later in this chapter, but it also requires equally intensive efforts to be made in successive stages of the process. As well as strengths in research and experimental development, it must have commensurate problem-solving talents in the design-engineering, pilot and prototype development and testing stages, and so on, often spread over a wide range of disciplines and skills. Because of the sparsity of precedence, problems may arise at perhaps the pilot/prototype production stage which cannot be resolved by the "rule of thumb" methods. If the innovation is incorporating new science and technology, even at this stage problem-solving may require recourse to "scientific first principles." Therefore, the firm must have ability to make this recourse. Examples of revolutionary innovations where this recourse was applied are Pilkington's development of the "float glass" process and Farben's development of polyvinyl chloride (see 4, p.175). Sparsity of precedence also applies to craft (or technician) as well as science and engineering (or technologist) skills. The demand for welders who can work to the exacting standards imposed by the deep-sea oil drilling program is a well-known example in this category.

Patents and Licensing

The same general considerations apply to what might be called the more peripheral functions in the corporate technological base. The company will probably wish to adopt a strong patent position to protect its technological leadership for as long as possible. This is not as self-seeking a charactertistic as first appearances might indicate. It is becoming clear that a company pursuing an offensive strategy must be expected to be technology (T)-intensive and invest a bigger proportion of its budget in R & D and related activities than the industry average. Despite its technological competence, it may also be expected to experience a high proportion of project failures, whether for technological or commercial reasons. Therefore, it will wish to protect its "winners" from the competition for as long as possible to ensure that they accrue the maximum profits possible, which may be invested in other projects (recall the analogy with horse breeding, training, and racing suggested in chapter 1). By adopting a strong patent position, the company is seeking to maximize the winnings of its successful "horses" so as to stay in the "breeding, training, and racing" business. Adopting a strong position implies filing a patent application as soon as feasibility allows, both for the primary invention and for any secondary and tertiary inventions that are developed as the innovation proceeds. These patents may then be used later as barriers or bargaining counters when competitors seek to enter the market. An astute licensing policy can also help maximize "winnings," particularly in off-shore markets which the company does not wish to enter directly or where the patent protection law is weak.

Marketing, User Education, and Services

The roles of the user needs and services are equally if not more important. A revolutionary innovation, such as radio and the computer, can be viewed as a technology-push-market-pull synergy because it seeks to satisfy an unmanifested but nevertheless latent user need. Often, as with radio and the computer, the innovations are both technologically and socially revolutionary. Since the offensive innovator is the "first to market" such innovations, it means that considerable effort must be made to ensure the reliability of the product in use and to educate users in their operation. A notable feature of the offensive innovators in the radio and computer industries is that they invested considerable efforts in setting up after-sales servicing networks and user training programs. Marconi and IBM are names associated with internationally famous schools for training wireless operators and computer programmers, respectively. Moreover, the process of training a user and servicing a piece of equipment leads to a more refined identification of user needs and of desirable incremental improvements in the equipment. During the "performance-maximizing" stage of the industry life cycle, the offensive innovator will be under pressure to maintain its technological and market leadership position by introducing incremental improvements in the innovation. Information feedback of user needs and problems to the R & D functions is an important role for the user needs and service functions, and training centers are usually located close to company R & D functions to encourage this communication.

This last sentence brings us to the last but by no means least point concerning the offensive strategy that should be emphasized—what Ansoff and Stewart call the need for *high downstream coupling,* which they illustrate in figure 4.4, which is essentially identical with the technology-push versus market-pull distinction we made in chapter 2. They argue that the implementation of an offensive strategy requires good "coupling" or communication between all the functions involved to ensure the swift identification and solution of problems, and they use a hydrological analogy to emphasize their point. Using our control engineering analogy, we express the same argument by suggesting that offensive innovation should be an open-system process replete with feed-forward and feed-back loops to ensure optimal goal-seeking behavior. That is, it is a cybernetic self-organizing system that has the ability to identify opportunities and threats in its environment and to exploit and adapt to them before

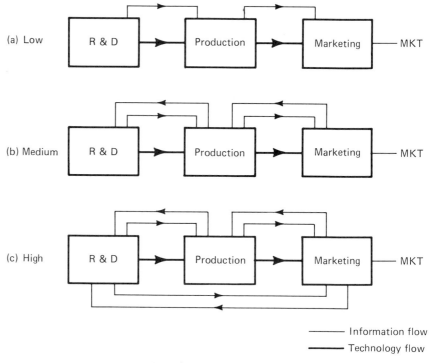

(a) Low

(b) Medium

(c) High

—— Information flow

—— Technology flow

Figure 4.4 Downstream Coupling.

its competitors. As control engineers know, this responsiveness is a function both of the loop-structures and of signal "delays" which occur in the system, since the system must exhibit a swift "reaction time" if it is to beat its competitors. Our earlier chemical process analogy or the innovation chain equation is also useful here. The offensive innovator seeks to complete the process before his competitors do, so he needs to accelerate the chain reaction process which occurs. In chemical engineering, this could be achieved by raising the temperature and pressure. The offensive innovator needs to achieve an analogous effect in the context of a managed institution, which is perhaps why such firms often have "pressure-cooker" climates.

Before concluding this subsection, a word must be said about one specialist type of innovator—the *opportunistic-offensive* innovator. All technological innovation may be viewed as opportunistic, since it involves seizing technicobusiness opportunities when they arise, and the successful technologically innovative company is one which effectively institutionalizes technological opportunism. However, not infrequently, quite radical if not revolutionary innovations are introduced by small companies set up specifically to seize the opportunity. Quite often, such companies are spun-off from parent organizations, such as government, university laboratories, or large companies where the invention to be innovated was made. These companies play an important role in the evolving technosphere since, if successful, they stimulate the adoption of the innovation often against the reluctance of larger companies and sometimes grow into large successful companies. This pattern of opportunistic-offensive innovation is notable in the field of scientific instruments.

4.6
Defensive Strategy— "Follow-the-Leader" Policy

The offensive strategy is fraught with risk and is rarely followed by a company for any period of time. Moreover, larger corporations are unlikely to encourage all their divisions to follow it at any one time, whether they are technologically homogeneous or otherwise. The industrial life cycle model suggested that relatively good profit opportunities occur in the performance-maximizing stage in the cycle when the innovation has been initially marketed, but the dominant design has yet to emerge. It also stressed that new product and company mortality rate can be high at this stage. These two considerations make a *defensive strategy* attractive.

The strengths of the defensive innovator are broadly identical with his offensive counterpart. Such an organization is likely to be equally technology-intensive as its "offensive" counterpart, but with differences of emphasis. The defensive innovator is averse to the risks of being first to market an innovation which is both technologically and commercially unproven. He reasons that if the innovation, when first marketed, looks like a "loser," the defensive innovator loses nothing. In contrast, if it looks like a "winner," provided the defensive innovator swiftly follows the leader with his own (probably improved) version, he stands to win much of the spoils. The successful implementation of the strategy requires an organization with a technological base which continually monitors technology-market opportunities, operates close to the state of the art in its successive phases, and is able to innovate swiftly. The differences of the defensive as against the offensive innovator are as follows:

1. To maintain his close proximity to the fundamental research phase of the state of the art, the defensive innovator may undertake some nondirected fundamental research and, certainly, directed applied research. This research will be of a "defensive" rather than "offensive" character, that is, on topics which match and perhaps duplicate the current research concerns of the offensive innovators. It ensures that the company has the autonomous scientific knowledge to exploit a new innovation once it appears to be successful.

2. Clearly the defensive innovator must be strong in experimental development and design engineering and successive functions in the technology base, since he needs to make up for "lost time" in marketing the innovation, and manufacture a product with superior performance. The only area on which the innovator may be able to place less emphasis is education, training, and advisory services. He may be able to "piggyback" on the success of the offensive innovator in this area because the market will already have had some experience of using the innovation. Such services must still be given emphasis, however, since the offensive innovators will have penetrated only a relatively small part of the market and substantial user education may still be required. Furthermore, dependent upon the technology, the first customers of the offensive innovator may have been "locked into" the latter's version of the innovation, forcing the new innovator to seek out and educate different customers. This has been a notable feature of the computer industry.

3. One area whose function differs significantly between offensive and defensive innovators is patents and licensing. We have already argued that the offensive innovator will seek to establish a strong patent position to protect his technological dominance, a position which the defensive innovator must seek to subvert. The latter will therefore be required to develop his own patents wherever possible and use them as bargaining counters to weaken the dominant position enjoyed by the competitor. Needless to say, a "cost" of delayed market entry to the defensive innovator may be a much lower licensing revenue than the offensive counterpart.

In practice, as was stated at the beginning of this section, most of the larger multiproduct companies are likely to "spread" the risks endemic to innovation and enjoy "economies of scale" in R & D by following a mixed technological strategy, that is, be defensive in some areas and offensive in others. Some may adopt a largely defensive stance but pursue opportunistic-offensive excursions when they spot particularly good opportunities. The European semiconductor industry has followed a largely defensive strategy, in deference to its stronger U.S. counterpart, since 1950. Over a similar period, the French chemicals industry has followed a defensive strategy in deference to its German counterpart.

The organizational structure and climate of a defensive innovator does not differ greatly from that of an offensive innovator. Both may operate with a comprehensive technological base as defined in chapter 3. Since such a firm is following the leader, by definition it has to demonstrate product superiority to "catch up." We have already argued that the offensive innovator must possess high downstream coupling to secure and maintain its competitive position. If anything, this argument is *more applicable* to the defensive innovator from experimental development onwards. His success is dependent upon the innovator's ability to identify and produce improved versions of the innovation—a requirement that places a high premium on superior technological product development, marketing intelligence, and responsiveness.

4.7
Imitative Strategy—"Me-Too" Policy

As an industry matures, a dominant design becomes established and it moves from a fluid to a transitional and then specific state, so other strategic options appear. As was stated earlier, the establishment of a dominant design tends to stimulate, delineate, and coalesce the market so that excellent opportunities are presented for incremental innovations or improvements in the dominant design, based more upon design, reliability and cost considerations than major technological differences.

The imitative company will be development design-, production-, and service engineering-intensive rather than R-intensive. It follows that its "costs" should be lower, except that it may well have to purchase the imitative technology through licenses and "know-how" agreements with the primary innovators. This strategy can be particularly attractive to domestic companies in countries that traditionally lag behind the leading countries (such as the U.S.) in adopting new technology. If a U.S. primary innovator has no corporate presence in a given country, it may prefer to license the innovation to a domestic manufacturer, rather than incur the costs, scarce resource investments, and risks associated with exploiting that market itself. The domestic manufacturer finds the arrangement equally appealing. Since it incurs no R & D expenditures, its direct manufacturing costs should be lower (dependent upon raw material, equipment, and labor costs in that country) and, particularly if it is protected by tariff barriers, profit opportunities may be excellent. Japanese companies have followed this strategy very successfully since World War II, although some could be said to have now moved to defensive or even offensive strategies, via absorbent strategies (see section 4.10).

The imitative company possesses a truncated technological base from design engineering onwards, as shown in figure 4.5. If it does not enjoy a protected market, it will clearly have to be very efficient at this truncated operation. The primary innovators may still be able to produce technological improvements (particularly during the transitional as against specific state of industrial development), based upon their R & D capabilities. The imitative company can compete with only design improvements and lower manufacturing costs. This implies a design-, production-, and service engineering-intensive

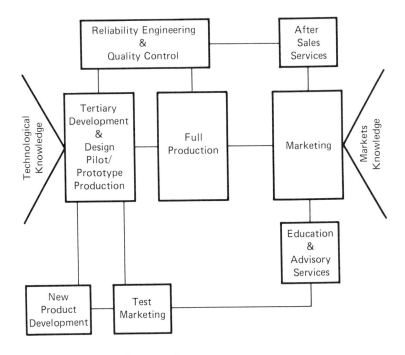

Figure 4.5 Imitative Strategy Base.

company as was stated above. This difference in technological emphasis imposes different requirements on the managerial style and organizational climate of the company. Tasks are typically more clear-cut, since a major emphasis is placed upon the efficient enactment of a specific design concept. Supervision is more directive, individual and group tasks are structured to ensure the efficient dovetailing together of the component activities, and established management techniques like PERT (Program Evaluation Review Technique) may be used. Thus the organizational climates favor efficiency rather than innovativeness.

4.8
Applications Engineering—
Interstitial Strategy

Given the complex and kaleidoscopic nature of the technosphere, the primary innovators are unlikely to seek to satisfy every potential application or occupy every potential market niche for the innovation. A judicious analysis of the primary innovators' strengths, weaknesses, and strategies, combined with a search for unrealized applications, will frequently identify specialist niches that can be

profitably exploited; that is, identify and develop incremental innovations for "new" markets. Such companies are analogous to organisms that adapt to specialized niches in the ecosphere.

In general, the strengths of a firm that identifies and exploits
specialist applications are similar to those following a "me-too"
strategy, with possibly less emphasis on development and more
emphasis on marketing and user services, to relate more closely to
user needs.

Most of the firms in this category emphasize design and development and can be properly described as applications engineering.
However, there is a subcategory of firms that warrants special mention, which may invest substantial effort in primary development
and follow an interstitial strategy. These companies also identify a
niche or interstice in the market, which is too small to interest the
larger primary innovators, but is still quite large and technologically demanding. One of the most notable examples of such a company
is Control Data Corporation (CDC), which was able to market computer systems tailored to users whose needs could not be satisfied by
systems available from the IBM product line. CDC were thus able to
compete effectively in a U.S. computer market dominated by IBM,
without confronting the latter until later when it had built up the
technology-market strength to do so. Its strategy at that time could
be viewed as opportunistic or interstitial. Similar strategies have
been followed by smaller companies in the aircraft industry that
have been unwilling and unable to compete with Boeing, MacDonell-
Douglas, and Lockheed. They may now be observed in the microelectronics industry. Such firms are likely to have managerial styles and
organizational climates that reflect the orientations of offensive and
defensive, as well as imitative innovators.

4.9
Dependent Strategies—
"Branch Plant" Policy

The company following a dependent strategy is characterized, in the
extreme case, by a technology base truncated to production and
marketing and their ancillary functions (see figure 4.6). It is typically a subsidiary or specialist department of a larger firm, or one of the
latter's suppliers. Often national subsidiaries of multinational corporations (MNCs) occupy such positions since this is one way for an
MNC to exploit an innovation in offshore markets. One of the major
preoccupations of Canada is its *branch-plant* economy, as much of
the country's technology is based in such companies, which are
subsidiaries of U.S. and European parent corporations.[8]

Figure 4.6 Branch Plant Base.

4.10
Absorbent Strategies

An impressive feature of Japan's industrial achievement after World War II is the way that nation has avoided developing a branch-plant economy through the pursuit of absorbent strategies by some of its companies. An offshore company, probably with a truncated technology base as shown in figure 4.6, acquires a license or licenses from (typically) an offensive-defensive innovator to exploit innovations in its domestic market. However, rather than acting merely as a passive agent for the manufacturing and marketing of the innovations, the licensee uses the surplus cash flow and market niche it thereby establishes to assimilate the technological know-how and build up its own R & D capability to launch its own performance maximizing and cost-reducing incremental innovations in both its domestic and offshore markets. The sustained successful implementation of this strategy enables a company to extend its technology from figure 4.6 to first figure 4.5 and possibly then figure 3.1, to become an offensive-defensive innovator.

We have now reviewed the repertoire of alternative technological strategies which may be pursued by a firm dependent upon its own strengths and weaknesses, the current stage of the industry in its life cycle, and the segments in the technology-market matrix the firm wishes to exploit. We have also seen that companies must establish technological bases congruent with their adopted strategies, and the profiles of these bases as a function of such strategies are summarized in table 4.1. Before concluding this chapter however, we will comment briefly on traditional and some other (non-technologically) innovative strategies.

Strategy / Function	Offensive.	Defensive.	Imitative.	Applied Engring. /Interstitial.	Branch Plant.
Nondirected Fundamental Research.	M	M	N	N	N
Applied Directed Research.	H	M	N	N	N
Experimental Development and Design.	H	H	N	L/M	N
Advanced Development and Design.	H	H	H	H	L
Pilot/Prototype to Full Production.	M	M	H	M	M
Quality Control/ Product Design.	M	M	H	M	M
Patents and Licenses.	H	M	L	M	N
"After-Sales" Services.	H	M	H	H	M
Education and Advisory Services.	H	M	L	H	M
Long-Term Planning.	H	M	L	H	N

Key: None.
Low.
Medium.
High.

TABLE 4.1 Strategies and Capabilities

4.11
Traditional Strategies

In section 4.2 we contrasted the roles of innovation versus operations managers because most management efforts in an established company are expended in manufacturing and marketing established products. Much business activity whether of low, medium, or high technology encompasses the "harvesting" of profits from established products (or old technology) in established (or old) markets. It is largely characterized by companies in industries which are in the mature stage of their life cycle. One could place the farming and wool-textile industries in this category. Both are very traditional craft-based industries that have displayed a continued capacity to absorb technological changes in work methods. The farming industry after World War II is an interesting example because it is sandwiched between suppliers and customers who have both introduced quite radical technological innovations. Pesticides and fertilizers have increased productivities for farmers, while the development of freeze-drying techniques represents a radical innovation by food processors.

4.12
Other (Nontechnologically) Innovative Strategies

In chapter 2 we pointed out that potential technological innovations must compete for support with other innovative opportunities which a firm may perceive. These other opportunities may not necessarily be based on new technology, but still be entrepreneurially rewarding and may profitably extend the duration of the mature product life cycle,[9] since they often require relatively little front-end investment in new manufacturing plant, marketing, distribution, and servicing efforts. As one very successful technological entrepreneur, Simon Ramo, implies, the successful profitable management of high-technology companies requires a prudent balance between the introduction of innovations and the maintenance and extension of ongoing operations.[10] Thus at any point in time, a company may have numerous opportunities for innovation, *not all of which are technological,* so it is useful to look at examples of these.

First, the sales of existing products may be increased by promotion innovations, distributing innovations, financial innovations, and so forth, as Ford and Ryan suggest.[11] Many such innovations are technologically *cosmetic,* involving only changes in product presentations to enhance customer appeal, with the products remaining unchanged. Such so-called cosmetic innovations are vital competitive tools in consumer expendable industries such as detergents, personal toiletries, food, and so forth, where significant long-term technological innovation does occur (Proctor and Gamble's development of Crest toothpaste incorporating a fluoride, for example), but far less frequently than the short-term cosmetic ones. It is significant to note that Telfer in his remarks quoted at the beginning of this chapter, did not distinguish between technological and nontechnological innovation. Because he is a senior executive in a corporation in the high-competitive food industry (Maple Leaf of Canada), it could be reasonably conjectured that his remarks reflect the views of enlightened management in that industry.

Second, many commercially successful innovations achieve this success through the judicious synthesis of technologically and non-technologically innovative features, in which the nature and role of innovative technology may vary. Two U.S. companies, MacDonald and Benihana of Tokyo, have achieved commercial success through marrying well-established and essentially noninnovative manufacturing management principles (but technologically innovative in their *own* industries) with innovative marketing approaches in the fast-food and restaurant industries, respectively.[12]

Third, the invasion of "new" markets with "old" technology can be highly profitable, provided the company "understands" the new market, and has the necessary access into it. Quite often a company perceives an opportunity for its own generic technological capabilities when technological and economic changes occur in the new market. For example, the Dowty Group, a developer and manufacturer of aircraft undercarriages, recognized that the introduction of face mechanization in the British coal mining industry created the need for mechanized roof support systems to replace pit props. They also recognized that the problems endemic to the design of mechanized roof support systems were generically similar to those endemic to the design of aircraft undercarriages. They therefore applied their generic technological expertise to the new problems, with considerable commercial success.

As we indicated in chapter 2, this process of invasion of one industry by another is, quite frequently, a mechanism for stimulating technological change in the host industry. Sometimes it occurs when companies in the host industry have the capability to exploit the technology themselves, yet view the innovation as being technologically feasible, but economically unattractive. By the 1950s Gillette had the capability of developing a stainless-steel razor blade, but declined the opportunity, because the longer life of the stainless steel blade would reduce the total razor blade market. At the same time Wilkinson Sword Edge had also developed the capability of making stainless steel blades from their generic technological capability in making stainless steel cutting edges. Because they were not then in the razor blade business they had nothing to lose from following a *maverick strategy* and invading this market with a stainless steel blade. Their new product was so enthusiastically received by shavers (including this author), who appreciated its superior performance every morning, that for a short time after its first introduction, supplies of these blades had to be rationed to retailers. The traditional manufacturers (Gillette and Schick) responded quickly with their own stainless steel blades, but by then Wilkinson had established a permanent niche in the razor blade market, through their "maverick" behavior.

The examples of technological invasion cited so far illustrate situations where the generic capability has been developed in one industry and later transferred to another. We have already stressed the pivotal importance of identifying the potential synergies between the technology and the marketplace during the innovation process. This frequently involves identifying a number of potential markets for a proposed new product. Quite frequently, the market finally chosen for the introduction of the new product was not one envisaged at the beginning of the innovation process. In this situation, it is difficult to distinguish between "old" and "new" markets because the introduction of an innovation may well trigger the invasion of one industry by another. This is illustrated by the invasion of the textile industry by the chemicals industry with the introduction of synthetic fibers and, more recently, the invasion of the watch industry by the electronics industry with the introduction of integrated circuits.

Finally, I will cite two examples illustrating that innovative thinking is not just confined to the technological rows of this matrix. In introducing a revolutionary innovation (xerography) into the marketplace, Xerox enhanced its market appeal and its profitability to the company through an innovative pricing approach. Rather than selling or leasing its copiers as fairly expensive capital cost units, it marketed them on a "charge-per-copy" basis. This overcame potential customer resistance because it enabled users to experience the performance benefits of the new technology without a substantial capital outlay. Xerox thereby achieved more unit "sales," and in many cases, a higher "sale-price" per unit, because the net revenue of copying charges derived from a unit exceeded the price that would have been charged in a direct sale. Both Texas Instruments and the Japanese solid-state electronics industry rather similarly exploited the introduction of transistor radios. Because transistors are smaller, cheaper, and more reliable than vacuum tubes, the transistor (as opposed to the vacuum tube radio) could be made smaller, cheaper, and more reliable (requiring little or no servicing). This meant that it could be sold as a "new" product (the personal as opposed to the portable radio) in a "new" market (as a pocket radio to individuals, notably "teenagers," rather than to "homes"). Furthermore, its improved reliability rendered the traditional dealer-retailing network of the radio and TV industry redundant because negligible "after-sales" servicing was required, and transistor radios could be distributed more cheaply through traditional nonspecialist retailing outlets such as drugstores. In both of these examples, the commercial benefits to be exploited from the introductions of revolutionary technologies were enhanced by a novel approach to marketing columns through market definition, distribution, and pricing.

All the examples cited in this section reinforce the point that technological innovations should be judged in the context of all the innovative opportunities available at any given time, and by business entrepreneurial *as well as* technological entrepreneurial yardsticks. Technological innovations (and their innovators) must be prepared to compete with other innovations (and their innovators) to receive shares of scarce company resources and must also be prepared to be judged by business criteria.

References

1. J. A. Telfer, senior vice-president, Domestic Operations, Maple Leaf Mills Ltd., in a presentation on Strategic Planning at the Innovation Canada—1976 Conference.

2. G. S. C. Wills, "The Preparation and Deployment of Technological Forecasts," *Long-Range Planning* 2, no. 3 (March 1970): 44–52.

3. H. Igor Ansoff and J. M. Stewart, "Strategies for a Technology-Based Business," *Harvard Business Review* 45, no. 6 (November-December 1967): 71–83; Christopher Freeman, *The Economics of Industrial Innovation,* 2nd ed. (London: Francis Pinter, 1982), chap. 8.

4. Freeman, *Economics of Industrial Innovation.*

5. Ansoff and Stewart, "Strategies for a Technology-Based Business."

6. Thomas Burns and C. M. Stalker, *The Management of Innovation,* (London: Tavistock Publications, 1961).

7. Tracy Kidder, *The Soul of a New Machine* (New York: Little, Brown & Co., 1981).

8. John N. H. Britton, and James M. Gilmour assisted by Mark G. Murphy, *The Weakest Link* (Ottawa, Ont.: Background Study 43, Science Council of Canada, 1978).

9. David Ford and Chris Ryan, "Taking Technology to Market," *Harvard Business Review* 59, no. 2 (March-April 1981): 117–26.

10. Simon Ramo, *The Management of Innovative Technological Corporations* (New York: John Wiley & Sons, 1980).

11. Ford and Ryan, "Taking Technology to Market."

12. Theodore Levitt, "Production Line Approach to Service," *Harvard Business Review* 50, no. 5 (September-October 1972): 41–52.

5

Forecasting Future
Technology Markets

Any firm operates within a spectrum of technological and market possibilities arising from the growth of world science and the world market....Its survival and growth depend upon its capacity to adapt to this rapidly changing external environment and to change it.

Christopher Freeman, **The Economics of Industrial Innovation**

5.1
Introduction

In the previous chapter I argued that a high-technology company should articulate and implement one or more technological strategies and discussed the alternatives that are available. I indicated that the rational implementation of a given strategy required the development and management of a technological base consistent with this strategy. Many of the leading companies in the high-technology industries pursue a mixture of offensive and defensive strategies over the range of technology-market niches that they occupy which, if they are to succeed, requires them to maintain strong coupling between the technological state of the art and the marketplace. In this chapter we will examine the implication of this requirement in the context of the technological innovation process and review the methods that may be used to perceive and assess the future technological markets for the company.

In chapter 2, I introduced the innovation chain equation, pointing out that commercially successful technological innovation is dependent on the occurrence of a particular chain of events, which reflect the matching of a technological capability to an acceptable social need by entrepreneurial and management actions. We also saw that the timescale of this innovation process can extend over a number of decades. Although this innovation chain was represented in isolation, it does, of course, occur in the context of evolving technological and social milieus. The evolving technological milieu or technosphere reflects a continous process of scientific and engineering inventions and developments embodied into economically competitive new products that satisfy social needs. The technological strategy of the company constitutes an overall framework for evaluating, selecting, and managing the individual innovations in this context. If it is to be effective, this framework must incorporate some perception of the technicoeconomic and sociopolitical environment of the organization over the timescale of the innovations being considered, that is, from a few years to several decades. Since the 1960s various approaches and techniques for perceiving and predicting such future environments have been developed under the general label of *technological and social forecasting* and, in this chapter, we will review some of these techniques that the company may employ to pursue its goals. Because some readers may be either unfamiliar with or have misconceptions about the nature of technological forecasting (TF), we will begin with an outline of its scope and development to date.

71

The Emergence
of Technological
Forecasting:
The Concept
of Future
History

5.2
The Emergence of Technological Forecasting: The Concept of Future History

TF or, more generally, futures studies (FS) is an approach to the study of the future that has burgeoned over the past twenty years although, as with many new management aids, its origins can be traced back much further in time. Both Leonardo da Vinci and Jules Verne, for example, speculated as to future technologies, and H.G. Wells proposed in 1902 that it should be feasible to develop a systematic study of the future, comparable to the sciences.[1]

> *And I am venturing to suggest to you that, along certain lines and with certain qualifications and limitations, a working knowledge of things in the future is a possible and practicable thing.*
>
> *I must confess that I believe quite firmly that an inductive knowledge of a great number of things in the future is becoming a human possibility. I believe that the time is drawing near when it will be possible to suggest a systematic exploration of the future....But suppose the laws of social and political development, for example, were given as many brains, were given as much attention, criticism and discussion as we have given to the laws of chemical combination during the last 50 years—what might we not expect?...*

During the interwar years, political scientists, sociologists, and economists published materials (particularly those warning of the dangers of the Nazi regime in Germany) that could be labeled futures research. However, it was not until after World War II that widespread interest was shown in the idea, initially for military reasons. The East-West cold war and arms race based on nuclear weapons technology made it imperative to explore possible future situations in which a war could break out, in order both to develop cost-effective weapons and (more important) to try to minimize the risk of a nuclear holocaust. These considerations dictated a requirement to explore the political and socioeconomic impacts arising from the development of alternative new weapons systems, as well as to predict developments in military technology itself. There thus existed a practical need to develop methods of technological and social forecasting. By the 1960s future thinking had been applied to other aspects of human affairs and became more "visible" through several notable publications. In 1964 (but based upon work which began there in 1948), Gordon and Helmer[2] of the Rand Corp. published their development of the Delphi method of forecasting (see section 5.10). Also in 1964, Gabor, who won a Nobel prize for his invention of holography, published his *Inventing the Future,* while in 1967 Kahn and Wiener published *The Year 2000.*[3] A body of techniques for systematically studying the future was identified by Jantsch in *Technological Forecasting in Perspective,* and De Jouvenal in *The Art of Conjecture* sought to provide an epistemological basis for the study of the future.[4]

This interest was reinforced by public concern for the harmful impacts of technological innovations on the natural ecosystem. This environmental concern led the Club of Rome to sponsor a global study, using systems dynamics techniques (see section 5.9), which forecasted worldwide food and other resource shortages, pollution buildups, population imbalances, and so forth, over the next hundred years. The controversial results of this study were published under the title *The Limits to Growth.*[5]

At first sight, it might be supposed that TF or FS really consists of traditional business forecasting methods in a new guise. Economists and statisticians have been forecasting the future behavior of important variables and parameters in the economic environment of the firm for a number of decades, and industrial operations research groups frequently perform similar exercises. Furthermore, technological forecasters do use statistical, econometric, and OR techniques to extrapolate technological and economic parameters into the future (as the Club of Rome study cited above illustrates), so TF does exploit techniques common to other disciplines. Despite these commonalities, it is broader in both scope and intent than traditional forecasting methods for the following reasons.

First, traditional forecasting methods are primarily concerned with forecasting key variables and parameters over the short and medium term, from perhaps a few months to five years or so. Because of the timescale of the innovation process, FS must be concerned with forecasting the environment of the firm over a much larger term—typically three to twenty-five years and maybe even longer.

Second, traditional methods are primarily based on statistical forecasting techniques that extrapolate historical data into the future, assuming no underlying economic, social, and technological change—an assumption that may be valid for forecasts up to five years, but is unlikely to be valid beyond that. Technological forecasters do use similar techniques when forecasting incremental and changes of scale in technological capabilities, but they are also concerned with forecasting the technological, socioeconomic, and political impacts of technological changes, either individually or in combination, which perturb the status quo, thus making an extrapolative approach invalid.

These requirements dictate that FS must first adopt a truly multidisciplinary approach (embracing expertise in the social and behavioral sciences, as well as the physical and life sciences, engineering and management), if it is to forecast future innovative opportunities. Second, because it seeks to predict technological changes based upon as yet undiscovered inventions or ideas, it must be able creatively to assess and evaluate technological trends to identify the characteristics, timings, and probabilities of occurrence of future inventions and innovations. These requirements are perhaps best described by using the term *future history* as a synonym for future studies or technological and social forecasting. Since the *Concise Oxford Dictionary* defines "future" as "of time to come,...describing events yet to happen" and "history" as "study of past events, especially human affairs," we must clearly justify the use of this apparent contradiction in terms!

The professional historian, after careful research and scholarship, presents his interpretation of the past events by some form of descriptive analysis. The factual data and other evidence available to the historian are clearly limited by the gaps and biases inherent in the historical recording process, so that his analysis must inevitably be based upon partial information. Moreover the historian is judicious and selective in the use of these facts in developing and propounding his interpretation of the past. As Carr puts it, "The facts speak only when the historian calls on them; it is he who decides to which facts to give the floor, and in what order or context."[6] Since history is a "seamless web" of unfolding events and trends, any competent historical analysis of the recent past should be amenable to extrapolation and informed speculative projection into the future. Such forecasts or projections will lack the formal mathematical rigor of the statistical methods of time series analysis which are used in the short-term forecasting of, for example, sales of an item and, because they must often accommodate the occurrence (or otherwise) of uncertain future events, they can rarely be expressed in terms of statistical confidence intervals. Nevertheless, the historical processes of technological, social, and political change can be projected into the future to provide a forecast of the future technological, social, and political environment for the company, and the opportunities and threats it may face in pursuing its corporate goals. It is in this sense that the term *future history* is used. It provides a "picture" or "scenario" of the future environment for the company, or the future technology-market matrices which the company may wish to exploit and, as such, is a required element in corporate planning. Any corporate planning activity must, either implicitly or explicitly, plan on the basis of some scenario of the future environment for the company. FS is an approach to building up this scenario in as critical and logically consistent a manner as the inherent uncertainties in the situation allow, rather than relying on an intuitive and impressionistic view of the future, or assuming it will be like the present.

For example, an intelligent observer of the semiconductor industry in the early 1960s could have predicted that there was a high probability that within ten to twenty years integrated circuit technology would provide a cheaper, more reliable, and more accurate replacement for the traditional mechanical watch. Had the Swiss watch industry studied "its" future in a critical systematic manner it would have discovered this "picture" and would have taken appropriate measures earlier to meet the threat of invasion of its traditional markets by the electronics industry. The development of scenarios is discussed in some detail later (sections 5.15 and 5.16).

This example illustrates an important observation on the purpose of business forecasting in general made by Peter Drucker.[7] As he pointed out, business forecasting is concerned not with the future itself, but with the *futurity of present decisions* taken by management *today*. It is in this spirit that a high-technology company should explore its futures and apply the TF techniques which we now discuss.

Technology Monitoring

One simple approach is to monitor "signals of technological change."[8] (Had the watch industry done so it could have anticipated the development of the electronic watch.) It is an approach that can readily be used in small companies that cannot afford to employ a full-time forecasting staff and may constitute a first step in the establishment of a TF group in larger companies. It is particularly suited to the technological and market gate-keeping roles described in chapter 12.

Technology monitoring involves reading or scanning a selection of publications that can be expected to provide signals of change. Martino gives a detailed description of the approach and stresses that all relevant environmental aspects (including sociocultural, political, and ecological factors as well as technological and economic ones) should be monitored, and Bright recommends maintaining a journal for this purpose.[9] The major difficulty with the approach is that if too many publications are monitored, continually updating the journal becomes too onerous an exercise, and the "signals" of change may be lost in the "noise" of the large amount of published material collected. Jones and Twiss therefore suggest a less formal method involving collecting xeroxes of extracts from relevant articles, and so forth, in a scrapbook which may be periodically pruned to discard redundant material.[10]

5.4
The "S"-Shaped Logistic Curve

The growth in a new technological capability typically follows an "S"-shaped curve (figure 5.1), which can be roughly divided into three stages. The first stage is of slow initial growth as the new technology has to prove its superiority over existing technologies. Then, once this superiority is demonstrated, a period of rapid or explosive growth follows. Finally, its growth is limited by technological or socioeconomic factors and levels off towards some upper limit.

The commercially successful exploitation of technology often depends upon the astute perception and exploitation of this growth, so forecasters have paid significant attention to extrapolating the growth of the S-shaped curve of a technological capability at some relatively early stage of its "life." In so doing, they have used mathematical functions or models which were originally developed to describe the growth of biological organisms. Thus they have explicitly exploited the conceptual similarity between biological and technological evolutions discussed in chapter 2.

One of these is the Pearl function (named after the American biologist and demographer Raymond Pearl, who extensively studied the growth patterns of organisms and populations):

$$y = \frac{L}{1 + ae^{-bt}}$$

where "y" is the dependent variable whose growth is to be forecasted, "L" is the upper limit to growth, and "a" and "b" are parameters.

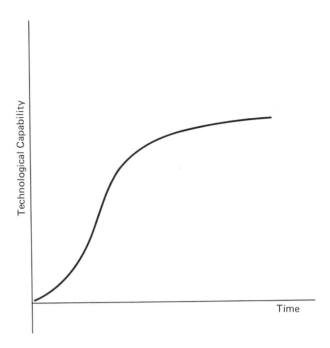

Figure 5.1 The "S"-Shaped Curve.

The Pearl function may be reduced to a linear form by the appropriate logarithmic transformation. If, at an intermediate stage in the "growth" of the technological capability, we have a historical record of "y" values for different years "t," we may plot a least-squares regression to determine the parameters "a" and "b." The Pearl function is then fully determined and may be extrapolated to forecast "y."

An alternative function that generates an S-shaped curve with a longer upper tail is called the Gompertz function (named after an English actuary and mathematician):

$$y = Le^{-be^{-kt}}$$

The Gompertz function may also be converted to a linear form by a double logarithmic transformation and forecasted values of "y" determined by least-squared regression. Both Pearl and Gompertz functions have been applied quite extensively in TF work, and many computer centers have standard software packages to perform the required logarithmic transformations and regression analyses. Thus, provided a reliable data base is extant, the goodness of fit of the data to each model can be readily tested.

Apart from ensuring that one is working with a homogeneous data base (see section 5.8) the practical utility of these curves fitting techniques for generating forecasts is limited by two factors. There is difficulty in setting a value to the upper limit "L" particularly when the technological capability under examination is a synthesis of several subtechnologies. Clearly this is often the case. Martino points out that aero-engine technology is a synthesis of several subtechnologies, so that developing an accurate performance forecast for a given engine type is difficult.[11]

In addition, the inherent "scatter" in the data points may place relatively wide confidence limits on the parameter estimates and subsequent extrapolated forecasts.

5.5
Envelope Curves and Trend Extrapolation

Technological evolutions typically progress through successive substitutions of new approaches. Whereas individual technological capabilities (A,B,C, and so forth) follow the S-shaped curve, each capability is superseded by a technologically superior successor (A by B, B by C, and so forth) so that overall functional performance continues to rise along an *envelope curve* generated by successive technologies (see figure 5.2). The changeover points $(X_1, X_2,...)$ may occur when the performance of the new innovation is *significantly* better than the old one. This is because of the economic, social, and political factors that tend to resist technological change. An industry may have so much fiscal and human capital invested in a given technology that it is unwilling to change to a new one, even when the latter is markedly superior. Indeed, often such a change is imposed by a company outside the industry.

A plot of the envelope curve over a lengthy period of time may yield a statistical trend which again may be extrapolated forwards to obtain a forecast.

Martino (11, chapter 5) cites numerous examples of logarithmic measures of given capabilities against time including:

1. Efficiency of light sources since 1850.
   y = log (lumens/watt) against time t.
   y = -128.71511 + .06851t.

2. Top speed of US combat aircraft since 1909 (bombers & fighters)
   y = log (speed mph) against time t.
   y = -118.30568 + .0640t
   Innovations occurring were closed cockpit, monoplane, all-metal airframe, and the jet engine.

3. Capacity of Random Access Storage/Random Access Time for computer since 1951.
   y = log (random access/storage time, bits/microsec).

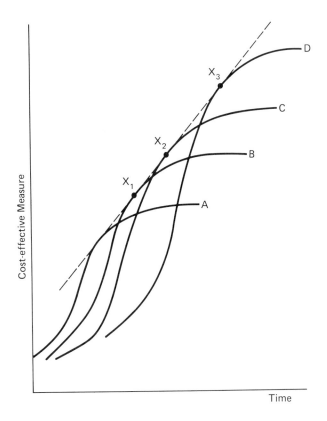

Figure 5.2 Envelope Curve.

$$y = -1080.81958 + .55125t$$

Innovations occurring were vacuum tubes, transistors and ferrite cores, and integrated circuits.

The above results plus the others given in Martino show remarkably high correlation coefficients (over 0.95) so it is hardly surprising that trend extrapolation has been used quite extensively to derive a performance forecast. However, we stress that trend extrapolation must be used intelligently and not naively. This caveat can be illustrated with a fairly obvious example.

As we shall see in the next section, the log-linear trend to combat aircraft speed (example 2 above) is replicated in passenger aircraft transport cruising speeds and a naive extrapolation of trends in the 1960s would have forecast fleets of supersonic transports (SSTs), flying in the 1970s. This "forecast" would have been proven invalid, for the following reasons:

1. Environmental pollution concerns made the SST socially and politically unacceptable.

2. The "energy crisis" and escalating costs of aviation fuel made it less attractive commercially.

3. Cruising speed is not an appropriate performance measure for passenger air transports. Passengers are primarily interested in getting from A to B as quickly and as comfortably as possible. Clearly, for the shorter haul flights, any reduction in flying time will be masked by the time to access and egress the airplane from/to a "downtown" location. Thus "total trip time" is a more appropriate performance measure and an SST will only offer a worthwhile reduction in this measure over the longer transcontinental and the intercontinental routes.

5.6
Precursor Trends

We have just referred to the fact that both combat and passenger aircraft speeds follow log-linear trends, as is illustrated in figure 5.3. More important, the passenger aircraft trend appears to parallel its combat aircraft counterpart, but is displaced about ten years in time. That is, passenger aircraft performance matches combat aircraft performance ten years later. The reason for this crosscorrelation between the two trends is fairly obvious to identify. Combat aircraft performance is achieved by the aerospace industry based upon R &

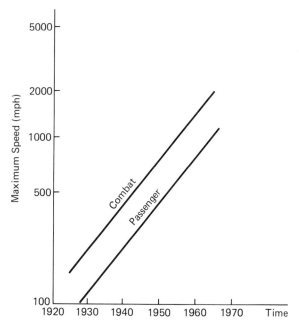

Figure 5.3 Precursor Trends.

D funded by government. This technological know-how can be readily transferred to the development of civilian transport aircraft so two parallel envelope curves are produced.

Although the application of precursor trends to forecasting civilian aircraft speeds broke down in the 1970s for the reasons cited here, it is an approach that can be used in other contexts. Schimdt and Smith found similar associations between aircraft and marine technology, while Kovac used the performance parameters of racing car tires to forecast those for passenger car types.[12] Precursor trends can also be used to forecast the adoption of new geographical as well as application areas. In many instances, U.S.-based multinational businesses develop and sell new products in their domestic U.S. markets before selling and possibly manufacturing them offshore. Mumford found a consistent lag between the growth of sales and aerosol products in the United States and corresponding sales in the United Kingdom which could be used to forecast the latter.[13]

5.7
Fisher-Pry Substitution Model

While S-shaped curves describe the growth of an individual technological capability and envelope curves describe the growth of an overall performance through a succession of technologies, it is often of considerable commercial interest to focus on the specific rate of substitution of technology A by technology B in figure 5.2 over time, that is, the rate and total extent to which technology A surplants technology B in its various product and geographical applications. Many such studies are described in the literature, and a useful review is provided by Linstone and Sahal.[14]

Because technology B substitutes for technology A by a Darwinian competitive process, the growth of adoption of technology B follows the familiar S-shaped logistic curve. A number of workers have suggested explicit models for generating these curves and that due to Fisher and Pry appears to have enjoyed the most widespread acceptance.[15]

From the evolutionary treatment, we can argue that technology B is a new and superior mutation of technology A and that B invades and occupies A's environmental niche(s) systematically "killing" A in the process. Fisher and Pry start with an assumption consistent with this analogy, that the rate of adoption of technology B is proportional to the fraction of the market (environmental niche) still using technology A.

Thus, if "f" is the fraction of the market *already* captured by technology B, "t" is time and "b" is a constant, we have:

$$\frac{1}{f} \quad \frac{df}{dt} = b \ (1 - f)$$

where $\frac{df}{dt}$ is the first derivative of f with respect to t.

Now if t is the time at which technology B has been substituted in half a market (that is when f = 1/2), the above first-order differential equation yields the following solution:

$$f = \frac{1}{1 + e^{-b(t-to)}}$$

Bearing in mind that if f = 1/2 corresponds to y/2 (since the upper limit to f is 1) and writing a = e^{bto}, we find that the Fisher-Pry model yields the Pearl function of section 5.4.

The Fisher-Pry model (and its extensions) is possibly the most widely used of the extrapolative TF techniques. Also it can be combined with precursor trend identification to produce useful forecasts. For example, Jones and Twiss report that a Swedish TV manufacturer successfully used this substitution model together with an analysis of the lag between U.S. and Swedish adoptions of TV innovations to predict the changeover from 21-inch to 27-inch tubes in their market.[16] Their forecasts enabled them to avoid the overproduction and excessive stocking experienced by their competitors.

5.8
Some Pitfalls of Extrapolative Techniques

The extrapolative techniques described above and others described in the now voluminous TF literature have proven useful to technological planners in both government and business. Indeed, as Sahal argues, they may well reflect some underlying laws of technological evolutions.[17] There are, however, some common pitfalls to be avoided if they are to be used effectively. We will now briefly discuss these before moving on to other techniques.

It is important to identify and clarify a common phase in the innovation process when measures are observed. Historical data may well be drawn from a variety of sources, which may report performance measures at different phases of the innovation process. One source, for example, might report the speed of a passenger air transport at the prototype trials phase when it is being tested for air worthiness, while another may report the speed of another air transport when it is in widespread use. The performance specifications of new capabilities are of course quite frequently changed during the successive phases of the innovation process, in the light of the experience gained, so that it is important to use the same innovation phase for all the observations. Also, a forecaster must test the robustness or sensitivity of his forecasts to changes in model parameter values. As we indicated in section 5.4, the extrapolated values obtained using the above models are typically quite sensitive to the parameter values which, in turn, may be subject to quite large statistical standard errors. Thus the forecaster should develop a good "feel" for the robustness and limits of the forecasted values. Needless to say, as with other extrapolation techniques, the data should not be extrapolated further forward than the total time base warrants. A lengthy extrapolation increases the statistically based error, but more important, it is unlikely that the underlying core assumptions will be valid over a longer period of time. This important issue will be discussed next.

This pitfall was illustrated above when we referred to SST forecasts. Implicit in all extrapolative techniques is the *ceteris paribus* assumption—other things being equal or unchanged. Clearly the forward extrapolation of historical trends is only valid so long as the underlying assumptions continue to hold. Apart from the fact that higher airspeeds do not produce commensurate reductions in total journey time, the trend forecast of speeds in the 1970s proved invalid, because the underlying assumptions broke down. The forecaster must never apply extrapolative techniques in a mechanical "unthinking" manner, but must always examine the underlying assumptions to ensure that they remain valid, or modify the approach in light of any new exogenous factors that must be considered. Indeed, Ascher, in an extensive review of the accuracy of many technological and social forecasting studies, argues that many such studies have proved inaccurate through choice of "wrong" underlying assumptions, rather than the use of "wrong" techniques.[18] He argues that the choice of a correct set of core assumptions is crucial to the forecasting task.

5.9
Systems Dynamics Models

The crucial importance of developing forecasting models based upon valid core assumptions suggests that better forecasts might be obtained from models that explicitly simulate the behavior of the underlying innovation process. Given the complexity of this process such simulations must be performed on computers.

Such models seek to represent and quantify the interconnections or causal paths between the various technological, economic, and sociopolitical factors in the innovative process. This typically involves developing a system or set of equations describing the process, identifying critical parameters and variables and their values. The equations and underlying logic of the model, together with the deterministic and stochastic properties of the parameters and variables (including time), are incorporated into a computer program. The program may then be run repeatedly under different assumptions to simulate the dynamic behavior of the process under different conditions. Computer simulations may be performed sequentially over successive time periods (incorporating past-present-future) to generate a forecast.

The most common of these approaches was originally pioneered by Professor Jay Forrester of Massachusetts Institute of Technology and was known originally as Industrial Dynamics and now more generally as Systems Dynamics. Forrester's approach is to view any socieconomic system as a hierarchy of closed-loop feed-back control system (he views the feed-back control loop as the "atomic" building brick of socioeconomc systems), which may be described as a series of mathematical equations. From this basis he developed simulation programs first for individual companies, "Industrial Dynamics," then for cities, "Urban Dynamics," and finally for the earth as a whole in "World Dynamics."[19]

Systems Dynamics leaped to prominence in the TF context through the "Limits to Growth" study performed by Meadows et al. for the Club of Rome.[20] These authors used Forrester's approach to simulate and forecast the impact of population growth, resource consumption, and pollution over the next 100 years. Their forecasts, which predict a virtual breakdown of human society within the next century, aroused widespread controversy in the early 1970s. Much of the Limits to Growth controversy centers on the validity of the core assumptions made in the study, and, particularly as we are concerned with TF in the corporate context, need not concern us here. Systems Dynamics has, however, also been applied to TF at the corporate as well as global level.[21]

Computer simulation approaches not using Forrester's systems dynamics paradigm have been developed by other works. Some have developed models for specific high-technology industries. For example, Allenstein and Probert describe a long-range planning model of the telecommunications business developed for the British Post Office.[22] Although this is a "corporate planning" rather than TF model, one module within the model is explicitly concerned with technological forecasts, since microelectronics and fiber optics or optronics will have radical impact on the telecommunications business over the next few decades.

5.10
Delphi Method

Some of the most widely used and most accurate of the TF approaches are those based upon the Delphi method and its extensions. It was originally developed by Helmer and coworkers for the Rand Corporation.[23] Since the original publication of the method, it has enjoyed widespread use and has been subjected to considerable extension and modification. It derives its name from the famous Oracle of Delphi who was the prime source of future forecasts in classical Greece.

The Delphi method is based upon the not unreasonable premise that the "best" sources of technological forecasts are the opinion of experts in the given technology. That is, the simplest way of making a forecast is ask the experts in the field to do it.

It is undesirable to base a forecast on one "oracle" or expert, however distinguished, so the opinion of a sample or committee of experts is sought. The considered judgment or consensus of a committee of experts provides a viable approach to deriving a technological forecast, but suffers the disadvantage that this consensus may be biased to the opinions of its dominant members. The Delphi approach avoids the disadvantage by requiring members to participate anonymously.

The approach is iterative with each "iteration" called a *round*. In each round a *panel* of experts is interrogated *individually* and *confidentially* (usually by a form containing a series of questions) for their views on the likelihoods and timings of the occurrences of certain hypothetical future events. Lateral interactions between panel members are forbidden, for the reason indicated above. The whole process typically takes four rounds. Thus, the "structured interactions" have the following characteristics: anonymity between panel members, iteration with controlled feed-back, and statistical group response. The Delphi procedure which we now outline is usually conducted by one individual (with whatever support staff that is required) known as the *director*.

Selection of Panel

The selection of panel "experts" must usually be performed on a pragmatic basis. Factors that must be considered follow:

1. Individual "expertise" in area of study. This may require organizational as well as technical knowledge.

2. Commercial and military security. This may limit the organization(s) from which panel members may be selected.

3. Individual availability and willingness to participate—Delphi can be very time-consuming.

4. Avoidance of bias.

Bearing in mind these factors, panel members can usually be selected from peer judgments, literature citations, honors and awards, patents and professional society status. Typically a panel will have from ten to fifty members. We will now discuss the rounds in turn.

Round 1

The first questionnaire is as unstructured and open-ended as possible. It requires panelists to provide a forecast of developments over the period and within the area under study. Panelists may respond in whatever form they think fit, such as by suggesting a sequence and chronology of key events or by providing a narrative or scenario (see section 5.16).

Round 2

From the results of the first round, the director compiles a list of key events that are presented to the panelists as the round 2 questionnaire. They are asked to suggest their estimated dates when each event will occur (including "never" or after a given date). They are asked to provide their reasoned justification for each estimate. From these results, the director prepares a consolidated statistical summary of the distribution of dates given for each event. The median, lower-, and upper-quartile dates for each event are calculated.

Round 3

The third questionnaire presented to the panel includes the above statistical data plus the reasons advanced for the extreme "event dates" and after reviewing this information, panelists are asked to provide revised event dates. Specifically if they suggest dates outside the interquartile range (that is, outside the "group consensus") they are asked to provide the arguments for doing so. Thus they are required to cite facts or factors that they think the other panelists may be neglecting or of which they are unaware. The director then repeats the consolidation of round 2, producing revised statistical distributions and arguments, particularly the "pros and cons" of extreme dates.

Round 4

This round repeats round 3, and panelists provide their *final* estimates of event dates, with full knowledge of the pros and cons that have been advanced.

The director then consolidates the results of this final round which constitute the reported forecast. (More than four rounds may be required before opinions have converged sufficiently to constitute consensus, but usually this number is sufficient). The panel's forecasts are usually presented in the form of the median date and interquartile range for each of the events considered. Figure 5.4 presents a typical "output" from one study.[24]

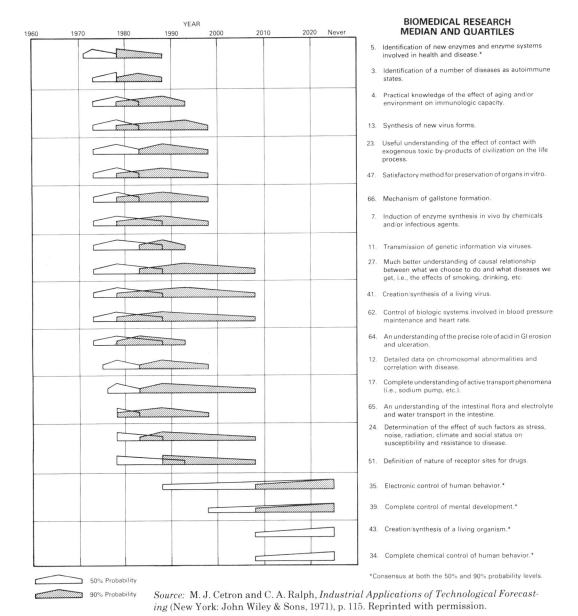

**BIOMEDICAL RESEARCH
MEDIAN AND QUARTILES**

5. Identification of new enzymes and enzyme systems involved in health and disease.*

3. Identification of a number of diseases as autoimmune states.

4. Practical knowledge of the effect of aging and/or environment on immunologic capacity.

13. Synthesis of new virus forms.

23. Useful understanding of the effect of contact with exogenous toxic by-products of civilization on the life process.

47. Satisfactory method for preservation of organs in vitro.

66. Mechanism of gallstone formation.

7. Induction of enzyme synthesis in vivo by chemicals and/or infectious agents.

11. Transmission of genetic information via viruses.

27. Much better understanding of causal relationship between what we choose to do and what diseases we get, i.e., the effects of smoking, drinking, etc.

41. Creation/synthesis of a living virus.

62. Control of biologic systems involved in blood pressure maintenance and heart rate.

64. An understanding of the precise role of acid in GI erosion and ulceration.

12. Detailed data on chromosomal abnormalities and correlation with disease.

17. Complete understanding of active transport phenomena (i.e., sodium pump, etc.).

65. An understanding of the intestinal flora and electrolyte and water transport in the intestine.

24. Determination of the effect of such factors as stress, noise, radiation, climate and social status on susceptibility and resistance to disease.

51. Definition of nature of receptor sites for drugs.

35. Electronic control of human behavior.*

39. Complete control of mental development.*

43. Creation/synthesis of a living organism.*

34. Complete chemical control of human behavior.*

*Consensus at both the 50% and 90% probability levels.

50% Probability

90% Probability

Source: M. J. Cetron and C. A. Ralph, *Industrial Applications of Technological Forecasting* (New York: John Wiley & Sons, 1971), p. 115. Reprinted with permission.

Figure 5.4 Sample Delphi Output.

Creativity Stimulating Techniques

The techniques considered so far for forecasting technological futures have essentially been *passive*. That is, they have sought to estimate the dates of occurrence of future technological events either by the objective analytical extrapolation of the "past" into the "future," or by the subjective performance of a similar exercise by "experts." In none of these approaches has the forecaster sought *actively* to influence the occurrence of the innovation under consideration. In practice, the realization of a new technological capability (such as the successful completion of a lunar mission with astronauts) requires the sequential completion of a number of inventive problem-solving steps that can often be approached in a number of different ways. Clearly the time taken to achieve a new technological capability is a function (among other factors) of the efficacy of attack on this problem-solving sequence. Thus another approach to making a technological forecast is to identify the time scale of the sequence of alternative problem solutions to be derived to achieve the technological capability to be forecasted. That is, the forecasting process is viewed as a creative problem-solving process and techniques which can be used equally well to stimulate group creativity (see chapter 8, section 8.5) can be used in TF. We will now consider two of the more important of these techniques.

5.12
Morphological Analysis and Mapping

Like Delphi, Morphological Analysis is a TF approach which derives its name from the classical Greek world—*morphē* or the Greek word for "Form." As Wills points out, the origins of the approach can be traced back to Plato and Aristotle, particularly in descriptions of the animal kingdom.[25] Goethe, in the late eighteenth century, continued this tradition when he described the study of the structural interrelationships between living things as *morphology*. Darwinian biological evolutionary theory essentially continues this theme. Biological evolutionism is based upon the continued improvement of a species through the retention of advantageous genetic mutations that offer survival advantages to the organism as a whole. Since a technological innovation is achieved in the context of a comparable evolutionary process, it is hardly surprising that the systematic examination of the morphology (or form) of a technological or functional capability can also identify new "mutations" that offer potential performance improvements in the capability.

Like an advanced biological organism, a given capability (for example, an airplane) often consists of a synthesis of a number of separate subcapabilities (engines, fuselage, wings, and tail plane), each of which may be realized in a number of different ways. Morphological analysis performs a systematic exhaustive categorization and evaluation of the possible alternative combinations of subcapabilities which may be integrated to provide a given functional capability. We use the illustrative example of building bricks based upon one developed by Wills.[26]

For purposes of illustration, we will assume that building bricks may be manufactured by various combinations of material and process technologies, producing bricks of different properties and shapes. In this example we have five subcapabilities, each of which may be realized in up to five different ways (see table 5.1)

	1	2	3	4	5
A Material	Natural Clay	Metal	Plastic	Waste Materials	
B Forming Process	Extrude	Mould	Press		
C Bonding Process	Heat	Chemical	Molecular		
D Properties	Opacity	Thermal Insulation	Elasticity	Aesthetic	
E Form	Rectangular	Spherical	Interlocking	Cubical	Aesthetic

TABLE 5.1 MORPHOLOGICAL ANALYSIS OF BUILDING BRICKS.

Given the number of alternative realizations of each of the five subcapabilities we have $4 \times 3 \times 3 \times 4 \times 5 = 720$ different combinations or realizations of building bricks. Thus, a relatively simple product will readily generate quite a large number of alternative realizations. More complex technological capabilities rapidly generate large numbers of combinations of realizations. By viewing table 5.1 as a two-dimensional matrix, we may define a given realization as a sequence of matrix cells: Thus A1 – B2 – C1 – D1 – E1 is a conventional house brick. The National Coal Board in the United Kingdom manufactures house bricks from "discard" or noncarbonaceous waste materials, and this realization is represented by A4 – B2 – C1 – D1 – E1. The thermal insulation brick is represented by A1 - B1 - C1 - D2 - E1. These are three examples of known realizations of technologies. Configurations which have yet to be achieved in practice constitute technological suggestions, such as the combination A3 - B2 - C2 - D1 - E2 , an opaque spherical plastic brick.

Zwickey applied the approach to an evaluation of the jet engine.[27] He defines eleven subcapabilities or parameters, each of which could be realized in two, three, or four different ways. He thus found there were 36,864 possible combinations. Not all these combinations are technologically compatible or feasible, but there remained 25,344 technologically feasible potential realizations. Not all of these have yet been tried out or "invented."

By exhaustively examining all possible combinations, morphological analysis ensures that all feasible (as well as some infeasible) approaches to producing the capability are considered. Also, it often reveals promising unexpected combinations or "new inventions" which otherwise would have remained undiscovered. Wills outlines an extension of the approach in the concept of morphological mapping and explores its potential applications in marketing to identify unsatisfied needs (or unoccupied niches in the technology-market matrix).[28]

5.13
Relevance Trees

At the beginning of section 5.12, we pointed out that the realization of a new technological capability requires the sequential completion of a number of inventive problem-solving steps which can often be tackled in a number of alternative ways. The Relevance Tree provides another conceptualization of the sequence which can complement morphological analysis. This approach essentially involves the drawing of one or more tree diagrams which structure the sequence of technological problems that must be solved in order to reach some overall objective. The basic concept is very simple and its application to the domestic automobile is shown in figures 5.5 and 5.6. The form and detail of a tree are dependent upon the objective(s) sought. Figure 5.5 is a "descriptive tree" which identifies the components comprising an automobile. On the other hand, figure 5.6 represents a "solution tree" in which, in this particular example, some of the possible alternative engine systems are shown. Corresponding solution trees could be developed for other components.

A relevance tree is drawn up in the context of the pursuit of some mission or objective which in turn requires the identification and solution of specific technological problems or tasks. As with morphological analysis, when developing a relevance tree, care must be taken to ensure that

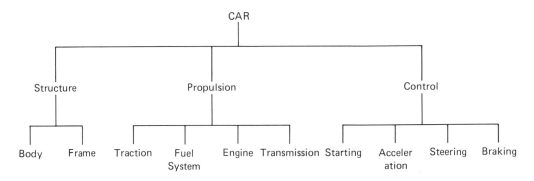

Figure 5.5 Descriptive Tree.

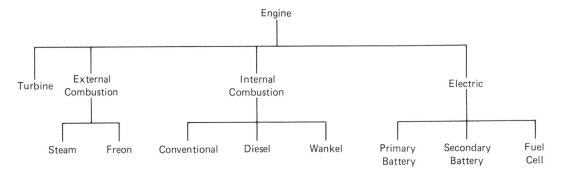

Figure 5.6 Solution Tree.

1. The overall objective/mission is clearly defined and rigorously broken down into sub-missions and tasks.

2. All possible ways of performing sub-missions and tasks are considered exhaustively.

3. A detailed hierarchical tree is developed which rigorously relates tasks and sub-missions to the overall objective. It is vital to perform this analysis in a rigorous and exhaustive manner to ensure that no tree levels and possible problem-solving approaches are excluded.

Honeywell Inc. has extensively applied relevance trees to identify future health care needs and objectives. Given that specific objectives have been set, resources must be allocated to the branches to achieve them. Because such resources (particularly in health care) will almost certainly be "scarce," the question arises as to how best to allocate them between alternative solutions at different levels. Although this allocation may ultimately be done subjectively, it can

90

be facilitated by the development of relevance numbers of weighting factors which may be assigned to the alternative tasks and approaches required to achieve objectives.[29] Relevance numbers is one way one can weight the relative importance of R & D projects, so is of potential value in R & D project selection. Also the comparison of relevance trees with Critical Path Method/Program Evaluation and Review Technique (CPM/PERT) networks and the notion of predecessor or precursor events leads naturally to a role for them in R & D management and control.

At the end of the previous section, we indicated that morphological analysis is valuable in exploring for potential synergies in the technology-matrix. The relevance tree approach can also be used to integrate R & D planning and marketing and indeed overall corporate operations through developing *binary relevance trees*. Richards describes such an approach to integrating a company's marketing strategy into its overall policy, and Hubert describes its application to related corporate and R & D activities in Unilever.[30]

5.14
Cross-Impact Matrices

Until now we have considered the forecasting of capabilities individually. In practice, the probability of development of a given capability may well be dependent on developments in contiguous technological areas. When relevance trees are developed it is readily evident that commonalities can exist between technological solutions of realizations developed in different branches of the tree. The process of rigorously and exhaustively developing relevance trees usually involves developing relevance matrices that identify the commonalities or interactions between activities in different branches of the tree. Beastall[31] developed many matrices to identify the commonalities between differing subsystems and activities in the British Post Office, and Hubert also developed similar matrices in the Unilever example cited above.

Gordon (who helped pioneer the Delphi method) and Haywood,[32] recognized that such cross impacts should be accommodated in Delphi exercises and independently developed the concept of cross-impact matrices for this purpose. Cross-impact matrices are similar to relevance matrices but seek to identify the impact that one technological innovation may have on the probability and time of occurrence of another.

Technological interactions may be direct and of the first order or indirect and of second or higher order (two forecasted events may interact through others rather than directly). The interactions between two forecasts can occur in several ways:

Mode of interaction. One event may enhance or diminish the likelihood of another event. It may advance or delay the second event. It may necessitate or obviate the second event. It may enable or prevent the second event.

Force of interaction. Strong ... weak.

Time lag. Immediacy of influence, length of influence.

A cross-impact matrix (figure 5.7) may be constructed as follows:

1. The events under consideration (E_1, E_2, E_3) are arranged in their expected chronological order in both rows and columns.

2. Consider cell $E_1 E_2$. This represents the impact of the *occurrence* of E_1 on the likelihood of occurence of E_2.

3. Consider cell $E_2 E_1$. This represents the impact of the *nonoccurrence* of E_1 on the likelihood of occurence of E_2. Since events are arranged in expected chronological order E_2 cannot impact on E_1).

Each cell includes the mode (positive or negative), strength, and lag of the interaction. For example, the entries in cell $E_2 E_1$ indicate the nonoccurrence of event E_1 on the likelihood of occurrence of E_2. They indicate that the impact will be five years later and will diminish the probability of E_2 occurring by twenty percent. It is also of interest to compare $E_2 E_3$ and $E_3 E_2$. Comparing these entries indicates that E_2 is largely a precursor event to E_3.

Martino (11, pp. 271–281) describes some conceptual and practical difficulties, as well as applications of the method. Cross-impact matrices must be developed with the same rigor and coherence as other approaches if they are to aid the development of reliable forecasts. Even quite a small list of events can generate a large number of interactions. A five-event matrix (E_1, E_2, ...E_5) can generate twenty

Event occurrence ↓	Impact on:		
	E_1	E_2	E_3
E_1		+ .20 Immediate	+ .70 5 years
E_2	− .30 3 years		0 Immediate
E_3	− .20 Immediate	− .90 Immediate	

Figure 5.7 Cross-Impact Matrix.

cells (excluding the leading diagonal cells) of first-order interactions apart from any second order (say the interaction of event E_1 on event E_3 through event E_2) or higher order interactions. Clearly it is very difficult to evaluate mentally all possible cross impacts with even a quite small number of events, and formulas for doing this have been suggested.[33] They have also been employed fairly widely in Delphi and other TF studies. They can be incorporated into computer simulation models as part of a scenario generation process (see the next section). Jones and Twiss report a variety of cross-impact studies ranging from the original one by Gordon and Haywood used to evaluate the Minuteman missile system, through studies on the future development of Canada to specific forecasts of farm tractor and industrial chemical technology.

5.15
Composite Forecasts—Scenarios

Cross-impact matrices represent one way of evaluating the interactions between events when combining forecasts. One may also wish to make a composite set of forecasts of numerous diverse events over a wide-ranging field. This situation first arose in "social" rather than "technological" forecasting and is particularly associated with the work of Herman Kahn of the Hudson Institute. In *The Year 2000* (coauthored with Anthony J. Wiener) the authors suggest a number of scenarios for the "postindustrial society" of the turn of the millennium.[34] Some of their features are shown in Wills.[35]

In section 5.2 we introduced the notion of "future history" and, in their book, Kahn and Wiener were seeking to write a future history for the turn of the century. As we stated, the professional historian presents his interpretations of the past events by descriptive narrative or "work picture" and the forecaster may also present his conjectures and interpretations in a similar manner. In forecasting, this form of presentation of composite forecasts is known as a *scenario,* or "a word picture of some future time, possibly including a discussion of the events which lead to the situation depicted."[36] To readers who question this reasoning it is pointed out that the reliability or "accuracy" of a historical interpretation of past composite events or a "historical scenario" is clearly dependent upon the extant data and information on the period concerned. Even if the extant data are ample, historians frequently disagree in their individual interpretations or scenarios. However, when one reads interpretations of prewritten and prehuman history (that is, as history shades into anthropology and biology), one cannot fail to be amazed at the extrapolations and interpolations made on the sparse extant data, to produce interpretations or "scenarios" of the distant past.

This suggests that the epistemological credence for a "future" history is at least as valid as that of some "past" history. Bearing in mind that *no attempt* to forecast the future is, in effect, an implicit forecast that the status quo will continue, it is of course impossible to avoid either explicit or implicit forecasting. Therefore, it can be argued that a careful and rigorous composite forecast is the "best" picture of the future available. It also opens up the possibility of "inventing the future" as we shall see shortly.

We have already stated that the pioneering work of Kahn and Wiener was concerned with broad socioeconomic and political forecasting at an international level and so it does not directly concern us here. However their concept and approach of scenario generation was quickly adapted by others to produce forecasts of value to individual firms. MacNulty, who has developed scenarios for numerous corporations in both North America and Europe, is probably the most experienced forecaster in this field and her approach will be discussed shortly.[37] Linneman and Klein report that, in 1977, 150 of the Fortune 1,000 industrial corporations were using multiple scenario analysis in their planning.[38]

The basic approach is to develop several rather than only one scenario:

1. The surprise-free scenario, which we will call scenario A, essentially involves the extrapolation of current trends based upon the assumption that they will continue indefinitely into the future without major disturbance or turbulent events occurring. This can be viewed as the "standard world" for the organization.

2. Alternative scenarios are then developed using a set of internally consistent assumptions which will generate possible plausible outcomes with extreme variations in either direction from the surprise-free scenario. These we will label scenarios B and C and represent two extreme alternative futures for the organization.

3. In practice, it is unlikely that either of the extreme alternative futures will obtain. On the other hand, because of the ubiquity and accelerating rate of technological and social change, it is likely that future actualities will progressively differ from the surprise-free scenario toward one or other of the alternatives, as time progresses. Therefore forecasters generate a number of canonical variations upon the surprise-free scenario, that is, alternative futures which incorporate elements of scenario A with scenario B or scenario C in different combinations. This is also illustrated in figure 5.8.

The organization can then incorporate these alternative or multiple scenarios or futures in its overall technological planning and strategy formulation process. Recalling Peter Drucker's maxim (quoted in section 5.2) that forecasting is concerned with the futurity of present decisions, the organization can develop its technological

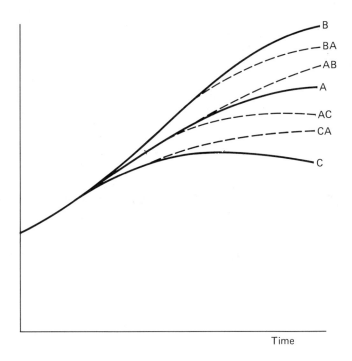

Figure 5.8 Alternative Scenarios.

planning in light of these scenarios and test the robustness of alternative technological strategies against the alternative futures which have been generated.

5.16
Generating Scenarios

As we stated in the previous section, MacNulty has considerable experience in generating scenarios for individual corporations, so I will describe her approach here. Her overall approach is broken down into a number of steps.

Step 1—Develop a Data Base
In all forecasting work it is generally agreed that the historic data should extend at least as long into the "past" as one wishes to project a forecast into the "future." In the long-term forecasts required in scenario generation, MacNulty reasons that this may not always be possible. She does stress, however, that if one wishes to produce a scenario for ten to twenty years hence, one should work with *at least* a ten-year historical data base. This base is both quantitative and qualitative, reflecting the technological, economic, social, cultural, demographic, political, and

market influences that may affect future developments in the technosphere. It also naturally includes internal organization performance measures and trends in production outputs, sales levels, financial measures, employment and turnover levels, and so forth. Quantitative measures are clearly "quantified," and qualitative measures are expressed in a narrative form.

Step 2—Select Organizational Objectives

Recall that in chapter 2, I stated that a potential innovation should be evaluated against a yardstick of the objectives and goals of the organization. Since we also develop scenarios implicitly to provide yardsticks for evaluating such innovations, the scenario generation process explicitly incorporates the consideration of the objectives and goals of the organization.

Steps 3 and 4—Selection/Evaluation of Organizational and Environmental Variables

The organizational and environmental variables whose interactions determine the performance of the organization in pursuit of its objectives are identified. If possible, mathematical and logical models expressing relationships between these variables are formulated.

Step 5—Scenario Selection

Once steps 1–4 have been completed (which is likely to require a fairly substantial effort), appropriate scenarios as discussed in section 5.16 are then selected for detailed development as elaborated below.

Step 6—Development of Scenarios

The "surprise-free" scenario or scenario A is first developed. A matrix (typically with 60–200 rows and/or columns) is drawn up with the row constituting organizational and the columns environmental variables. A team of three to seven people, probably from corporate and senior functional managements, is brought together. They are chosen because they will take any decisions arising from the scenario generation/technology planning exercise. Thus, in effect, they constitute the innovation management steering function of the organization.

A monitor then conducts a lengthy structured "question and answer" session. Each question is of the following form: If organizational row variable "x" increases, will environmental column variable "y" increase, decrease, or remain unchanged? Respondents have a choice between seven answers—high, medium, or low increases; no change; low, medium, or high decreases. If there is a disparity in these responses, discussions follow until a consensus is reached. This consensus (or spread of responses, if no consensus is reached) is recorded. If scenarios are to be generated for more than five years ahead, respondents are asked to provide their answers over future five-year time intervals. All matrix elements of interest are evaluated in this manner, and it typically takes a two-day session for the team completely to evaluate the matrix.

The forecasting team takes the results of this exercise and combines it with the data bases to generate internally consistent and coherent forecasts for the variables of interest. Quantifiable variables are forecasted, then qualitative variables. The output of the exercise is a narrative surprise-free scenario plus graphical forecasts of the quantifiable variables and a statement of the assumptions on which it is based.

The alternative scenarios (scenarios B and C and any canonical variations) are developed by the forecasting team alone, without the top management team. Apart from reducing the demands on scarce top management time, this procedure ensures that the internally consistent assumptions which premise the generation of scenario B and C, and so forth, are maintained throughout the matrix analysis process. The top management team has been less intimately involved in the forecasting exercise, and there is a greater risk that some of its members may unconsciously violate these assumptions in their responses. Otherwise these scenarios are generated in a narrative form as for scenario A.

Step 7—Analysis of the Implications of the Scenarios

The most likely outcome of the above exercises will be that the organization mainly meets its objectives in the surprise-free scenario, but fails to do so with one or more of the alternatives. In this case the objectives and underlying assumptions are reexamined to identify why the particular objectives will not be met. It may then be possible to modify the objectives so that they should be met in their modified form. Alternatively, contingency plans are devised that enable the firm to modify its objectives if one of the alternative scenarios appears likely to occur. At the same time, a procedure for monitoring the key variables which differentiate between the surprise-free and alternative scenarios is initiated. These contingency plans are then implemented if, in the future, changes in these variables indicate that an alternative scenario is emerging.

MacNulty's framework provides a realistic, if time-consuming, procedure for scenario generation, but Linneman and Klein indicate that corporations condense it in practice.[39] Most importantly, it requires a sustained commitment from top management if it is to be effective, and this feature is briefly discussed again in section 5.20. By employing her procedure (possibly in an abbreviated form), the organization can develop its technological strategies and test their robustness against the alternative futures generated as suggested at the end of section 5.16.

5.17
Scenarios and Inventing the Future

In section 5.2 we pointed out that one of the books that stimulated an interest in future studies was Denis Gabor's *Inventing the Future*,[40] in which he introduced the notion of alternative futures or scenarios for a social group from which it should select or "invent" the future it desired. Once a firm has developed a number of alternative scenarios, it can pursue technological plans or strategies that it seeks to create or "invent" the alternative future or scenario that it prefers. This notion is illustrated conceptually in figure 5.9.

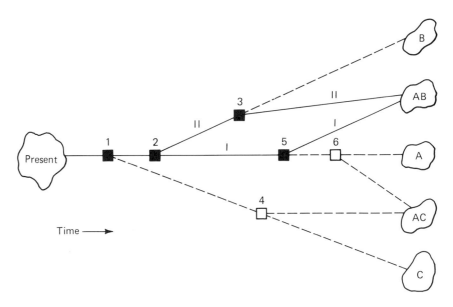

Figure 5.9 Inventing the Future.

The present state and environment is depicted at the left-handed side of this figure with the alternative scenarios or futures depicted on the right-hand side. The continuous and dotted lines notionally represent the alternative paths through time that the firm must follow in order to realize each of these alternative futures. These futures may have common paths initially, but as time progresses nodes are reached where they diverge. These nodes represent decision points where the firm must choose which path it wishes to follow or which future it wishes to invent. If the firm chooses scenario AB, that is, invent that "future," it must follow one of the continuous paths shown in the figure and make the appropriate decision at the shaded nodes to ensure that it does so. On each of the continuous paths the sequence of decisions made at the shaded nodes can be viewed as a technological plan or strategy. Note that we have drawn two paths (I and II) which the company may follow to invent its preferred future, that is it has two alternative technological strategies, I and II. The strategy chosen will depend upon the company's estimates of the probabilities of each of the scenarios occurring, the expected benefits which each offers, and the company's attitude towards risk.

Readers may object to this view of inventing the future on the grounds that it is both naive and utopian. Clearly most companies can exercise only a very limited influence on the future environment in which they operate, which is continually subjected to uncertainties and turbulences over which they have little control. Running most businesses is more like sailing a small boat, in which continual adjustments and changes may have to be made in the light of changing sea and wind conditions, rather than "driving" a large ocean-going ship. However, provided the organizational and environmental variables are judiciously selected in the process of scenario development described in the previous section, the company may define a comparative few which are relevant to its operations, and it is the alternative "states" of these that constitute its alternative scenarios. (Rather like a biological organism which evolves a set of senses which are relevant to its behavior and survival in the environmental niche it has "selected"). It may then be able to exercise a sufficient influence on these variables significantly so as to influence future developments in the direction it seeks. Most readers will accept that Texas Instruments has been one of the most successful high-technology organizations since World War II, and that corporation provides a graphic illustration of the notion of inventing the future. In the 1950s when the semiconductor industry grew rapidly based upon the manufacture of successive generations of discrete semiconductor devices, most companies concentrated on germanium technology. Although it has inferior properties to silicon, most companies chose to develop and manufacture germanium devices because to do so was technologically easier. Texas Instruments, however, committed itself to silicon technology because its technological management believed the advantages of silicon outweighed its disadvantages. Thus they sought to invent a "silicon" future for the company. The developments in integrated circuit technology from the late 1950s onwards did produce a semiconductor industry primarily based upon silicon.

5.18
The Validity of Technological Forecasts

We have now completed our review of the more frequently employed TF techniques. This overview does not claim to be exhaustive, and it is recommended that readers study the specialist texts and journals on the subject if they wish to pursue it in more depth. However, all readers will be concerned with the accuracy and efficacy of TF efforts to date, so it is worthwhile to comment briefly on this matter.

First of all, it is important to distinguish between efficacy and accuracy in TF. The purpose of TF is to aid technological planning and decision making and should be viewed in that context. Its ultimate purpose is to promote "efficacious" forecasts in the sense of promoting "good" technological decision making, rather than "accurate" forecasts. Thus it is possible to distinguish between what can be labeled "passive" and "active" forecasting studies.

1. In a passive study the criteria of efficacy and accuracy coincide. Its purpose is to forecast the behavior of a technological function or capability as accurately as the situation allows.

2. In an "active" study, the forecasting process seeks to promote a "self-fulfilling" or "self-defeating" prophecy. Ideally, a self-fulfilling forecast will stimulate a sequence of technological decisions which will "make" the forecast "come true" (particularly if it is performed as an attempt to invent the future). In contrast, a self-defeating forecast will stimulate a sequence of technological decisions which will ensure that the forecast fails to come true. In both situations the value of the forecast must be judged by its efficaciousness in triggering the desired sequence of technological decisions. The well-known *Limits to Growth* study quite dramatically illustrates this argument. The assumptions and methodologies used in that study have been criticized so that its "accuracy" is, to say the least, questionable. It could, however, be viewed as an efficacious example of self-defeating forecasting. The prestige of its sponsoring body, The Club of Rome, and the dramatic "doomsday" scenario published, focused worldwide attention on the dangers of untrammeled energy consumption, population growth, and pollution and helped make them "political issues." The public concern which it aroused has stimulated governments to be more actively concerned with these dangers than they would have been had it not been performed and published.

Having pointed out that the criterion for evaluation is dependent upon its contextual use, I must also note that credibility of a forecast (whether extrapolative, self-fulfilling, or self-defeating) in the eyes of the technological decision makers is dependent upon the credibility of the methodologies and techniques used to generate it. One of Ascher's overall conclusions is that a major determinant of forecast inaccuracies is the validity of the core assumptions of the forecasters; that is, the validity of their insights into the underlying social, economic, political, and technological factors and trends which are interacting to produce the outcomes being forecast.[41] He suggests that forecasts based upon Delphi exercises which identify the consensus of expert opinion may be superior, because "experts" are, by definition, capable of predicting from the best set of core assumptions.

5.19
Organizational Location of the TF Function

At the beginning of this chapter it was suggested that a TF function should perform the technological futures studies required for corporate technological planning in the light of the characteristics of the innovation process and the alternative technological strategies discussed in chapter 4. It was also pointed out earlier that TF was a significant corporate activity in 1966 when Jantsch reported that 500-600 of the largest U.S. firms were engaged in such activities.[42] The effective deployment of TF in corporate technological planning depends upon the way it is set up and its location in the organization. Jones and Twiss devote a whole chapter to this issue and we shall only make some very general comments here.[43]

Fusfield and Spital cite GE, Monsanto, Whirlpool, Cincinnati Milacron, and Goodyear as firms that have successfully developed technology forecasting and planning programs.[43] They suggest the following three requirements for effective technology planning and forecasting.

The first is a formal integrating mechanism. TF requires the part-time participation of diverse members of the corporate technological base, and some such individuals may assign it a low priority as compared to their primary work roles. It is vital, therefore, to assign specific responsibility and accountability for planning/forecasting to an integrative group. This group may well constitute the nucleus of the innovation management function discussed in chapter 4.

The second requirement is that TF must be designed and geared to support planning. This implies three distinct but interrelated requirements. First, forecasts must be derived and presented in a format that satisfies planning needs. The success of Monsanto's planning efforts is partially attributed to its forecasters' ability to identify potential planning problems and to provide forecasts that explicitly address these problems. At Whirlpool, results are presented in the form of "what this means to Whirlpool." Second, the cycle time of forecasting must be synchronized with the cycle time of the planning process which, in turn, is dependent on the lead time of the innovation process. This planning horizon can, of course, vary from six months for companies following an "applications engineering" strategy to upwards from a decade for companies following offensive/defensive strategies. In some industries, characterized by a very rapid rate of technological change, forecasting and planning effort may be required over several increasing time cycles from a year upwards to match this rate to ensure that such companies can respond rapidly to environmental changes. Third, forecasting must be comprehensive and supportive to planning efforts and must include assessments of competitors' future technological capabilities.

The third requirement is that TF must enjoy top management support. The successful applications of TF in Monsanto, Whirlpool, Cincinnati Milacron, and Goodyear have been particularly attributed to support from top management. This is best achieved by involving senior management in the forecasting process, as illustrated in MacNulty's approach to scenario generation described in section 5.16.

Given the characteristics of the innovation chain equation and the above considerations, we see that the effective use of TF requires participation and support throughout the corporate technological base. In chapter 4, we suggested that firms incorporate innovation

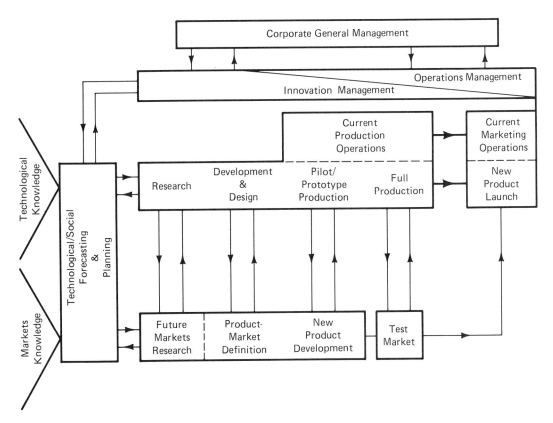

Figure 5.10 Location of TF Function.

34. Kahn and Wiener, *The Year 2000.*

35. Wills et al., *Technological Forecasting,* pp. 163-64.

36. Martino, *Technological Forecasting,* p. 267.

37. C. Ralph MacNulty, "Scenario Development for Corporate Planning," *Futures* 9, no. 2 (April 1977).

38. Robert E. Linneman and Harold E. Klein, "The Use of Multiple Scenarios by U.S. Industrial Companies," *Long-Range Planning* 12, no. 1 (February 1979): 83-90; Robert Linneman and Harold E. Klein, "The Use of Scenarios in Corporate Planning—Eight Case Histories," *Long-Range Planning* 14, no. 5 (October 1981): 67-77

39. Ibid.

40. Denis Gabor, *Inventing the Future.*

41. Ascher, *Forecasting an Appraisal for Policy-Makers and Planners.*

42. Jantsch, *Technological Forecasting in Perspective.*

43. Jones and Twiss, *Forecasting Technology Planning Decisions,* chap. 4.

44. Alan R. Fusfield and Francis C. Spital, "Technology Forecasting and Planning in the Corporate Environment: Survey and Comment," in *Management of Research and Innovation,* ed. Burton V. Dean and Joel L. Goldhar (TIMS Special Studies in the Management Sciences, 15, 1980), 151-62.

6

Environmental
Concerns and
Technology Assessment

*If you talk to enough people about technology assessment,
the concept begins to sound as marvelous as motherhood.
Except in this case, nobody knows how to get pregnant.*
Nina Laserson, "III—Technology Assessment at the
Threshold," **Innovations,** *no. 27 (January 1972)*

6.1
Introduction

The publication of Rachel Carson's *The Silent Spring* in 1962 could
be said to have first aroused public concern for environmental pollution.[1] This concern grew throughout the 1960s and early 1970s,
stimulated by the 1973 energy crisis and by publication of other
books such as *Limits to Growth* discussed in the previous chapter,
and Barry Commoner's *The Closing Circle.*[2]

The cumulative impacts of successive technological innovations
can be viewed in the total context of technological evolution. Until
the advent of the first industrial revolution in the late eighteenth
century, human culture (including technology) was primarily based
upon agriculture and had relatively little impact upon the "natural"
environment or the global ecological system or ecosphere. The cumulative effect of the growth of industrialization, human population,
and the combustion of fossil fuels, and so forth, over the past two
hundred years is now believed to be modifying the ecosphere in a
potentially dangerous and irreversible manner. One dramatic manifestation of such danger is the greenhouse effect—the impact that
growing concentrations of carbon dioxide in the upper atmosphere,
produced by the combustion of fossil fuels, may have on the amount
of solar radiation reaching the earth's surface and/or the amount of
heat radiated back into space. Depending upon the overall balance

106

between these two conflicting effects, the temperature of the earth's surface may be slowly rising or falling, and quite small changes in either direction could induce radical changes in the global climate patterns, sea level, and so forth.

The notion of environmental regulation can be traced back some time. The burning of sea coal was banned in English cities in the thirteenth century,[3] and public health authorities have been concerned with environmental health hazards for many years, certainly since the last century when improvements in public water supply and sewage considerably reduced morbidity and mortality from infectious diseases. Furthermore, during the 1950s and early 1960s, "clean air" legislation in Britain, which banned the use of house coal in domestic fireplaces and restricted the amount of smoke that could be released into the atmosphere from industrial plant chimneys, did much to reduce "smog" in British cities. These regulations were enacted before the "environment bandwagon" developed, and serious political and legislative attention was paid to the broad assessment of the impact of technological innovations.

First, it should also be stressed that technology assessment in a traditional sense has always been performed by companies in evaluating new capabilities out of commercial prudence. However, such assessments have been based on the *primary* impact of the innovation on the company's operations and its markets. Until recently, companies have been under limited legal responsibility to assess the *secondary and higher order* impacts (particularly on third parties) of innovations. The environment was treated as a "free good," like the "commons" in medieval agrarian society. Polluting this environment was not viewed as antisocial, because it was merely exercising a citizen's right to "free grazing" on the "commons." It has now been realized that the natural environment, like the commons, is a semirenewable resource that can be destroyed by overpollution, just as common land can be destroyed by overgrazing.

Second, it should be stressed that (up to the present time at any rate) the primary and secondary benefits of most technological innovations far outweigh their secondary disadvantages. Human social and technological evolution is inextricably linked and it is platitudinous to observe that our present high state of material well-being is primarily based on technology. However, since a law of "diminishing psychic and social returns" may apply to technological innovations, society must critically examine further innovations to ensure that benefits outweigh problems. In fact we can adapt Maslow's hierarchy of needs in this context. This implies that a society's evaluation of the potential benefits and disadvantages of a given technological innovation is subjectively dependent upon its unsatisfied needs. Thus a "poor" economy might tolerate an innovation that significantly increases environmental pollution because these disadvantages are outweighed by the economic development benefits and the increase in employment it provides. In contrast, a "rich" economy might reject the same innovation on the grounds that the environmental pollution outweighs the marginal contribution of the innovation to the economy and employment opportunities.

On several occasions already, we have quoted the example of the fate of the Concorde and the SST to illustrate this fact. Although this is an extreme example, nowadays, some evaluation of the environmental impacts as well as the commercial expectations of a prospective technological innovation must now be made in many industries, before substantial sums are invested in its continued developement. These environmental impacts and technology assessments must typically be integrated within the technological planning and R & D project evaluation procedures of organizations, and so we will consider such assessments in this chapter.

6.2
The Technology Assessment Process and Environmental Impact Statements

Some high-technology industries, such as the pharmaceutical and aeronautical industries, have traditionally been subject to formal requirements for testing potential innovations. Moreover, the public environmental concern is only one facet of a wider concern for the safety of new technological products. The thalidomide tragedy and Ralph Nader's *Unsafe at Any Speed*[4] cast doubts upon the reliability of drug and automobile testing and safety standards, so that from the 1960s onwards, governments in most developed countries enacted new laws to assess and regulate prospective technological innovations to reflect public concern for safety standards, environmental pollution, and energy conservation. Some potential innovations may have quite diffuse impacts and unexpected side effects which affect various interest groups in society-at-large. Therefore, the assessment process must be a political process which admits adequate avenues of expression to all parties or constituencies which may be affected. Also, once such innovations are launched, their impacts must be monitored for any unexpected undesirable side effects, analogous to those used before and after the introduction of a new drug.

Technology assessment (TA) should be performed at two levels. First at societal level, where its broadest impacts should be evaluated. This exercise often requires resolutions of conflicts of interests between various constituencies, so it becomes a political process. It should be performed by one or more public agenices with no vested interest in its outcome. Second, the technology should be evaluated at the company level, based upon technioeconomic factors, but giving due consideration to the above broader impacts. Both levels of assessment should be performed interactively with the regulative agencies and company collaborating in joint evaluation and testing programs to assess actual and potential impacts.

109

The Technology
Assessment
Process and
Environmental
Impact Statements

Thus TA should be viewed as a technicoeconomic process embedded in a broader political process. Several writers have treated TA as political process. For example, Gibbons[5] views it as an extension of Etzioni's "mixed-scanning" paradigm of societal decision making,[6] while Bozeman and Rossini[7] view it in the context of Allison's analysis of the resolution of the Cuban missile crisis.[8] Although the Science Council of Canada has sponsored several studies which recognize TA as a political process, many researchers place more emphasis upon its technicoeconomic aspects. This is apparent from reading one edited text on the subject[9] where, for example, Strasser exemplifies the latter approach in his assessments of automobile emission controls, computer-communications networks, enzymes, sea farming, and water pollution.[10] Also all of the above citations focused on TA at the societal rather than corporate level and, since we are concerned with the latter, they are not our direct concern here. Most Western countries have enacted laws that require individual companies to prepare an Environmental Impact Statement (EIS) before certain technological initiatives are approved. The terms *Technology Assessment* and *Environmental Impact Statement,* among others, have tended to be used loosely and interchangeably in the literature. For exposition purposes here we view EIS as a product of the company level TA process. This is broadly consistent with a distinction made by Waller.[11]

The EIS constitutes the basis for the consultative process between company, regulative agency, and other parties which is integral to the assessment procedure. It takes the form of documentation examining the full ramifications of the proposed innovation. Its scope and detailed content are obviously dependent upon the specific technological project under consideration. Typical projects in the private sector might be the building of new chemical, oil, power, or steel complexes which would require different detailed assessments from a water resources project in the public sector. Assessments would also be different for the development of a new product such as a pesticide, plastic, or food additive as opposed to a new plant. Such differences are no more marked than in TF, so it seems reasonable to expect that, as with TF, generally applicable techniques for preparing EIS's should have been developed. Moreover, any comprehensive technological forecast must include an impact assessment, so corporate TA is really part of corporate TF. For example, the Business Planning Group of Bell-Canada, which is the futures studies function for that organization, performs both activities.[12] One can also expect TF and TA to share a common body of techniques. This is indeed so. Coates and Linstone et al. have written extensive reviews of TA modeling which include TF approaches and are published in a TF journal.[13]

It is not surprising, therefore, that the most common TA approach reported is to generate an environmental impact matrix derived from the cross-impact matrices of TF, as suggested by Gordon and Becker.[14] A major difficulty in preparing a comprehensive and exhaustive environmental impact matrix is its herculean nature, given the innumerable possible ramifications of technology impacts from its primary, secondary, and higher-order effects. This can be seen in the approach of Leopold et al.[15] which generates a matrix with 8800 cells. Their basic approach has been applied quite extensively in the U.S. and is reflected in California state planning requirements. It was also applied in Canada in a preliminary study of the environmental impacts of the James Bay Development Project. Both the Battelle Institute[16] and the British government's Department of the Environment[17] have reduced the matrix sizes to under 1000 cells, but these still remain unduly large for corporate use. Fischer and Davies[18] suggest a TA framework that reduces the sizes of impact matrices though the initial screening out of relatively minor impacts, and the MITRE corporation provides a broadly similar approach.[19]

The MITRE framework is a seven-step process that is illustrated in figure 6.1:

1. Define the assessment task
2. Describe the relevant technologies
3. Develop the state of society assumptions
4. Identify the impact areas
5. Carry out a preliminary impact analysis
6. Identify possible action options
7. Carry out a complete impact analysis

Day[20] and Feldman[21] describe the application of this framework to an assessment of the impact of computer-aided instruction (CAI) in colleges, performed by the Business Planning Group of Bell Canada. The group used the Delphi method and trend analysis. After identifying a larger preliminary impact matrix, they were able to condense it to that shown in figure 6.2. Impacts are measured on a seven-point scale from -3 (large negative impact) to +3 (large positive impact).

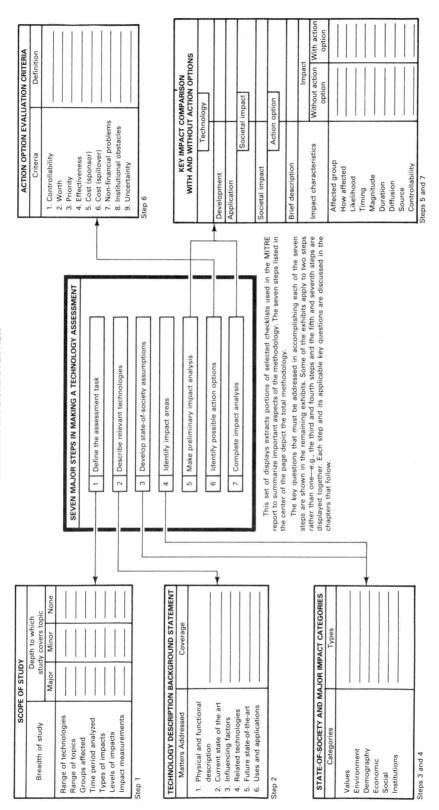

**TECHNOLOGY ASSESSMENT:
A METHODOLOGICAL OVERVIEW**

SCOPE OF STUDY

Breadth of study	Depth to which study covers topic		
	Major	Minor	None
Range of technologies			
Range of topics			
Groups affected			
Time period analyzed			
Types of impacts			
Levels of impacts			
Impact measurements			

Step 1

TECHNOLOGY DESCRIPTION BACKGROUND STATEMENT

Matters Addressed	Coverage
1. Physical and functional description	
2. Current state of the art	
3. Influencing factors	
4. Related technologies	
5. Future state-of-the-art	
6. Uses and applications	

Step 2

STATE-OF-SOCIETY AND MAJOR IMPACT CATEGORIES

Categories	Types
Values	
Environment	
Demography	
Economic	
Social	
Institutions	

Steps 3 and 4

SEVEN MAJOR STEPS IN MAKING A TECHNOLOGY ASSESSMENT

1	Define the assessment task
2	Describe relevant technologies
3	Develop state-of-society assumptions
4	Identify impact areas
5	Make preliminary impact analysis
6	Identify possible action options
7	Complete impact analysis

This set of displays extracts portions of selected checklists used in the MITRE report to summarize important aspects of the methodology. The seven steps listed in the center of the page depict the total methodology.

The key questions that must be addressed in accomplishing each of the seven steps are shown in the remaining exhibits. Some of the exhibits apply to two steps rather than one—e.g., the third and fourth steps and the fifth and seventh steps are displayed together. Each step and its applicable key questions are discussed in the chapters that follow.

ACTION OPTION EVALUATION CRITERIA

Criteria	Definition
1. Controllability	
2. Worth	
3. Priority	
4. Effectiveness	
5. Cost (sponsor)	
6. Cost (spillover)	
7. Non-financial problems	
8. Institutional obstacles	
9. Uncertainty	

Step 6

**KEY IMPACT COMPARISON
WITH AND WITHOUT ACTION OPTIONS**

		Impact	
		Without action option	With action option
Technology			
Development			
Application			
Societal impact			
Action option			
Societal impact			
Brief description			
Impact characteristics			
Affected group			
How affected			
Likelihood			
Timing			
Magnitude			
Duration			
Diffusion			
Source			
Controllability			

Steps 5 and 7

Figure 6.1 MITRE Framework. Reprinted with permission

If this action occurs... ...then this impact is more (+)* or less (−) likely to occur:	Prior probabilities	Actions of government		Actions of colleges		Actions of professors		Actions of industry	
		New types of copyright protection are legislated	A common author language is developed	Disadvantaged students without normal qualifications are admitted	Professional incentives for programmed materials are created	CAI is used mainly as a supplement to instructors	CAI is used mainly as a substitute for instructors	CAI programmer consultant services are instituted	Cooperative projects involving colleges and industry are instituted
		.7	.5	.7	.5	.7	.3	.7	.7
Impacts on colleges — There will be a general improvement in the quality of education	.5	0	+1	0	+2	+1	−1	+1	+2
There will be more dissemination of CAI programs to many colleges	.7	+1	+2	+1	+2	+1	+1	+1	+1
The actual cost of CAI software will increase	.5	+2	−1	0	0	0	0	+1	0
The actual cost per student of instruction will increase	.7	+1	−1	+1	0	+1	0	0	0
Impacts on professors — The ability to cope with poorly—prepared students will increase	.7	0	+1	+1	+1	+1	0	+1	+1
The student–instructor relationship will improve	.5	0	0	0	+1	+1	−2	0	+1
Faculty will use the CAI concept more efficiently	.7	0	+1	+1	+2	+1	0	+1	+1
Segments of published material will be directly quoted in CAI programs	.7	+1	0	0	+1	0	+1	0	0
Impacts on students — Gifted students will be able to finish their studies more quickly	.7	0	+1	0	+2	+1	+1	0	+1
There will be a decrease in the drop-out rate of disadvantaged students	.5	0	+1	0	+1	+1	0	0	+1
The education process will become more impersonal	.3	0	0	0	0	0	+2	0	0
Student unrest will develop to a greater degree	.3	0	0	0	0	0	0	0	−1
Impacts on industry — The future work force will be better suited to industry needs	.5	0	+1	+1	+1	+1	0	+1	+2
There will be new CAI education—related business opportunities	.7	+1	+1	+1	+1	+1	+2	+1	+2
There will be a lack of patent and copyright protection	.5	−2	0	0	0	0	0	0	0
There will be a change in the relationship between schools and industry	.5	+1	0	+1	+1	0	+1	+1	+3

*A positive sign (+) indicates the effect of the occurrence of the development will be to increase the likelihood of occurrence of the subsequent development and a negative sign (−) indicates the converse. The strength of the impact is indicated by the following code: 0 = No impact, 1 = Minor impact; 2 = Strong impact; and 3 = Very strong impact. (Source: Enzer).

Source: Lawrence H. Day, "Technology Assessment in Bell Canada." Testimony before the Technology Assessment Board, United States Congress, Washington, D.C., 10 June, 1976. ☉ 1976 Bell Canada. Reprinted with permission.

Figure 6.2 Bell (Canada) Impact Matrix for CAI.

6.3

113
Organizational
Framework
for Conducting
Technology
Assessments

Organizational Framework for Conducting Technology Assessments

We have reasoned that some procedure for conducting technology or environmental assessments should be incorporated into the innovation management process of companies developing potential hazardous and/or polluting products or processes. We now consider how such a review might be incorporated into this process.

Unilever is a typical example of a large offensive/defensive innovator, which operates a central research and engineering function and about 500 constituent companies, including some with product development functions. Its major products are foods, detergent, and toilet preparations; paper; plastics; chemicals; and animal feeds, and so it generates potentially hazardous and/or polluting products and processes. Unilever's approach to protecting the consumer and the environment from potential hazards is outlined in Philp.[22] Corporate responsibility for ensuring the safety of its products is held by the director of research, to whom reports an environmental safety officer supported by an Environmental Safety Division (ESD). This division employed forty scientists and 170 support staff in 1974 and also has access to the full support services of the major Unilever Research Laboratories.

The corporation has an established procedure for examining proposed innovations and clearing them as environmentally "safe," which is integrated into the R & D process. When a research project has proceeded to the stage of an intended new product, the intention is committed to paper and the tentative proposal with technical specifications, support data, and analyses is submitted to the "clearance group" of the research division for a "clearance forecast." This group consists of an independent team of scientists, uncommitted to the project. It is vital that a proposal should be subject to an independent internal review and this consideration could be readily incorporated in the evaluation process described in chapter 9. This independent team considers the impact of the product on the environment and the extent to which the physical and chemical attributes of the proposed product could interact in biological systems relevant to consumer safety. Their scrutiny is largely theoretical at this stage and, if the outcome of this scrutiny is favorable, the "clearance forecast" will indicate the possible constraints that might affect the successful completion of product development and then specify the testing program required. This testing program then becomes an integral part of the overall project so its costs and outcomes must be incorporated in further project evaluations. Such programs are stringent, and costs and time durations may be high.

It takes at least three years to test for the carcinogenicity of a potential new food additive, and possible new detergents must be tested thoroughly to ensure they will not produce allergic dermatitis in a small minority of their users. Thus, test programs are not undertaken unless the probabilities of technical and commercial success for the proposed product appear to be high. Also some projects which offered high technical and commercial promise have been aborted because they failed to pass a stringent testing program. If the project survives this scrutiny and evaluation, the original team will continue development and testing work in collaboration with the ESD.

If the project fulfills this earlier promise, it will be adopted by an operating company to develop as a new product. It will then pass through a formal clearance process made in writing. The operating company will submit the results of the testing program together with a manufacturing and marketing plan for the product, for clearance. This plan is first reviewed at the central corporate level to compare it with past products which were similar and, assuming this review is satisfactory, the plan is then passed on to the clearance area of the research division. If the results of the test program are satisfactory, and it is unlikely that the project would be adopted by an operating company otherwise, this clearance process should be straightforward. The ESD will repeat its review of the proposed product, now including the results from the test program and the manufacturing and marketing plans. If formal clearance is given, the ESD will monitor outcomes during the early stages of the manufacturing and marketing of the new product. For product proposals which originate in an operating company, as against the R&D function, it would appear that the clearance forecast and formal clearance procedures are collapsed into one.

Challis[23] provides a valuable review of the problems and remedies in environmental management in the chemicals industry, written from the perspective of Imperial Chemicals Industries (ICI). His review will not be discussed here (and since his and Philp's paper appear together in the same issue of the *Proceedings of the Royal Society,* it is convenient to read both together), but he provides a useful visual illustration of the registration process of new chemicals in the United Kingdom. His illustration of the process for an agrochemical product is shown in figures 6.3 and 6.4 (adapted from his figures 6 through 9). Figure 6.3 shows that a sequence of tests from "laboratory screening" to "commercial studies" is performed by the firm as an integral feature of the innovation process, and the regulatory body (the U. K. Ministry of Agriculture in this case) is continually kept informed of progress. Figure 6.4 shows the time durations and cost breakdowns of the process.

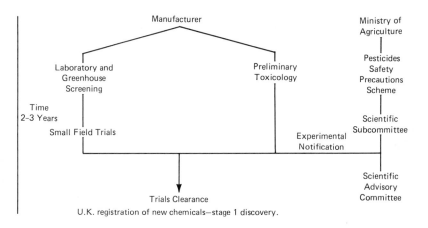

Trials Clearance

U.K. registration of new chemicals—stage 1 discovery.

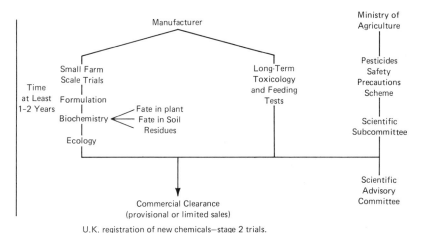

Commercial Clearance
(provisional or limited sales)

U.K. registration of new chemicals—stage 2 trials.

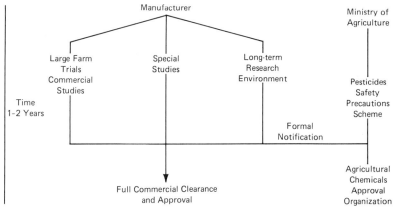

Full Commercial Clearance
and Approval

U.K. registration of new chemicals—stage 3 commercial development.

Source: E. J. Challis, "The Approach of Industry to the Assessment of Environmental Hazards," *Proceedings of the Royal Society,* series B, no. 1079 (1974), 183-97. Reprinted with permission.

Figure 6.3 Clearance Processes.

TIME	STAGE	% COST
Year 1 to Year 3	Discovery ↓ Trials Clearance	20%
Year 4 to Year 5	Trials Clearance ↓ Limited Commercial Clearance	30%
Year 6 to Year 7	Limited Commercial Clearance ↓ Full Commercial Clearance	50%
Total 7 Years	Approval	£2–4 million

Development of an agricultural chemical. It takes as long as it does to grow an apple tree to fruit and costs up to £4M.

Source: E. J. Challis, "The Approach of Industry to the Assessment of Environmental Hazards," *Proceedings of the Royal Society,* series B, no. 1079 (1974), 183-97. Reprinted with permission.

Figure 6.4 Time-Scale.

References

1. Rachel Carson, *Silent Spring* (Hardmondsworth: Penguin Books, 1972).

2. Donella H. Meadows et al., *The Limits to Growth* (New York: Universe Book Publishers, 1972); Barry Commoner, *The Closing Circle* (New York: Alfred A. Knopf, 1971).

3. Jean Gimpel, *The Medieval Machine: The Industrial Revolution of the Middle Ages* (Harmondsworth: Penguin Books, 1977), p. 82.

4. Ralph Nader, *Unsafe at any Speed* (New York: Grassman Publishers, 1965).

5. Michael Gibbons, "Technology Assessment: Information and Participation," in *Directing Technology Policies: Promotion and Control*, eds. Ron Johnston and Philip Gummett (London: Croom Helm, 1978), pp. 175-91; Michael Gibbons and Roger Voyer, *A Technology Assessment System: A Case Study of East Coast Offshore Petroleum Exploration* (Ottawa: Background Study no.30, Science Council of Canada, March 1974).

6. Amatie Etzioni, "Mixed Scanning. A 'Third' Approach to Decision Making," *Public Administration Review* 27, no.5 (December 1967): 385-92.

7. Barry Bozeman and F. A. Rossini, "Technology Assessment and Political Decision-Making," *Technological Forecasting and Social Change* 15, no.1 (September 1979): 25-35.

8. G. T. Allison, *Essence of Decision: Explaining the Cuban Missile Crisis* (Boston: Little, Brown Co., 1971).

9. M. J. Cetron and Bodo Bartecha, eds., *Technology Assessment in a Dynamic Environment* (New York: Gordon & Breach Science Publishers, 1973).

10. G. Strasser, "Methodology for Technology Assessment—Case Study Experience in the United States," in Cetron and Bartecha, eds., *Technology Assessment in a Dynamic Environment*, pp. 905-40.

11. R. A. Waller, "Assessing the Impact of Technology on the Environment," *Long-Range Planning* 8, no.1 (February 1975): 43-51.

12. Laurence H. Day, "Technology Assessment in Bell Canada," testimony before the Technology Assessment Board United States Congress, 10 June 1976, Washington, D.C.

13. Joseph F. Coates, "The Role of Formal Models in Technology Assessment," *Technological Forecasting and Social Change* 9, no.1/2 (1976): 139-90; Harold A. Linstone et al., "The Use of Structural Modeling for Technology Assessment," *Technological Forecasting and Social Change* 14, no.4 (September 1979): 291-327.

14. J. J. Gordon and H. S. Becker, "The Cross-Impact Matrix Approach to Technology Assessment," *Research Management* 15, no.4 (July 1972): 73-80.

15. L. B. Leopold et al., *A Procedure for Evaluation Environmental Impact* (Washington, D.C.: U.S. Geological Survey Circular no.645, 1971).

16. I. L. Whitman et al., *Final Report on Design of an Environmental Evaluation System* (Columbus, Ohio: Battelle Memorial Institute, 1971).

17. *Assessment of Major Industrial Applications: A Manual* (London: Research Report no.13, Department of the Environment, 1976).

18. David W. Fischer and Gordon S. Davies, "An Approach to Assessing Environmental Impacts," *Journal of Environmental Management* 1 (1973): 207-27.

19. Martin V. Jones, "The Methodology," *Technology Assessment* 1, no.2 (1973): 143-53.

20. Day, "Technology Assessment in Bell Canada."

21. Philip Feldman, *A Technology Assessment of Computer-Assisted-Instruction Use in Colleges*, rev. ed. (Montreal: Business Planning Group, Bell Canada, July 1974).

22. J. McL. Philp, "A Multi-National Company: The Public and the Environment," *Proceedings of the Royal Society*, series B, 185, no.1079 (1974): 199-208.

23. E. J. Challis, "The Approach of Industry to the Assessment of Environmental Hazards," *Proceedings of the Royal Society*, series B, 185, no.1079 (1974): 183-97.

Part 3

The R&D Setting

Part 3 begins with some general considerations of R & D organization and planning. Effective R & D management requires the judicious balancing of behavioral and technicoeconomic considerations. These considerations are discussed then in chapters 8 through 12.

7

R & D Management: Some General Considerations

A technological corporation's R & D program must be balanced between needed profitable improvements in existing products and other short-term projects, on one hand, and longer-term or strategic efforts on the other.... Generally, it is well to arrange a deliberate separation of these two kinds of R & D programs, because the results obtained need to be rated so differently and the mechanism of use of results and the organizational control are both different.
Simon Ramo, **The Management of Innovative Technological Corporations**

7.1
Introduction

In chapter 4 we saw that the nature and climate of the R & D function are dependent upon the technological strategy of the firm, so that at first sight it might be supposed that we should discuss R & D management sequentially in the context of each of the strategies described there. Such a repetitive discussion is unnecessary, however, because the sequence of strategies from offensive to traditional reflects a decrease in R-intensiveness and a sequential truncation of the technological base of the company. As stated in chapter 4, the most visible high-technology corporations generally follow mixed offensive/defensive strategies over product ranges and time, so we may focus our discussion on such firms. The extent to which the discussion applies to companies that follow imitative, interstitial, and so forth, strategies is dependent upon the scopes of their technological bases.

7.2
Organizational and Geographical Locations of R & D

Both the organizational and geographical locations of the R & D functions can be expected to be functions of the historical evolution

119

and size of the individual firm. A small- to medium-sized company which exploits a comparatively narrow technological base will typically operate with a single R & D facility often physically associated with one of its manufacturing plants, with this original location determined historically. In contrast, medium-to-large corporations that may well be offensive/defensive innovators exploiting a broad technological base and operating several manufacturing divisions operate between the extremes of pure *concentrated* or *distributed* R & D functions.[1]

In a concentrated function, R & D is performed in a corporate laboratory for the whole corporation, probably under the executive direction of a corporate vice-president of R & D, who also coordinates R & D efforts with divisional requirements. By concentrating its R & D capabilities in one function, the corporation creates a central pool of expertise that can be flexibly deployed on changing R & D needs and can establish an outstanding reputation in its fields. In a distributed function, R & D is performed in individual divisions, with much looser coordination at the corporate level. Each division exercises more autonomy over its "own" R & D activities which should therefore be more responsive to divisional needs.

In practice, the R & D activities of many larger corporations are performed at both corporate and divisional levels, possibly with corporate laboratories being oriented towards "research" and longer-term innovations, while divisional laboratories are oriented towards "development" and shorter-term innovations. Moreover, such organizations are often multinational corporations that have grown through a process of merger and acquisition. They may therefore duplicate both corporate and divisional R & D laboratories in different countries, especially if they grew from the international merger of individual national corporations, each with its own laboratories. These days, some MNC's rationalize their R & D activities across national boundaries by encouraging each national component to focus its capabilities within specific technology—market segments to develop "world product mandates."

The issue of the geographical location of R & D activities also deserves a brief mention. Divisional laboratories are typically located close to a manufacturing function as we stated at the beginning of this section. In contrast, corporate laboratories may be located so as to be geographically separate from individual divisions for two interrelated reasons. First, corporate R & D laboratories are usually competing with universities and government R & D establishments in their professional R & D staffing requirements. They therefore need to provide organizational climates comparable to academic and government institutions, free from the shorter-term pressures of divisional needs (see chapter 8). Second, if the corporate function is physically located with a manufacturing division, there is a risk that it will become the R & D satellite of that division in the long-term and lose sight of its broader corporate mission. Given this organizational and geographical distribution of R & D efforts, longer-term innovations are typically initiated in the corporate R & D facility and then transferred to the appropriate divisional facility when the secondary development phase is reached, or later. Shorter-term innovations may be initiated in divisional facilities, often in response to specific requests from operations managers and customers. This arrangement provides response mechanisms to both long- and short-term innovation opportunities.

Regardless of its organizational form, there are seven inter-related requirements to be considered in the management of the R & D efforts of a firm. First, the overall level of expenditure or the pro-portion of the total corporate budget to be invested in R & D activi-ties. Second, the R & D plan to be developed to implement the corporate strategy (or strategies) congruent with the future sce-narios envisaged for the firm. Third, the appropriate linkage be-tween R & D planning and the budgetting process. Fourth, the development and maintenance of a creative organizational climate in the R & D function. Fifth, the institution of effective and realistic procedures for the evaluation of individual R & D projects and their integration into an overall project selection process. Sixth, the institution of ongoing evelation and control procedures for projects, once they have been approved. Seventh, the reconciliation of the personal development needs of professional R & D staff and the requirements for effective project management.

We will examine the first three of these requirements in this chap-ter and will consider the remainder in chapters 8 through 12.

7.3
R & D Budget Setting—Pragmatic Aspects

The scope of R & D activity is obviously dependent upon the amount of money spent on it. This budget size can be expected to be related to the technological strategy and goals of the company but, in practice, annual budget allocations to R & D are also likely to be highly depen-dent on other criteria. The process of formulation of a corporate pol-icy and the setting of growth, profit, and market share, and so forth, goals in the technology-market matrix are likely to evolve historically over a time period, during which there is a corresponding evolution of R & D activities. Although the scope and nature of these activities should (and usually do) reflect corporate policies and goals, annual budgetary allocations have traditionally often been based upon more pragmatic criteria, and Clarke provides a useful review of current practice.[2] Therefore, it is useful to begin our discussion by exam-ining these traditional criteria for fixing the size of the R & D budget.

122

R & D
Management:
Some
General
Considerations

We shall then examine the formulation of an R & D program in the context of corporate goals, and so forth, in the next section.

Broadly speaking, there are three potential sources for R & D funding in a company. First, internal funding from the corporate budget, which is discussed below. Second, income from selling or licensing R & D inventions to other manufacturers. Given the relative prolific capacity of a good R & D laboratory to generate inventions (recall chapter 1, section 1.1) to some extent serendipitously, companies frequently find that they have invented and patented potential product concepts that do not readily "fit" into their strategies and goals in the technology-market matrix. Moreover, because R & D costs represent a relatively small proportion of the total costs for developing many innovations, companies are likely to lack the corporate resources to exploit all the inventions that they discover. In such circumstances, companies frequently find it advantageous to recoup these expenditures by the selling of licensing patents to others.

The third source for R & D funding is contract R & D. Most (if not all) companies at least partially support their R & D effort through contract research from outside bodies, notably government. Moreover, some national governments seek to promote domestic technological innovation by supporting corporate R & D through specific grants and tax incentives.

The extent to which a company seeks dependence upon each of these three sources for its R & D funding is really a policy decision, but companies following offensive/defensive strategies are likely to pay for a substantial proportion of their effort from internal funds. Therefore, we will focus our discussion on alternative pragmatic criteria for setting this funding level.

Percentage of Sales Turnover

One crude yardstick is to spend a certain percentage of the previous (or a number of recent) year's sales. Ramo quotes the 1978 figures for a sample of thirty-nine U.S. high-technology companies ranging in size from Beckman ($338 million turnover) to GM ($63 billion turnover).[3] R & D spending as a percentage of sales ranged from 0.5 percent for Arco and Exxon to 9.4 percent for Fairchild (see table 7.1). These figures reflect industry and size factors, but also suggest that R & D expenditures do vary widely. For example, this author knows one small high-technology company that spends 15 percent of its turnover on R & D. Quite possibly this is the most common yardstick used, but it suffers from two obvious disadvantages.

The first disadvantage is that in the absence of an industrywide recession, declining sales might suggest that the majority of products in a company's product range are in the declining years of their lifecycles. Therefore, it would be more appropriate to spend *more* rather than less money on R & D, to restore a flagging product range. However, increased expenditures will only have significant impact if the lead time between R & D expenditures and consequent sales is relatively short. We have already seen that the time scale of the innovation process is often long, maybe decades, for revolutionary innovations. When such lengthy lead times apply, the impact of increased R & D expenditures will be too delayed to satisfy the shorter-term need to restore a sagging sales record. Increased R & D expenditures are only likely to increase sales in the relatively short term (that is, next year) if the money is spent on advanced development, where the payoff is obtained quickly. This suggests that it offers the most benefit when the industry has reached maturity so that innovations are incremental and cost reducing. This applies particularly if a company follows an "applications engineering" strategy and is strong at providing incrementally improved products for specialist applications.

The second disadvantage is that if, conversely, sales are increasing, it does not follow that R & D expenditures should be increased. Firstly, there is no need to increase expenditures to restore flagging sales. Secondly, if the increased money were spent effectively, it would lead to more R & D inventions being available for potential commercial exploitation. If the company is already enjoying buoyant sales, it may lack the downstream capacity (in product, sales, and servicing) and management willpower to manufacture and promote further innovations. Admittedly, patents on surplus R & D inventions may be sold or licensed, but that may not always be feasible or congruent with corporate policy. Furthermore, increased expenditure may lead to the hiring of more R & D staff than can be usefully employed by the company in the long term, and this can create "redundancy" problems later.

Percentage of Profit

This measure is also shown for Ramo's thirty-nine corporations in table 7.1 and range from Arco (8 percent) to Fairchild (202 percent). This positive feedback rule suffers from the same objections as using a percentage of sales turnover, with the added implication that R & D is a "luxury" on which surplus cash can be squandered, but is unrelated to profitability in either the long or short term. The criterion does have some relevance in the context of "tax management," however. Judicious temporary increases in R & D expenditures may be a worthy and convenient means of reducing the tax liability on the increased profits, particularly if they are spent on the purchase or construction of improved equipment or facilities, which entail no long-term employment commitments.

Firm	% Sales	% Profits	Firm	% Sales	% Profits	Firm	% Sales	% Profits
AT & T	2.1	16.0	Ford	3.4	92.1	Motorola	6.0	106.6
Arco	0.5	8.0	GE	2.6	42.4	NCR	5.3	71.3
Beckman	7.9	121.5	GM	2.6	46.6	Polaroid	6.3	73.0
Bendix	1.4	39.1	GTE	1.5	20.3	RCA	2.1	50.5
Boeing	5.1	85.5	Goodyear	2.0	67.1	Rockwell	1.1	29.0
Burroughs	5.9	56.3	HP	8.9	100.7	Sperry-Rand	5.3	110.5
DEC	8.1	81.4	Honeywell	5.3	103.1	TI	4.4	79.1
Dow	3.4	40.2	IBM	6.0	40.3	TRW	1.4	30.7
DuPont	3.6	47.9	ITT	2.4	56.1	Union Carbide	2.0	39.5
Eastman-Kodak	5.5	43.1	McD'L Douglas	4.1	104.8	United Tech.	7.0	187.9
Exxon	0.5	10.5	Merck	8.1	52.5	Westinghouse	2.3	48.8
Fairchild	9.4	202.3	3M	4.4	36.2	Xerox	5.3	66.9

Note that profits are defined as the differences between sales or revenues and costs (including depreciation + taxes + interest).

TABLE 7.1   R & D SPENDING.

Interfirm Comparisons

Another commonly used yardstick is to spend the same percentage of turnover as similar companies in the same industry. A difficulty with this "keeping up with the Joneses'" approach is that firms resemble people in that no two are sufficiently alike to make direct comparison valid. In companies large enough to have a substantial R & D effort, there may be a multiplicity of product lines, each with different R & D requirements, so that each firm will have a unique technological profile. Furthermore, differences in accounting systems, both between and within companies, may make the disentanglement of meaningful comparative costs from public records virtually impossible. Finally of course, peer group firms used for comparison may have set their expenditures by other yardsticks such as those given above, so that an interfirm comparison criterion becomes a surrogate for another measure.

A more important role for interfirm comparisons is in technology monitoring, as discussed in chapter 5. By monitoring its competitors' R & D expenditures, activities, recruiting policies, and so forth, a company can anticipate potential innovations coming from the opposition and recognize potential threats and opportunities. Despite reports to the contrary, U.S. companies probably continue to exercise technological leadership in many fields and maintain a fairly "open" attitude in their R & D activities. Consequently, companies in Canada, Japan, and Europe can still often anticipate future technological developments by observing the activities of their U.S. counterparts. In so doing, they can take note of the expenditures of these counterparts in fixing their own.

Costing on a Project-by-Project Basis

It might be argued that each R & D project should be costed and budgeted and the total appropriation determined, based upon aggregate requirements. In practice, given the fecundity of a good R & D function, such aggregate costs are likely to exceed any financial appropriation which could be even remotely made available. Therefore, projects must be eliminated to satisfy the constraints of the overall budget set by some other criterion.

Advocacy Process Based upon Historical Allocation

In reality, the current level of R & D expenditure, based upon whatever methods have been used historically, is likely to be the major determinant in fixing future allocations, since they are unlikely to change radically from year to year. This being so, annual appropriations are likely to be determined by a bargaining process between senior R & D and corporate managements. As Steele[4] describes the process, the R & D director may argue a case for an appropriation based upon the following considerations:

1. Percentages of sales, or profits or industry "norms" as discussed above.

2. Maintenance of a historical growth rate for R & D in real monetary terms.

3. The degree of esteem or disfavor with which corporate management views the R & D functions, and their satisfaction or dissatisfaction with its recent efforts.

4. The overall business and economic climate.

In practice, the exact size of the budget is likely to depend upon the R & D director's skill and judgment in advocating his function's case at corporate cabinet level. On the one hand, the director will not wish to ask for too much money, lest he be viewed as being naively unaware of what money can be realistically made available and unsympathetic toward the problems of corporate management. On the other hand, the director will not wish to make too low a bid, lest he be viewed with disdain by corporate management as a weak advocate for his functional interests. If the director has the necessary grasp of corporate politics *plus* sound technological and business judgments (and he should not occupy this position if he does not), it is likely that he will secure a realistic financial appropriation for R & D.

126

R & D
Management:
Some
General
Considerations

7.4

R & D Planning and Budget Setting —
Normative Aspects

The foregoing discussion has focused on rather pragmatic approaches to budget setting, whereas a more normative approach could be deemed desirable. The financial performance measures discussed in the previous section, together with the overall economic climate within which the company currently operates, as well as other competitive investment opportunities, will almost certainly set an upper limit to the budget allowed for R & D activities. However, the criteria for determining actual R & D expenditures within this upper limit should be based upon the considerations examined in chapters 4 and 5.

In chapter 4, I argued that responsibility for innovation should rest with an innovation (in contrast to operations) management function spanning both corporate and functional management areas. In chapter 5 we examined the techniques that enable this function to identify and define the alternative future scenarios in prospect for the company and match its technological capabilities to envisioned market opportunities to constitute a technological strategy. Pearson describes specifically how relevance trees might be used in R & D long-range planning.[5] Because the R & D function should provide the base for the implementation of this strategy, the total R & D budget and its division between competing projects should be based upon these considerations, that is, to pursue a technological strategy that seeks to invent the company's preferred scenario or future. This overall planning and budgeting process is schematically illustrated in figure 7.1.

Senior corporate management can be expected to set overall R & D expenditure in the light of this overall technological strategy, and then review and approve (or reject) individual project fundings in the context of the implementation of that strategy. More detailed reviews are likely to be performed by differing groups acting as "innovation managers" using criteria which vary through the innovation process. The broad division of R & D expenditures between the activities that lie within the span of responsibility of the R & D function (say nondirected research, applied research, primary or experimental and secondary or advanced development) is likely to be made at the senior R & D management level, perhaps subject to the approval of corporate management, depending on overall corporate practice. At first sight, this reasoning might be taken to imply the R & D planning should be entirely "top-down," but this is not so. Able R & D staff who are monitoring the state of the art in their respective technologies and markets should and will wish to partici-

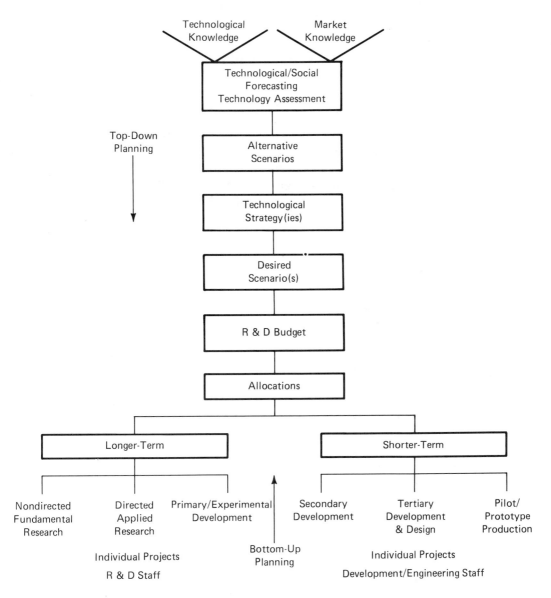

Figure 7.1 R & D Planning.

pate in both scenario generation/selection and R & D planning and budgeting activities. This consideration is illustrated by the two arrows in figure 7.1. Like its overall size, the allocation of the overall R & D budget between different departments and individual projects within the overall R & D function will be significantly influenced by the political advocacy skills of the interested parties from bench-level staff upwards. Thus R & D planning should be both a "top-down" and "bottom-up" process.

127

128

R & D
Management:
Some
General
Considerations

Despite the behavioral considerations, the criteria for dividing the overall budget between these broad divisions, and evaluating individual projects within each of them, should be based upon rational considerations and vary through the succeeding phases of the innovation process. It will depend upon the phase the project has reached in the process, the resources required for its initiation or continuance, and the *immediacy of the impact on the business of its successful completion.* Given this "immediacy of impact" consideration we may divide the successive phases into two broad groups—longer-term R & D, covering nondirected and applied research and experimental development, as opposed to shorter-term R & D, covering advanced development onwards.

Longer-Term R & D

If a company is following an offensive/defensive strategy, it is R-intensive and invests significant effort in nondirected research which, by definition, offers payoffs that cannot be quantified. Similarly, although directly applied research and experimental development may be focusing on a specific product concept, again by definition, it is questionable whether the technology will be sufficiently developed, or the product concept sufficiently delineated, to admit even remotely realistic cost projections. It should be recognized that since the technological innovation process can extend over a relatively long time scale, projects should not be subjected to the hazards of "premature economic assessment"[6] or the ROI and marketing criteria which must be applied later in the process. The ability to reconcile the paradoxical requirements of commercial stewardship and the nurturing of a climate for sustained technological innovation, is one of the most difficult and vital roles for senior innovation management. Thus corporate management should perhaps support these activities as acts of *faith and hope,* trusting in the quality of staff they have hired in these functions, rather than expecting explicit monetary rewards from each project. The size of budget for such activities is therefore a matter of largely subjective management judgment based upon the general financial health of the company, competitors' expenditures on probably rather similar efforts, and the maintenance of continuity of the activities (assuming previous efforts have been adequate). This approach should produce a fair stability without violent and disruptive fluctuations in efforts. As Twiss suggests, it does not disallow for the occasional budgetary cutbacks should the company's financial position impose such a requirement.[7] As Twiss indicates, provided it is humanely rather than viciously applied, a cutback can have a salutary effect and force the weeding out of unproductive equipment, facilities, and possibly people.

129

R & D Planning
and Budget
Setting—
Normative
Aspects

We have already seen (chapter 4) that firms following an offen-sive/defensive strategy may perform nondirected fundamental re-search and directed research in an R-intensive environment, working in close proximity to the state of the art, with low stability and predictability, scarcity of precedent, and so forth. The commercial outcome of such research is almost certainly long term, with little immediate business impact. Therefore, once corporate management has set the overall funding for these activities, it should delegate the detailed allocation of these funds between competing projects, as an internal and parochial R & D responsibility. The R & D director can then operate peer group internal evaluation and review procedures similar to those used in R-intensive laboratories in universities and government. Given that high-quality research staff are being re-cruited and retained, the quality of research output should be congruent with the company's long-term objectives and goals. In the event that this is not so, it is likely to be due to inappropriate R & D management and staff recruitment policies, rather than inappro-priate project evaluation and selection procedures.

Shorter-Term Development, and So Forth

Once one moves into the secondary or advanced development/ design innovation phase of R & D activities, different considerations apply. By now the product concept of a proposed innovation should be more clearly delineated, and some crude but reasonable estimates could well be possible. This means that the project can be assessed in terms of its congruence with corporate objectives and goals using evaluation methods discussed later in chapter 9. The extent to which a proposed innovation is congruent with corporate goals can be assessed in terms of the technology-market matrix and what, in corporate planning, is called *gap analysis*.

Recall that figure 4.1 schematically illustrated the requirement of a high-technology firm to generate a continued sequence of inno-vations in order to survive. The ordinate in figure 4.1 represented a measure of cash flow, but it could have readily represented another fiscal measure such as profit. Recall also that figure 4.3 schemati-cally represented the technology-market matrix in which such a firm might operate. The process of formulating a technological strategy and corporate objectives and goals should identify which cells in the technology-market matrix the company wishes to occupy and specific goals (in terms of profit, ROI, market share and so forth) it seeks, over time, in each of these occupied cells. Thus, the sche-matic representation of figure 4.3 may be disaggregated to represent (perhaps) the profit growth goals to be sought from a particular cell in the matrix. Given the finite life cycle of many technological prod-ucts, the projected profits from currently existing company products

130

R & D
Management:
Some
General
Considerations

in a particular cell might be sufficient to satisfy current profit requirements, but be insufficient to satisfy future profit goals. This
is illustrated in figure 7.2. The products currently being manufactured are at varying stages of their life cycles. The sum of their financial contributions are sufficient to satisfy profit requirements until
time X on the abscissa, but beyond that point there is a growing gap
between profits which these products can contribute and the requirements of the business plan. This gap is required to be filled by new
product concepts at present in the R & D phases of the innovation
process.

From the viewpoint of corporate management, the role of R & D
is to implement a technological strategy by generating a continuous
flow of new product ideas which will fill such gaps. So once a proposed innovation has reached the secondary development/design
engineering phase, it should be judged in this context, that is, in
terms of its potential to fill the company's future "profit gaps."

As was stated above, the detailed discussion of project selection
is deferred until later, and for the present, it is sufficient to note three
points. The first point relates to the vertical axis of figure 7.2. Given
the uncertainties of outcomes associated with potential innovations
at this stage of the process, the probability of any one project yielding a profitable product innovation is quite low. Therefore, corporate
management should ensure that the R & D budget is large enough
to enable a sufficient number of innovation avenues to be pursued,
to provide a *high degree of insurance* that future gaps in profitability
goals (in all technology-market cells) will be filled. That is, the budget should be large enough to ensure that enough "eggs" (projects)

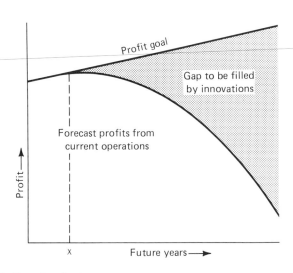

Figure 7.2 Gap Analysis.

are placed in enough "baskets" (product-market segments) for a
sufficient number of "eggs" to prove fertile and "hatch out" to achieve
corporate goals.

131

R & D Planning
and Budget
Setting—
Normative
Aspects

The second point relates to the horizontal axis of figure 7.2 or the
"timing" of new product developments. The cost associated with
the R & D phase in the innovation process is typically no more than
10 percent of the total cost, and since (as this cost proportion implies)
a new innovation places increasing demands on all company resources as it progresses further downstream, it is important to ensure that the project portfolio consists of a set of projects which are
"balanced" in time. This protects the R & D function from harmful
fluctuations in its budgets, ensures that promising projects are not
aborted later through lack of (perhaps) production capacity, and
that the "timing" of product entry into the market is congruent with
company goals.

"Timeliness" of product introduction is also very important from
a marketing viewpoint. As indicated in chapter 2, it is by no means
a rare occurrence for products to fail because they enter the market
too soon rather than too late. The Canadian government's development of the "Bras d'Or" hydrofoil, a product concept invented by
Alexander Graham Bell, is an ironic example of this phenomenon,
and others are readily apparent. Therefore, it is important to exercise *family planning* in project selection to ensure that the fertile
eggs placed in the baskets hatch out in orderly sequence.

Third, given the need to ensure future profitability by backing
more projects than would be required if all were successful, even with
astute "family planning," it is probable that the fecundity of the
R & D function will generate "successful" outcomes, which must be
aborted through lack of downstream resources and commitment.
It is important to recognize that such successful projects *do not*
represent bad investment decisions. The best way of ensuring
long-term profitability is to produce an *embarras de richesse* of new
product ideas, and those that remain unexploited may still have
tangible or intangible economic values. The patents filed for unexploited inventions may be either sold or licensed to other companies, or used as bargaining counters to derive patent concessions
from competitors.

As was stated earlier, once one moves into the secondary development phases of R & D activities and projects, different considerations apply. If budgets are set by (what might be called) an Innovation Management Steering Committee, based upon an aggregation
of costs of a selected project portfolio in the context of "gap analysis," it follows that such projects might be evaluated by that same
committee or its subcommittee(s). Although such an arrangement
might appear negatively bureaucratic, and could indeed become so,
it should be recognized that the majority of technological innovations fail for nontechnological reasons. Once a project reaches the

132

R & D
Management:
Some
General
Considerations

secondary development phase, it generally begins to absorb a significant amount of money and other corporate resources. It therefore becomes managerially irresponsible not to involve the critical evaluation of non-R & D personnel whose enthusiam and support will substantially influence the final success, or otherwise, of the proposed innovation anyway. The success of a project evaluation and review procedure of this kind, both in terms of screening out undesirable projects and, at the same time, in not dampening the morale of R & D staff, could critically affect the innovation climate of the company and the transfer of an innovation from the R & D laboratory to further downstream. It will be dependent upon the network of interpersonnel relationships between R & D staff and other members of the organization. These issues are discussed in subsequent chapters.

Overall Comments

The foregoing discussion suggests that the size of the budget can be determined by a process of rational negotiation between the R & D function and corporate management with the participation of the corporate planning/technological futures group of the organization. Indeed, it is possible that the Innovation Management Steering Committee made up of representatives from the above and senior line management functions could produce joint budget recommendation to a corporate board of directors. Obviously, at a certain stage in the innovation process (usually by the pilot/prototype stage), responsibility is transferred from R & D to operations management. Clearly, a project is more likely to be successfully innovated if it enjoys the support and sponsorship of senior line management. Therefore, given that a climate of mutual goodwill exists between R & D and operations management, the latter participation in R & D planning is desirable and may involve operations management financially patronizing some projects, thus augmenting the former's budget.

Whatever mode of budget setting is adopted, the foregoing discussion suggests that the budget level of the R-intensive activities of the R & D functions is best set by faith and hope, whereas the D-intensive activities are best set by a process of gap analysis. The latter process does, of course, provide more rational support to our objections to using fixed percentage of sales or profits. If sales or profits are falling, it is likely that the projected gap between future profit goals and forecasted profits is widening; therefore, a "gap analysis" would suggest increased spending. Conversely, increasing sales or profits should suggest a narrowing gap to which an aggressive corporate management should perhaps react by raising future profit goals or exploring other innovation opportunities, rather than reducing spending. In fact, what is being suggested is a Keynesian approach with R & D (albeit with a longer lead time) rather than public works spending, being used to stimulate demand during corporate recessions.

The above considerations, together with the innovation chain equation, provide us with a framework for a review of the problems and issues in the management of the R & D function. These problems and issues will now be reviewed in the following chapters.

References

1. M. F. Usry and J. L. Hess, "Planning and Control of Research and Development Activities," *Journal of Accountancy* 124, no. 5 (November 1967): 43-98.

2. Thomas E. Clarke, "R&D Budgeting—The Canadian Experience," *Research Management* 24, no. 3 (May 1981), 32-37.

3. Simon Ramo, *The Management of Innovative Technological Corporations* (New York: John Wiley & Sons, 1980), App. B.

4. Lowell W. Steele, *Innovation in Big Business* (New York: American Elsevier Publishing Company, 1975), chap. 9.

5. Alan Pearson, "Planning of Research and Development," *Long-Range Planning* 5, no. 1 (March 1972): 56-61.

6. M. W. Thring and E. R. Laithwaite, *How to Invent* (London: Macmillan Press, 1977).

7. Brian C. Twiss, *Managing Technological Innovation*, 2d. ed (London: Longman Group, 1980), p. 36.

8

Creative Thinking

Creative activity could be described as a type of learning process where teacher and pupil are located in the same individual. Creative people like to ascribe the role of the teacher to an entity they call the unconscious, which they regard as a kind of Socratic daemon—while others deny its existence, and still others are prepared to admit it but deplore the ambiguity of the concept.

Arthur Koestler, **The Drinkers of Infinity: Essays 1955-1976**

8.1
The Need for Creative Thinking

At the beginning of chapter 1 I stressed that innovation is much more than invention. But since the innovation process is based upon invention, innovation cannot occur without it. That is, the innovative organization must be an *inventive* organization. Although already on several occasions we have indicated that a good R & D function is prolific in generating such inventions, the innovative company will still be concerned with enhancing both their quality and quantity. Furthermore, *inventiveness* is a quality usually required and always desirable *in all phases* of the innovation process. Recall that the Japanese "invented" the personal radio, following the American R & D invention of the transistor. Later, in chapter 13, I shall argue that entrepreneurship is another manifestation of the human creative urge.

The *Concise Oxford Dictionary* defines "to invent" as to *create by thought*. In this chapter we explore the nature of this creative thought process, how it may be stimulated, and how organizations may identify people and foster a climate which stimulates their creativity.

The Creative Process—
The Socratic Daemon

As the quotation at the beginning of this chapter implies, there is no irrefutable means of defining the creative process, and although psychiatrists and psychologists have developed numerous approaches to "explaining" the phenomenon, the approaches appear to be of limited practical value. Perhaps the most important inference to be drawn from these approaches is that creativity is not markedly dependent upon intelligence (as measured by IQ tests), but is correlated with a number of other personality attributes. These inferences will be discussed later in section 8.4. Therefore, managers may most fruitfully view the process as a "black box" which has had its input-output characteristics fairly well documented, even though the nature of its Socratic demon is shrouded in obscurity.

Arthur Koestler is probably the most eloquent and perceptive writer on this subject. In the first two works of a trilogy concerned with the relation between reason and imagination, *The Sleepwalkers* and *The Act of Creation,* he explores in some detail the nature of scientific discovery and the creative act.[1] The first includes biographical and psychological sketches describing how some of the major figures in modern science, ranging from Copernicus to Kekule, "discovered" their ideas. The second book examines the process of creative thinking in humor, science, and art. To illustrate the creative process we quote two examples from the latter book. Koestler is almost exclusively concerned with human creativity, but he also postulates its capability in animals. Other primates, as well as humans, have developed tools or technological artifacts so I will first quote an example of "technological invention" in the animal kingdom which Koestler extracts from *The Mentality of Apes* by Wolfgang Kohler.[2]

> *Nueva, a young female chimpanzee, was tested 3 days after arrival (11th March, 1914). She had not yet made the acquaintance of the other animals but remained isolated in a cage. A little stick is introduced into her cage; she scrapes the ground with it, pushes the banana skins together in a heap, and then carelessly drops the stick at a distance of about three-quarters of a metre from the bars. Ten minutes later, fruit is placed outside the cage beyond her reach. She grasps at it, vainly of course, and then begins the characteristic complaint of the chimpanzee: she thrusts both lips—especially the lower—forward, for a couple of inches, gazes imploringly at the*

*observer, utters whimpering sounds, and finally flings her-
self on to the ground on her back—a gesture most eloquent of
despair, which may be observed on other occasions as well.
Thus, between lamentations and entreaties, some time
passes, until—about seven minutes after the fruit has been
exhibited to her—she suddenly casts a look at the stick,
ceases her moaning, seizes the stick, stretches it out of the
cage, and succeeds, though somewhat clumsily, in drawing
the bananas within arm's length. Moreover, Nueva at once
puts the end of her stick behind and beyond her objective.*[3]

Our second example is the possible apocryphal but usefully illustrative example which every schoolboy knows—Archimedes in his
bath.

*Hiero, tyrant of Syracuse and protector of Archimedes, had
been given a beautiful crown, allegedly of pure gold, but he
suspected that it was adulterated with silver. He asked Archimedes' opinion. Archimedes knew, of course, the specific
weight of gold—that is by its weight per volume unit. If he
could measure the volume of the crown he would know immediately whether it was pure gold or not; but how on earth is
one to determine the volume of a complicated ornament with
its filigree work? Ah, if only he could melt it down and measure the liquid gold by the pint, or hammer it into a brick of
honest rectangular shape, or...and so on. At this stage he
must have felt rather like Nueva, flinging herself on her back
and uttering whimpering sounds because the banana was
out her grasp and the road to it blocked.... One day, while
getting into his bath, Archimedes watched absentmindedly
the familiar sight of the water-level rising from one smudge
on the basin to the next as a result of the immersion of his
body and it occurred to him in a flash that the volume of
water displaced was equal to the volume of the immersed
parts of his own body—which, therefore, could simply be
measured by the pint. He had melted his body down, as it
were, without harming it, and he could do the same with the
crown.*[4]

Both of these examples (and numerous others that Koestler quotes)
have four stages in common:

1. Perception of the problem
2. Frustration at the inability to solve it
3. Relaxation
4. Perception of solution—EUREKA!

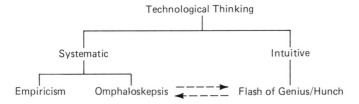

Figure 8.1 Creative Thinking A.

Green,[5] who wrote from the perceptive of an R & D management background (he was vice-president, Bell Telephone Laboratories when he wrote the article), essentially expands on the above stages in his description of the creative process. First of all, he proposes that creative thinking in science can be typified as in figure 8.1. On the horizontal or lateral dimension, there is a continuum of styles of thinking from systematic thinking to intuitive thinking, as he puts it: "The one is a deliberate act of the conscious mind, the other the gracious gift of the subconscious in return for the previous labors of the conscious mind."

Green suggests that systematic thinking is combination of *empiricism* and *omphaloskepsis*. The first term is self-explanatory and the second was devised by a British officer in India to describe oriental meditation. Green uses it to describe the contemplative and meditative process of the rational and analytical formulation of a theory, tested by guided empiricism, sometimes supplemented by serendipitous outcomes. Intuitive thinking yields the sudden flash of insight or genius or "hunch" as in an "output" of nonconscious though processes. This flash of insight or genius will be defined further shortly.

Green formulates the creative process as shown in figure 8.2, which though essentially self-explanatory, benefits from further elaboration.

1. The "problem" faced by the individual arises by the nature of the R & D process. The individual develops at least one preliminary conception of the problem and moves to:

2. *Accumulation* of data, ideas, concepts through literature reading, discussion, investigation, and experiment.

3. *Incubation* when the conscious and nonconscious mind assimilate and digest the information gathered.

4. *Intensive thinking* next occurs, whereby the individual seeks to solve the problem by weaving ideas in new combinations, but— despite the intense effort—without success. This leads to:

5. *Frustration*, dissatisfaction, and fatigue, so in the face of this psychological blockage, the individual abandons conscious consideration of the problem, takes some:

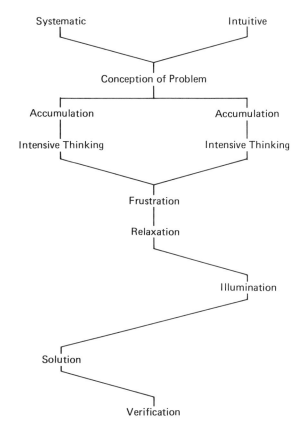

Figure 8.2 Creative Thinking B.

6. *Relaxation*, and "sleeps on it," leading later to:

7. *Illumination*, or sudden inspiration, the so-called "flash of genius" or EUREKA, and (in science and engineering anyway) finally:

8. *Solution*, verification, and embodiment.

Green's treatment, provides a useful outline of the creative process, and later we examine the techniques that may be used to enhance the effectiveness of stages 1, 3, 4, in his process. Before doing so, however, we elaborate upon the progression through the process, returning to Koestler's writings on the subject.

Although the explicit mental mechanics of the creative process are as yet largely inexplicable, it is important to avoid viewing it in purely passive terms. The Socratic daemon, like the poet's "muse," may provide inspiration, but *not in an autonomous, arbitary, or capricious manner*. R & D inventions, like good poetry, are not generated spontaneously in the inventor's or poet's conscious mind.

Inspiration and illumination only occur after the inventor has undergone the *perspiration* of extensive and frustrating conscious deliberation and meditation. Unfortunately, although lying on a beach on the Cote d'Azur may lead to a flash of genius, it will only do so *after* you have exhausted yourself in preparation. Therefore, the Socratic daemon hypothesis cannot be used as excuse for plain idleness!

Medieval thinkers described the flash of genius as a "flash of lightning" or *fulguratio*, implying a sudden thunderbolt from Zeus or intervention from God above. Contempory intellectual fashions make us look for mystical explanations from depth psychology rather than Hellenic or Judeo-Christian theology, and Koestler offers such a description. He points out that mental activity, or mentation, occurs with or without personal awareness. He postulates that there is a continuum or gradient of states from being totally unaware or unconscious (say from being knocked "senseless" by a hard blow on the head), to (as he puts it) the "laser-beam of focal consciousness." Some physiological control activities are performed entirely unconsciously, while others such as breathing can be performed unconsciously or consciously. Similarly, some complex consciously learned skills such as tying one's shoelaces or driving a car may be performed either unconsciously (absentmindedly) or consciously. Equally, some activities (both unconscious and to lesser degrees conscious) are nonverbal and nonrational and cannot be articulated in the framework of language and logic. Furthermore, these activities or "thoughts" are characteristically exposed to one level of awareness in dreams or dreamlike states of mind, which are rejected by the linguistic and logical articulations of the focal consciousness. Indeed, the preservation of human sanity requires the permanent repression of some of these archaic and prelinguistic "thought forms" to nonconscious levels of mentation.

Now the concrete realization of an R & D invention must be verified and embodied in the linguistic and logical framework of the scientific body of knowledge; therefore, it must be finally *articulated in this form*. However, Koestler provides massive anecdotal evidence to support the postulate that the mental act of creation or invention *does not occur in this way*. The creative process appears to occur at a prelinguistic level of mentation.

He suggests that creative people soak themselves thoroughly in the subject matter of the problem and then cogitate. The verb *to cogitate* is derived from *co-agiture,* meaning to shake together two or more previously separate entities. Now this process of cogitation can occur at two levels of mentation. First, at the linguistic and logical level of systematic thought in Green's treatment; second, at the prelinguistic and dreamlike or fantasy level, which provides the "flash of genius" in Green's treatment.

At the conscious level, this cogitation involves exploring new associations, combinations of data, and so forth, within the existing linguistic and logical framework of the state of the art, that is, within an existing framework of assumptions and rules. This framework constitutes a dominant perceptual pattern within which the problem is viewed. A characteristic of many creative acts in science is their formulation of new associations or patterns which overturn the existing framework through a gestalt switch of perception. Systematic and rational mentation does not readily make these novel associations because its perception is dominated by the existing pattern. Thus systematic and rational thinking leads to failure and frustration. The individual is frustrated by his failure to solve the problem so he "abandons" it and relaxes or does other work.

Now what happens when the individual "abandons" the problem? Up to this moment the process of mentation may have been occurring at both a conscious linguistic and rational level *and* at a preconscious, prelinguistic, and nonrational level. From now on, further cogitation is performed at the latter preconscious level, untrammeled by the "rules and regulations" of rationality. Further associations of ideas, concepts, data and so forth, however irrational and fantastic, may be combined and associated which never reach the threshold of awareness. Finally, a novel association of new perceptual pattern is "found" which has credibility at the rational level. It is, therefore, permitted to emerge into conscious awareness through a sudden thought in the bath or in a dream or dreamlike state.

Koestler postulates that the act of creation consists of the novel association of two previously unrelated concepts or ideas by *bisociation* to form a new perceptual pattern. He represents this act of bisociation as shown in figures 8.3 and 8.4. Both component concepts of "frames of reference" are represented by two-dimensional matrices of association M_1 and M_2—one in a horizontal and the other in a vertical plane. The individual, prior to the act of creation, does not associate them with each other, and may reject their association as irrational, fantastic, and outrageous. Thus, prior to this act, he searches for a problem solution within one matrix of association (the arrowed line in M_1), experiencing increasing frustration (point F) because the solution cannot be found there. The person relaxes, but preconscious cogitative mentation continues exploring novel matrix associations which his conscious mind would reject. The preconscious mind "discovers" a novel association or bisociation between M_1 and M_2 which holds promise of solution. Because of its promise, the bisociation is allowed to pass through the threshold of awareness and becomes a "flash of genius" or *eureka* in the conscious mind of its creator, leading to the problem's solution T (figure 8.4). This intuitive solution can then be proved using systematic, logical, and rational thinking.

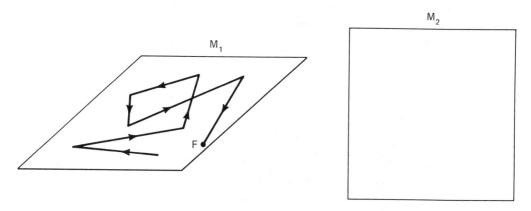

Figure 8.3 Separate Matrices.

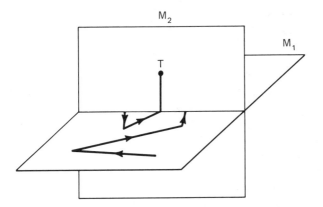

Figure 8.4 Bisociation.

This treatment may be readily applied to the two creative acts described earlier. The chimpanzee Nueva's problem was to obtain the bananas outside her cage. Her previous method of doing this—matrix M_1—had been to squeeze an arm or a leg through the bars of her cage and rake the fruit towards her until she could grasp it. The method is unsuccessful in this instance as the fruit is too far away. She reaches point F in figure 8.3 expressing despair. The other matrix M_2 is the stick which she associates with pushing and scraping objects on the floor of her cage (including banana skins). There is no recognition of the potential commonalities between the two matrices. A few minutes after her lamentations she makes the bisociation, grasps the stick, and uses it to reach the bananas, reaching the solution point T.

Archimedes's problem was to assay the king's crown which he defined as the problem of determining its volume. The methods available for doing this, based upon his current scientific knowledge,

141

required its destruction—matrix M_1. He was unable to devise a non-destructive method within this framework and so experienced frustration (point F). He probably enjoyed the relaxation and sensual pleasure and relief offered by a daily bath, and doubtlessly frequently observed the rise in water level as he sat down in it. Unfortunately, this observation was perceived within the pattern of "the taking bath" frame of reference—matrix M_2—and was assigned no significance. Again, during relaxation, Archimedes's mind made the bisociation between the two matrices, reached the solution point T, and (if legend is true) jumped out of the bath and ran naked down the street crying *Eureka*. As every schoolboy knows, his verbalization of this act of bisociation was bequeathed to us as the Principle of Archimedes.

In *The Act of Creation* (p. 121), Koestler provides numerous other examples of bisociation in the references cited earlier, including Gutenberg's invention of the printing press by the bisociation of the coin or seal stamp and the wine press. Perhaps his most charming example is that in which Kepler bisociates "Gravity and the Holy Ghost" (p. 124). To justify his laws of planetary motion, Kepler (who was educated initially as a theologian) postulated that a force emanates from the sun which influences the planets in an analogous manner to God the Father influencing the Apostles through the Holy Ghost. This postulate provided a basis for Newtonian gravitation and the integration of astronomy and physics. One might conjecture that classical physics could have developed differently had there been a different outcome to the conflict concerning the doctrine of the trinity in the early Church!

The creative bisociation of two frames of reference is not simply an additive association. Rather it exhibits an ontological discontinuity whereby the bisociated "whole" is greater than the associated frames of reference "sum of the parts." In chapter 2 I suggested that an R & D invention might be viewed as a technological mutation in technological evolution. Lorenz points out that in biological evolution, through etymological accident, new organic systems or capabilities are created with properties that cannot be described in terms of the properties of the associated subsystems.[6] He draws an analogy between the creative process and this manifestation of creativity in biological evolution, using an example from elementary circuit physics for illustration. See figure 8.5. In the first circuit (a) a battery is connected to a resistor and capacitor in series, and as the capacitor is charged with electricity, the voltage across it rises exponentially. In (b) the battery is connected to a resistor and inductor in series, and voltage across the latter falls exponentially. In (c) the battery is connected across all three components in series, and the voltage across the capacitor and inductor exhibits oscillatory behavior with exponentially falling amplitude. Circuit (c) represents an

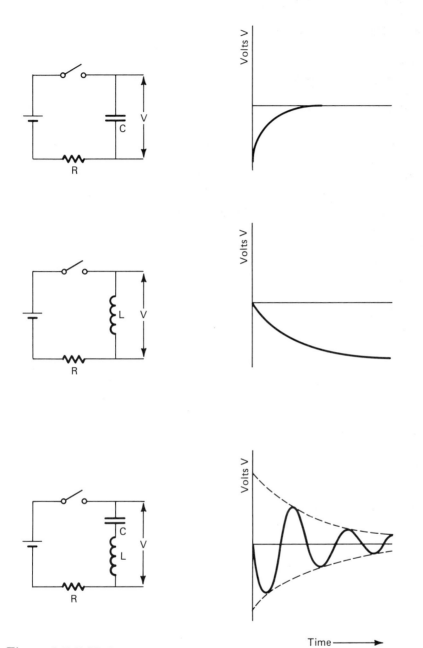

Figure 8.5 LCR Circuit.

association of circuits (a) and (b), but its phenomenological behavior is not readily expected from the phenomenological behaviors of circuits (a) and (b). It is in this sense that creative bisociation is more than simple association and the bisociative whole is more than the sum of the constituent parts. The unexpected emergent properties of systems synthesized from component subsystems is a well-known feature of cybernetics and general systems theory.

Recapitulating and summarizing this section, we can say that it provides a description of the creative process which displays the following salient features:

1. The creative act typically requires the synthesis or bisociation of two frames of reference or conceptual patterns, previously not associated.

2. Although *a posteriori*, such an act may appear obvious (why did I not think of it before?), because of its revolutionary novelty, it often constitutes a feat of some psychological courage on the part of its creator.

3. The mentation leading to bisociation involves cogitation or the shaking together of nonassociated conceptual patterns. Because these perceptions frequently occur at a preverbal and irrational level, the process requires a regression to preverbal, prerational, and preconscious levels of mentation and to childlike fantasy, without a loss of ultimate rational control and evaluation.

4. Since cogitation involves combining nonassociated matrices, perhaps on a random trial-and-error basis, the more matrices there are to combine, the more likely is a "creative" combination to be found. Therefore, a creative individual is likely to be one with wide as well as specialist scientific interests, someone with a childlike eclectic intellectual curiosity who is persistently asking "the why" of everything and delighting in the answers which he obtains. This person is an indefatigable idea and fact collector, is immersed in the literature of his own field, and is curious about everything else too. This way he stores a large repertoire of matrices in his mind from which a bisociative combination is more likely to be found.

5. Finally the individual must demonstrate the enthusiasm, persistence, and tenacity that is to generate the requisite perspiration that precedes inspiration.

8.3
Identifying Creative People

Although "creativity" (like "intelligence" and "physical strength") may be a trait possessed by everyone, just as some people are manifestly more intelligent or physically stronger than others, it can be tritely conjectured that some people are more creative than others. High-technology companies, particularly those following offensive/defensive strategies, are in the competitive business of exploiting technological creativity for profit. Consequently, it is axiomatic that such organizations must recruit and retain a cadre of "creative" individuals if they are to maintain their business competitiveness. In this section, we will identify the personality attributes which, according to the current state of the art in the field, are indicators of a person's creative ability.

The fairly detailed discussion of the creative process given in the previous section broadly identifies these indicators. Creativity requires the ability to "think" at the nonverbal level without a decline in the capacity for abstract thought and logical analysis. This viewpoint strongly suggests that it requires a mixture of Hudson's convergent and divergent thinking.[7] Indeed, Kuhn argues that the interplay between the convergent and divergent thinking provides the "essential tension" for the creative talent of the creative scientists.[8] Good academic performance in science subjects at the high school and undergraduate levels mainly requires convergent thinking. Therefore, it is important to look for other attributes as well as academic performance when seeking creative R & D staff.

This observation is possibly reflected in the results of Getzel and Jackson who, in a study of U.S. high school students, found that a threshold IQ score of 120 was required for creativity, but that there was no correlation between IQ and creativity beyond that score.[9] These findings were supported in research on teachers by Torrance and Hansen.[10]

Getzel and Jackson also found that "creative" students did as well as their more intelligent classmates, suggesting that they were "overachievers." This observation, which can be expressed as commitment to internal standards of performance, rather than those imposed by an external peer group, is supported by other research. MacKinnon conducted extensive research on architects.[11] He chose architects because they must combine elements of the artist and the scientist and also require business skills. From his research MacKinnon concluded that highly (as opposed to less) creative architects stressed the importance of performing to an inner standard of excellence, rather than that required by the profession. More important here, Roe[12] and Chambers[13] obtained similar results in their studies of eminent scientists versus other scientists, and scholars versus administrators. Again the creative individuals set their own internal standards which they adhered to religiously, being capable of considerable self-discipline and perseverance. Barron also obtained similar results with writers and mathematicians.[14]

Given that creative thinking requires both systematic and intuitive thinking—skills that are popularly supposed to be the fortes of men and women, respectively—it is hardly surprising that creative individuals possess what might be called complementary gender attributes. That is, highly creative men tend to get high scores on femininity scales. They are also more sensitive than most men and have higher aesthetic interests. On the other hand, highly creative women tend to score highly on masculinity scales. They are more interested in things and ideas than most women.

Their implacable commitment to high standards and goals, coupled with breadth of interest and curiosity, is reflected in the interpersonal behavior of creative individuals. They seek personal autonomy and are indifferent to group standards and control, but although "independent" they are also "dependable." They have good communication skills and are eager to discuss ideas with others. However, the purpose of such discussions is to promote a two-way traffic in the ideas themselves, rather than human fellowship, so they are unconcerned with policing their own or other people's feelings. On the other hand, although primarily interested in communicating ideas, they are sensitive to interpersonal aggression and prefer to repress or avoid interpersonal conflict. The commitment to internal, rather than group standards and control, does present a difficulty to the creative child. The educational system imposes continued external evaluative criteria upon a person from the age of six onwards and generally disapproves of nonconformist personality traits. In their study, Getzel and Jackson found that the teachers preferred the more intelligent to the more creative pupils. A person seeking an R & D career is likely to have "survived" the external evaluative criteria of the educational system to the level of a master's or doctorate degree. Some people may have had their creative potential irreparably stifled in reaching this level, but persons who manage to identify and preserve this creative talent throughout this twenty-year process are worthy of recognition and encouragement!

Before concluding this section, one other "attribute" can be reported. Because of the nature of the creative process Koestler, in *The Act of Creation*, suggests that the creative person has "multiple potentials," that is, the latent capacity to be creative in a number of fields (Leonardo da Vinci is an outstanding example). Koestler quotes Dr. Johnson as saying that true genius "is a mind of large general powers, accidentally determined to some particular direction, ready for all things, but chosen by circumstances for one." In other words, chance and necessity determine the outlet for a person's creative potential. Kepler was going to be a theologian, but was offered a job as a mathematician. Darwin was expecting to become a curate when invited to join the voyage of the Beagle. Therefore, probably the best indicator of an individual's creative potential is evidence of past creative achievement, not necessarily in the field of immediate interest to the prospective employer.

Creating a Creative Climate

A company will only derive the inventive benefits of having a potentially creative R & D staff if it provides an environment to support their potential. This implies "creativity-squared," that is, the creation of an organization climate in which personal creativity can thrive. In this section, we review the indicators of such a climate.

Although the research literature on criteria for creating a creative R & D climate is sparse, the findings of the few studies that do exist are both mutually consistent and congruent with "common-sense" expectations. Parmenter and Garber asked a sample of American scientists (whom their colleagues deemed creative) to rank order ten organizational factors that might encourage their creativity.[15] Gerstenfeld[16] conducted a broadly similar study in which he asked a sample of 122 scientists to rank order goal priorities from a list of eight goals, while Kaplan lists five important factors influencing organizational creativity.[17] Finally, the seminal study of Burns and Stalker provides an excellent background to the field.[18] The overall conclusions of these studies are as follows.

First, creative persons value autonomy and challenge. They desire freedom to choose challenging but realistic work assignments. However, provided they are involved in organizational research planning, they will seek to reconcile personal and organizational research goals. Some also welcome external "pressure to produce," possibly because it helps stimulate the psychic energy required for creative effort. To some, good R & D is like good journalism—"content under pressure."

Second, this means that the organization should be receptive and responsive to the individual's ideas, showing him appropriate recognition and appreciation. Despite a person's commitment to internally set goals and indifference to formal performance appraisal, his sense of equity or "natural justice" dictates that the organization should implicitly sanction his own sense of worthiness through remuneration and fringe benefits. Creative people will risk failure in order to explore novel ideas; therefore, the organization should not just reward "error-free" performance.

Third, puzzlement and wonderment, as well as errors, should also be nurtured in the continued search for invention. These are childlike traits which, as we have already seen, are typical of creative people. Furthermore, they are qualities that are encouraged in some creativity-enhancing techniques. Management convention tends to depreciate such traits as being symptomatic of indecisiveness, woolly-headedness, and immaturity—devasting management "sins"! Such traits are probably "sinful" for a person in a sensitive line management position, but for the researcher they may be "virtues" and should be accepted as part of the R & D climate. Indeed, tolerance of the nonconformist, provided he is productive in the long term, should be axiomatic in a creative climate.

Fourth, tolerance of conflict should accompany tolerance of eccentricity. Creativity thrives on conflicts of ideas and opinion, and it need hardly be stated that scientific advance is often carried forward on the wings of controversy. Provided the conflicts are substantively about ideas that generate creative problem solutions, and are not centered on personal territorial disputes, they should be accepted as one of the colorful features of the creative climate.

Fifth, we saw in the previous section that creative people are generally eager to discuss ideas with others. Indeed, the above conflict reflects a Hegelian approach to intellectual discourse in the pursuit of "truth." This consideration dictates that there should be minimal barriers to both intraorganizational and interorganizational communication between individuals, even at some risk to commercial security. It is believed that, in general, creative organizations gain more than they lose by the ready willingness to exchange ideas and information. This has implications for the staffing of the R & D function, which will be discussed further in chapter 12.

The research findings provide useful management guidelines for creativity-squared, but we should conclude the section with two brief caveats. First, although the psychological nature of the creative act may be invariant with respect to field, its contextual manifestation does vary throughout the innovation process. The creativity of an undirected research group in a company following an offensive/defensive strategy should itself be "undirected," since ideally its desired output is a continued but unspecified flow of novel inventive ideas. On the other hand, the success of a company following an applications engineering strategy is dependent upon the continued ability of its development engineers to provide creative solutions to particular user problems *in a timely manner*. The first group can afford to employ a Kepler or a Newton who can deliberate indefinitely before publishing a specific novel insight, although he may continually generate new insights throughout his creative career. In contrast, the second group cannot afford this approach. It must employ an Archimedes who can devise timely creative solutions to specific problems posed by an outside sponsor, especially when he knows that his personal survival may depend upon doing so! Although (as illustrated in the World War II careers of some eminent scientists) some creative individuals can be successful in either situation, it is important that the organization defines its manifest creative needs and seeks to recruit creative staff who can satisfy these needs. Too often companies are unsuccessful because they recruit the "wrong" sort of creative people (see Ansoff and Stewart[19]).

As was stated at the beginning of the chapter, management should also recognize the potentially critical role that creative thinking can play in all phases and aspects of the innovation process. Quite often the success of technological innovation is ensured through the timely creative solutions of "bugs" by technical and shop floor staff. R & D laboratories usually establish a cadre of good technical and craft personnel, but an innovation may be delayed or aborted through the absence of a similar cadre at the manufacturing stage.

Second, although creative staff are vital, they are obviously not the sole ingredients required for innovation to occur. Indeed, they might be viewed as the yeast which ferments the "innovation brew," because successful innovation requires the cooperative participation of all corporate elements. This remark is valid even in the R & D phases of the innovation process, because the successful development of a project through to the pilot or prototype production stage requires a mixture of skills. The important topic of R & D staffing for successful innovation is discussed in chapter 12.

8.5
Creativity Stimulating and Enhancing Techniques

Given that a company has hired creative R & D staff and provided a supportive organizational climate, the final issue which we will consider in this chapter is whether there are other means available for stimulating and enhancing individual and group creativity. Despite a commitment to internal standards, a creative individual is usually an active member of the scientific community in his field. Enthusiasm and curiosity make this individual ready to communicate ideas and information with colleagues and he sees this as a stimulus to his own as well as other people's thinking. This observation suggests that the creative process can be stimulated at the individual and group levels. This is indeed so.

In section 8.3, we distinguished between convergent and divergent thinking as being roughly equivalent to Green's systematic and intuitive thinking. We also quoted Kuhn's view that an individual's scientific creativity is based upon an "essential tension" between his convergent and divergent thinking. Thus the creative process may be viewed as dynamic or psychic interplay between convergent and divergent modes of thinking, and there are specific analytical and psychological techniques in use, which stimulate these modes of thinking at both the individual and group level. These may be conveniently classified as techniques for stimulating convergent and divergent thinking. Useful, detailed reviews of these techniques are given in Prince, [20] Rickards,[21] and Whitfield,[22] and we outline the more important ones in the remainder of this chapter.

8.6
Techniques for Stimulating
Convergent Thinking

In chapter 5 we stated that TF is often conducted as a creative problem-solving process using creativity stimulating and enhancing techniques. In sections 5.12 through 5.13 we discussed two such techniques—morphological analysis and relevance trees in the TF context. Both these techniques are analytical and logical and stimulate the creation of problem solutions by rational thought processes or convergent thinking, and other simpler techniques are also used.

Two of the simplest analytical approaches, which owe their origins to method study, are attribute listing and value analysis. Attribute listing requires the description of each component, the definition of the function of each component, and the evaluation of all possible ways in which each attribute of every component may be changed. Attribute listing may be viewed as an incomplete form of morphological analysis and also as the basis of another technique used in method study—value analysis. This latter technique compares the cost of each component with the value of the function it performs. Its application can often identify ways of improving the cost-effectiveness of a capability.

Earlier in this chapter we stressed that creative skills are vital throughout the innovation process. Given the nature of this process, particularly in the development phases onwards, workers face a problem requiring a "remedial" as opposed to an "inventive" solution perhaps before a critical time deadline. For example, the performance of an innovation may fail to meet technical specifications at development, pilot-prototype, or full production stages. Kepner and Tregoe are management consultants who pioneered an approach to analyzing management and organizational problems requiring the restoration of a performance attribute which has departed from some predetermined standard.[23] The deviation is precisely defined and alternative causes for this deviation are rigorously diagnosed or defined. Problem solving then focuses on remedying the most likely cause of the performance deviation.

8.7
Techniques for Stimulating
Divergent Thinking

Techniques for stimulating divergent thinking can be described as nonrational, in that they essentially seek to simulate an individual's unconscious and preconscious nonrational cognitions. These techniques are most effective when performed in a small group setting (rather than by a lone individual) and, in the light of Kuhn's stress on "essential tension," cyclically with the rational techniques we have already discussed. Because they seek to identify nonrational bisociations that normally occur at a prerational level of mentation, they may appear bizarre and frivolous to readers who have not met them before. Therefore, we stress that these techniques have been used effectively to stimulate creativity in many business organizations.

Brainstorming

Brainstorming is probably the technique most familiar to managers and the public at large. Indeed it is possibly too well known, since it is used loosely to describe any free-for-all group problem-solving or idea-generating session. The originator of the approach, Alex Osborn, points out that the verb *to brainstorm* means "to practice a conference technique by which a group attempts to find a solution for a specific problem by amassing all the ideas spontaneously contributed by its members." Osborn describes his brainstorming approach in *Applied Imagination,*[24] and the approach and applications are described in numerous publications. This outline of the approach is mainly based upon Rickard's description.

The approach, usually conducted by one or two group leaders or organizers, is divided into a sequence of sessions. In the present context, the topic to be considered is likely to be an R & D problem to be solved. The brainstorming organizer and researchers involved can be expected first to define the problem. Once the problem is identified, the next task is to select the group of six to ten people who will attempt to solve it, and Osborn's criteria for group selection follow.

1. Prior Experience. The group should include at least two people who have prior experience of brainstorming sessions and preferably, at least two people without prior experience.

2. Personalities. Members should be restricted to people who participate constructively in committees and are sympathetic to the brainstorming approach.

3. Disciplinary and Corporate Backgrounds. Membership should embrace the disciplinary and functional backgrounds potentially relevant to the problem, subject to the overall size limitation. Seating arrangements should be informal, preferably a semicircle.

A sequence of several sessions is conducted, each with a specific purpose.

1. Warm-up. The organizer may encourage members to give some preliminary thought to the problem, but brainstorming will begin with a warm-up session to encourage personal interaction and group awareness. The brainstorming approach will be described, and the session may end with a discussion and (possibly) a redefinition of the problem.

2. Idea Generation. This roughly corresponds to Koestler's cogitation process and seeks to encourage the *uncritical* generation of as many ideas as possible, while postponing critical evaluation and judgment. Each member is encouraged to suggest an idea that might be relevant, however "crazy," as it "comes into his head." The organizer lists ideas as they are generated, usually on a flip chart at the center of the semicircle, so that they may be readily viewed by the whole group. Members are encouraged to improve or "piggyback" on each other's ideas.

Typically, at the beginning, ideas are generated fairly rapidly as each member puts forth his "own" ideas, but pays little attention to those of the others. One session on increasing efficiency in an R & D contract research organization produced fifty-five ideas in ten minutes. There then follows a slump in ideas generation and possibly relatively long periods of silence. This period can be the most fruitful, because it encourages members to cogitate on each other's ideas and generate useful combinations of piggybacking. If the slump proves unproductive, continuing for too long so that members become bored, the organizer may stimulate the atmosphere by asking questions or introducing his own ideas from which to continue brainstorming. Rickard's quotes the example of a group session on improving the works canteen which appeared to have exhausted its capacity to generate ideas. Someone suggested the wild idea of topless waitresses which led to the practical suggestion to install beverage vending machines.

Specific techniques for generating ideas have also been used. "Trigger Sessions" is one approach in which each member works alone to produce an ideas list and then reads it out to the group as a stimulus. Another approach is "Recorded Round Robin," a variation on the party game of "consequences." Each member lists an idea on a card and passes it on to another member, who attempts to piggyback on it, then passes onto the next member and so on. If each of a six-member team generates three ideas on three separate cards, the "consequences" of eighteen ideas can quickly be explored.

3. Ideas Evaluation. The ideas generation session usually lasts up to two hours and produces a substantial list of ideas of varing usefulness. At least one member of the research team will have been included in the group. It is the researcher(s) who will be primarily concerned with the rational evaluation of the ideas generated. Further evaluation may be conducted with the participation of all group members, or just the research team. This rational analysis will further explore the ideas that look promising and may incorporate rational techniques such as morphological analysis. Other techniques for evaluating ideas are discussed under synectics below.

Synectics

"The word Synectics, from the Greek, means the joining together of different and apparently irrelevant elements." With this sentence, William Gordon, its inventor, begins the introduction to his book describing the approach.[25] The words "Synectics" and "Bisociation" are virtually synonyms, and Gordon's approach presents a more explicit procedure than brainstorming for the group stimulation of nonrational cogitations. Indeed, it is fruitful to compare Koestler's and Gordon's treatments of the creative process by reading both books together.

Synectics is quite similar to brainstorming in procedure in that an organizer or chairperson conducts a small group (typically four to six people) through a sequence of different sessions in an informal social setting. The criteria for group membership are also similar. The sequence of sessions is summarized as follows:

1. Making the Strange Familiar. Although we may suppose that most team members are already cognizant of the problem being considered, given the differences between individual perceptions, a group discussion is likely to uncover unexpected problem features and ramifications. This process of clarification or "familiarization" of the group with the "strange" problem yields the *problem as understood*. Group members may also suggest problem solutions at this stage which are evaluated or *purged* and, if one is acceptable, the procedure may be terminated. The term *purging* is used to encourage individuals to forget unacceptable solutions and not allow them to create psychological blockages to future constructive thought. Purged solutions are recorded, however, in case the group wishes to reconsider them later.

2. Making the Familiar Strange. Gordon contends that by making the familiar strange, or by viewing the problem in an familiar manner, creative solutions can be generated. He believes that such solutions are generated by metaphorical (as opposed to analytical) thinking and the development and exploration of analogies, and the synectics process is based upon a person's capacity to learn this metaphorical cast of mind. Gordon views biological analogies as being generally the most fruitful. He defines four types of analogy[26] for making the familiar strange and we consider each, in turn.

First, personal analogy involves the individual in an empathic personal identification with the elements of the problem; that is, he thinks of himself as a dancing molecule, or in the case of Kekule when inventing the benzene ring, as a snake swallowing its own tail. Gordon also quotes the writings of Faraday and Einstein to show that they exploited personal analogies in their creative thinking. If the problem concerns the development of a tire material for use on vehicles in a desert terrain a person may imagine himself as a tire traveling at fifty miles per hour over hot sand. The exploration of personal analogies enables the individual to get "inside" the problem and provides a novel perspective from which a solution may be generated.

Second, direct analogy involves the direct comparison of parallel observations or systems. We have already indicated Gordon's emphasis on biological analogies. He quotes several examples, including Kettering's addition of a red dye (tetraethyl lead) to kerosene to make it vaporize quickly, after observing that a wild flower (the trailing arbutus) blooms early in spring because its red-backed leaves more readily absorb heat. Twiss quotes the use of a direct analogy with the homing pigeon to develop a method for a novice brick-layer to build a straight brick wall.[27]

Third, symbolic analogy is impersonal and objective and aesthetically pleasing rather than technically accurate. Gordon says that James Clark Maxwell used it extensively and quotes a synectics group using the symbolic analogy of the Indian rope trick to generate the solution to the problem of designing a jacking mechanism that fits into a four-inch cube, but would extend to three feet and support four tons.

Fourth is Fantasy Analogy. Freud viewed creativity as a perpetuation of childhood fantasy, and Gordon views fantasy analogies as applying Freud's wish fulfillment theory of art to technical problem solving. It is rather similar to the wildest idea approach in brainstorming, in which the group is invited to develop and explore any wild fantasy, which may incorporate the judicious suspension of the laws of nature and common sense, to the problem. Gordon quotes an example applied to the invention of a vapor-proof enclosure for space suits. The group developed fantasy analogies of trained insects closing and opening the seal to military-style drill orders and then a flea sewing a spider's thread in and out through closure holes! These fantasies led to a practical solution involving a wire pulling together overlapping steel spiral springs embedded in the two rubber surfaces which had to be "stitched" together.

The organizer encourages the group to explore potentially fruitful analogies by diverting its thinking (or taking a vacation) from the problem to another (apparently) unrelated subject. This subject will be conceptually distant from the problem but should, in the opinion of the organizer, offer scope for the development of fruitful analogies. Gordon recommends that a fantasy analogy is best tried first (presumably biologically based, given his preference for biological analogies), but the choice is probably best left to the discretion and judgment of the organizer. Further analogies may be introduced at appropriate stages in the discussion until an acceable solution is generated. The whole session may typically take sixty to ninety minutes before a solution is suggested which offers practical promise. This solution (or solutions) may then be explored further, possibly using the other techniques we have already described.

Lateral Thinking

The lateral thinking approach has been developed by Edward de Bono and described by him in several books.[28] His concepts of *lateral* versus *vertical* thinking roughly correspond to divergent and convergent thinking introduced earlier. Vertical thinking is "conventional" logical thinking in which knowledge progresses in a sequence of analytical and logical steps. De Bono uses the analogy of water flowing down well-defined channels for this thinking, noting that the more water flows down these channels, the more it will continue to do so. In contrast, lateral thinking focuses on overall patterns and interrelationships between the elements of a problem and seeks to discover new patterns of association, using nonlogical approaches, which will lead to a problem solution. De Bono argues that the deliberate departure from the conventional logical or vertical approach to the problem through lateral thinking is equivalent to the damming up of existing water flow channels and exploring the new flow patterns thereby generated, to see if they offer a more fruitful irrigation catchment. As de Bono expresses it, in vertical thinking logic is in control of the mind, whereas in lateral thinking the mind is in control of logic. It is clear that the lateral thinking process is broadly similar to the synectics process and Gordon's metaphorical thinking, and seeks out creative bisociations.

It follows that de Bono's approach to creativity stimulation is broadly similar to brainstorming and synectics and may be used in similar group settings. He stresses that first one should critically examine the ideas associated with the vertical thinking approach to the problem. One should identify dominant ideas and boundaries that unduly constrain and influence solution approaches, tethering factors and assumptions that are implicitly included (without re-evaluation) in every solution approach, and polarizing tendencies that discourage the adoption of a position between two extremes. This critical process can free the mind from the thought habits and preconceptions of previous approaches and set the stage for lateral thinking. As with synectics, specific techniques for stimulating lateral thinking are recommended.

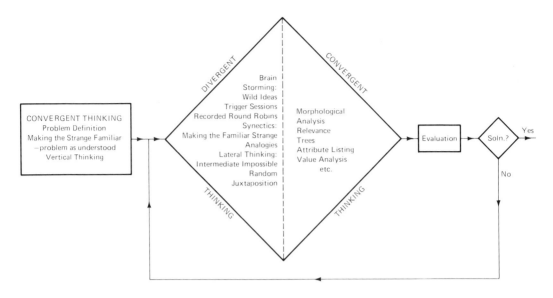

Figure 8.6 Convergent-Divergent Cycles.

The first is the intermediate impossible. This technique seeks to break logical constraints and resembles the "wildest idea" and "fantasy analogy" in brainstorming and synectics. Although rationally untenable, the intermediate impossible idea provides a start for potentially fruitful explorations. Whitfield (p. 91-2) considers the problem of making road transport more efficient and starts from the "intermediate impossible" idea of requiring road vehicles to be stationary.[29] The idea of stationary road vehicles leads logically to moving roadways onto which vehicles park, which has yet to be implemented. However, it also leads to practical solutions which are in widespread use: motorail services (in which cars are carried on flat rail wagons behind passenger trains), peak-hour traffic restrictions (which constrain the direction of traffic flow by time of day), "kiss-and-ride" schemes (whereby commuters are dropped off at suburban stations by their spouses), or tele/video-conferencing techniques (which obviate the necessity to travel from A to B anyway).

The second technique is Random Juxtaposition. This resembles the vacation approach in synectics, in which an apparently unrelated concept is introduced in the hope that it will trigger a sequence of associations which will lead to a problem solution. Whitfield considers the problem of providing a cheap domestic central-heating fuel.[30] He randomly selects the word "nonsense" which evokes the synonyms absurd, garbage, meaningless, trash, rubbish, waste. Garbage, trash and rubbish? Why not develop a domestic furnace which

156

will burn household garbage and trash? Waste? Why not control the decomposition of plant, animal, and human organic waste to convert it into fuel? (Whitfield reports that someone is running a car on chicken manure and that sewer gas is being used for street lighting.)

Clearly the brainstorming, synectics, and lateral thinking approaches all stimulate successive periods of group convergent and divergent thinking as illustrated in figure 8.6. Cycles of convergent and divergent thinking may be repeated several times, using these techniques (and perhaps others) until a solution is found.

Group creativity stimulating techniques are now used quite extensively in business and have been incorporated into overall systems for stimulating innovation and new product development in companies. Two such systems are described by Carson and Rickards[31] and Buggie.[32]

References

1. Arthur Koestler, *The Sleepwalkers* (Harmondsworth; Penguin Books, 1964); Arthur Koestler, *The Act of Creation* (London: Hutchinson & Co., 1969).
2. Wolfgang Kohler, *The Mentality of Apes* (London: Pelican Books, 1957), p. 35.
3. Koestler, *Act of Creation*, p. 100.
4. Ibid., p. 105.
5. E.I. Green, "Creative Thinking in Scientific Work," in *Research, Development, and Technological Innovation: An Introduction,* James R. Bright (Homewood, Ill.: Richard D. Irwin, 1964), pp. 118-28.
6. Konrad Lorenz, *Behind the Mirror* (New York: Harcourt Brace Jovanovich, 1977).
7. L. Hudson, *Contrary Immaginations* (Harmondsworth: Penguin Books, 1966).
8. Thomas S. Kuhn, "The Essential Tension: Tradition and Innovation in Scientific Research," in *Scientific Creativity: Its Recognition and Development,* ed. C.W. Taylor and Frank Barron, (New York: John Wiley & Sons, 1963), chap. 28.
9. J.W. Getzel and P.W. Jackson, *Creativity and Intelligence: Explorations with Gifted Students* (New York: John Wiley & Sons, 1962).
10. E.P. Torrance and E. Hansen, "The Questing-Asking Behavior of Highly Creative-Less Creative Basic Business Teachers," *Psychological Reports* 17 (July-December 1965): 815-18.
11. Donald MacKinnon, "Personality and the Realisation of Creative Potential," *American Psychologist* 20, no.4 (1965): 273-81.
12. Anne Roe, *The Making of a Scientist* (New York: Dodd, Mead, and Co., Inc., 1953).
13. J.A. Chambers, "Relating Personality and Biographical Factors to Scientific Creativity," *Psychological Monographs: General and Applied,* no.7 (whole no. 584, 1964).

14. Frank Barron, "The Disposition Towards Originality," *Journal of Abnormal Social Psychology* 51, no. 3 (November 1955): 478-85.

15. S.M. Parmenter and J.D. Garber, "Creative Scientists Rate Creativity Factors," *Research Management* 14, no.6 (November 1955): 478-85.

16. Arthur Gerstenfeld, *Effective Management of R & D* (Reading, Mass.: Addison-Welsey Publishing Company, 1970), p. 83.

17. Norman Kaplan, "The Relation of Creativity to Sociological Variables in Research Organizations," in Taylor and Barron, eds., *Scientific Creativity,* pp. 44-52.

18. Thomas Burns and C.M. Stalker, *The Management of Innovation* (London: Tavistock Publications, 1961).

19. H. Igor Ansoff and J.M. Stewart, "Strategies for Technology Based Business," *Harvard Business Review* 45, no. 6 (November-December 1967): 71-83.

20. G.M. Prince, *The Practice of Creativity* (New York: Harper & Row, 1970).

21. Tudor Rickards, *Problems Solving through Creative Analysis* (Epping, Essex: Gower Press, 1974).

22. P.R. Whitfield, *Creativity in Industry* (Harmonsworth: Penguin Books, 1975).

23. C.H. Kepner and B.B. Tregoe, *The Rational Manager* (New York: McGraw-Hill Book Co., 1965).

24. Alex F. Osborn, *Applied Imagination* (New York: Scribner's, 1957).

25. William J.J. Gordon, *Synectics: The Development of Creative Capacity* (New York: Collier MacMillian, 1961).

26. Ibid., pp. 38-53.

27. Brian C. Twiss, *Managing Technological Innovation*, 2d ed. (London: Longman Group, 1980), pp. 86-87.

28. Edward De Bono, *Lateral Thinking for Management* (New York: McGraw-Hill, 1971); Edward De Bono, *The Use of Lateral Thinking* (Harmondsworth: Penguin Books, 1971).

29. Whitfield, *Creativity in Industry,* pp. 91-92.

30. Ibid.

31. John Carson and Tudor Rickards, *Industrial New Product Development* (Epping, Essex: Gower Press, 1979).

32. Fredrick D. Buggie, *New Product Development Strategies* (New York: American Management Associations, 1981).

9

Project Evaluation

With the aid of this model and the key to unlock it, one can then invent further plants **ad infinitum** *which, however, must be consistent; that is to say, plants which, if they do not exist, yet could exist, which far from being shadows or glasses of the poet's or painter's fancy, must possess an inherent rightness and necessity. The same law applies to all the remaining domains of the living.*

Goethe

9.1
Introduction

We now turn to the subject of the evaluation of R & D projects as potential innovations. There is voluminous literature on this subject and it would be beyond the scope of this text to review it in detail. Up to (and including) this primary experimental development phase, a project is probably best subject to peer group evaluations as mentioned in chapter 7. However, prior to more advanced development, it should be subject to an increasingly comprehensive evaluation along the lines suggested in this chapter. Moreoover, this initial evaluation should mark the beginning of an ongoing process, which will continue until either the project is aborted or the innovation is launched and established in its market(s). In chapter 2, it was suggested that the evaluation of prospective innovations should incorporate a Popperian conjecture-refutation process. The approaches and the ongoing evaluation process described in this chapter, essentially reflect this process.

Before going further, it is useful to recall the distinction made between revolutionary and normal innovations in chapter 2. By definition, revolutionary innovations, like their scientific counterparts, often induce major changes in social as well as engineering practices and attitudes, sometimes creating new industries. Such innovations imply the application of new technology to old and often new markets, that is, the upper row cells of the technology-market matrix (chapter 4, figure 4.3). New technology must often displace old technology to establish itself, even when it has emerged from the same industry base. For example, the semiconductor electronics industry emerged from the vacuum tube electronics industry in the 1950s, following the invention of the transistor. Some electronics corporations found it expedient to set up separate semiconductor divisions to ensure that the growth of this embryonic technology was not stunted by persons with a vested interest in preserving the status quo of the old vacuum tube technology. These divisions were essentially semi-autonomous new business ventures. This consideration suggests that the evaluation of a revolutionary project or innovation requires a more extensive commitment of company resources and a more extensive evaluation process, than incremental normal ones. We, therefore, begin by discussing the evaluation of revolutionary projects or innovations.

9.2
Revolutionary Innovations—
White's Approach

One approach to the evaluation of the revolutionary innovations has been developed by Dr. George R. White, corporate vice-president and vice-president-Information Technology Group of Xerox Corporation.[1] He developed the approach while he was Carroll-Ford Foundation Visiting Professor at the Harvard Business School. He thus brought to it precepts and perceptions from his experience as a senior manager in a high-technology corporation which has itself been responsible for a revolutionary innovation, as well as those from the mecca of management academia.

White's approach evaluates the "credits" and "debits" of a proposed innovation in terms of its technology potency and business advantage. His overall conceptual framework (shown in figure 9.1) is developed in detail through a retroactive analysis of two fairly recent revolutionary innovations, the personal transistor radio and the Boeing 707/DC 8 passenger jets (see figure 9.2). He then applies the approach to two as yet (at the time of writing) unrealized innovations, the application of microprocessor technology to the control and operation of automobile engines and the SST.

The evaluation consists of asking the series of test questions shown on the left-hand side of figure 9.1 to determine the merits of the proposed innovation as shown in figure 9.2. We consider each of these in turn using the first two examples cited.

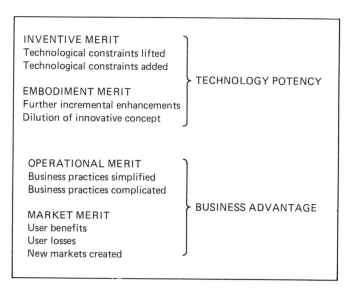

INVENTIVE MERIT
Technological constraints lifted
Technological constraints added

EMBODIMENT MERIT
Further incremental enhancements
Dilution of innovative concept

} TECHNOLOGY POTENCY

OPERATIONAL MERIT
Business practices simplified
Business practices complicated

MARKET MERIT
User benefits
User losses
New markets created

} BUSINESS ADVANTAGE

Figure 9.1 White's Approach.

	Transistor Radios	*Boeing 707/DC 8*
INVENTIVE MERIT	Solid state vs. vacuum tube amplification. Elimination of HT power source. Increased reliability and battery life. Reduction in weight and size. Increased initial costs, but cheaper later.	Gas turbine vs. piston engine. Increased power and speed. Increased reliability. Longer run-ways required.
EMBODIMENT MERIT	Ferrile rod antennas. Smaller loudspeakers and tuning capacitors. Power signal selectivity, tuning precision. and audio quality.	Swept back wings. Dutch roll problems.
OPERATIONAL MERIT	Distribution simplified and costs reduced.	Lower costs per seat-mile. Lower maintenance costs.
MARKET MERIT	Market outlets increased Creation of personal/pocket radio Initial higher cost/price	More comfortable speedier travel plus lower fares attracted new air passengers and others from surface transportation. Decline of luxury travel (e.g. Ocean Queens)

Figure 9.2 Transistor Radios and Boeing 707/DC 8.

Inventive Merit

Thus at the heart of a revolutionary innovation is an R & D invention which constitutes a new combination of scientific principles and new ensemble of technical elements which has the potential to displace existing technology. That is, in terms of our Lamarckian evolution analogy, it is a new technological mutation. In the case of the transistor it is amplication via positive hole and electron currents flow across semiconductor p-n junctions, displacing electron currents flow generated by thermionic emission in triode vacuum tubes. In the case of the 707/DC 8 it is rotary air compressor followed by a combustor and gas turbine, displacing the reciprocating piston crankshaft engine.

The new technology may eliminate or relax previous technological constraints (basic constraints lifted), but also may add or increase others (basic constraints added). The introduction of transistors into radios eliminated the need for a high-tension (HT) voltage and reduced weight, size, power-consumption, and hence battery life and increased ruggedness and reliability. At the same time, the low manufacturing yields plus high initial "start-up" investments in the semiconductor industry made transistors more expensive than vacuum tubes for some years. The jet engines were superior to piston and turbo-prop engines in power and speed, but were less efficient at takeoff so their initial use had to be on routes between airports which could justify runway extensions to two miles. On balance, the credits of these inventions outweighed the debits, so they could be assigned a high-inventive merit.

Embodiment Merit

R&D inventions are rarely, if ever, innovated in a technological vacuum, but are embodied in large technological products (radios or 707/DC 8 airframes). This process of engineering embodiment may offer opportunities for further incremental enchancements which were previously technologically or economically infeasible or unattractive, albeit with some dilution of the innovative concept. The potential reduction in the size and weight of radios offered by the transistor was enhanced by the Japanese electronic industry's introduction of ferrite rod antennae and smaller loudspeakers and tuning capacitors. However, this enhancement was achieved at the cost of poorer signal selectivity, sound reproduction, and tuning precision. The increase in speed offered by jet engines was enhanced by sweeping back the wings of the planes more sharply to an angle of thirty-five degrees (707) and thirty degrees (DC 8), an innovation which would be pointless on a propeller-powered airplane. This enhancement was diluted through the need to control an aerodynamic phenomenon called Dutch roll. The commercial success of both of these innovations suggests that the embodiment enhancements were overall meritorious.

High ratings on inventive and embodiment merits suggest technology potency, so attention is now turned to business advantage (the matching blade in Smookler's analogy, cited in chapter 2).

Operational Merit

By definition, a revolutionary innovation can be expected to change business practice. Therefore, it is necessary to identify the extent to which the innovation will simplify or complicate such practice. The introduction of transistors reduced the size and weight and increased the reliability of radios. Therefore, distribution costs could be reduced and, more important, the increased reliability made a franchised dealer service distribution system redundant. This provided the opportunity for new wholesale and retail sales outlets for the product. No complications of business practice followed. In the case of the jets, their limited introduction into long-range routes, plus their increased reliability, simplified practice.

Market Merit

With a revolutionary innovation which is displacing a prior technology in an existing market, the "market merit" questions in figure 9.1 are self-evident. Nonetheless, recollection of the technology-market matrix suggests that the "new market" aspects of these questions invite specific attention. The size and weight reductions of Japanese transistor radios, coupled with their increased reliability, opened a previously unrealized latent product-market opportunity for the radio industry. Previously, size/weight constraints had prevented anything smaller than portable or "pick-up, move, and put-down" radios to be manufactured. In return for a dilution of sound quality, and so forth, the concept of a reliable personal or pocket radio, which could be purchased from a neighborhood chain or drugstore, was introduced. This presented a new market which was steadily enlarged as the prices of personal radios dropped and the radio entertainment industry changed and increased its programming to cater to the teenage "pop" market which grew with increased personal radio sales.

The lower costs per seat mile of the 707/DC 8's meant that they were cost competitive with turbo-prop and the piston-engined airplanes over long-haul routes. Jet plane passengers quickly found that they offered a more comfortable as well as speedier journey, so airlines were forced to retire prior-technology airplanes prematurely to satisfy a final customer demand. Because costs per seat mile were lower and occupancy rates higher, fares could be lowered attracting new air travelers. The reduced fares coupled with reduced journey times also lured more passengers away from rail and sea travel.

The above historical analyses, which were used to illustrate the approach, are discussed in rather more detail in White's work. In the case of transistor radios, he argues that the U.S. radio industry failed to recognize the further engineering enhancements which could be achieved by corresponding miniaturization of other radio components to match the introduction of transistors, or the potential to develop a personal radio severed from the franchised dealer service network. Thus, the industry failed fully to exploit the innovative potential of the transistor and, at the same time, allowed its Japanese counterpart to establish a U.S. market presence. This example illustrates the importance of including engineering and marketing analysis and flair in the evaluation when assessing a potential new innovation.

This absence of "positive thinking" concerning engineering and marketing applies equally to Lockheed which developed the Electra, at the same time as its competitors developed the 707/DC 8. Lockheed failed fully to exploit the potential for engineering enhancement. They used turbo-prop engines and straight wings, thus maintaining some of the speed, noise, and higher maintenance costs constraints imposed by propellers, but retaining the facility for relative short takeoffs and efficient cruise which they offer. Passengers preferred the speed and comfort of the 707/DC 8 to the turbo-prop Electra, which was virtually driven off the market. Lockheed lost its position as "number two" to Douglas in the U.S. air transport industry and did not stage a comeback until the marketing of the Tri-star. In contrast, Boeing established a strong presence in the industry through its success with the 707, which it has reinforced through its later aircraft (727, 737, 747, and so forth).

The above two innovations are already "proven" so it is desirable to consider other potential revolutionary innovations which, at the time of writing, remained "unproven." White considers two—microprocessor technology in cars and the SST. The microprocessor is an R & D invention which offers the potential for improved operation and control of automobile engines in terms of lower fuel consumption and exhaust pollutants, and increased reliability. Because the invention is a digital microprocessor, engineering embodiment requires the development of sensors and actuators that will digitally encode and decode information from and to the engine assembly. Its introduction offers the prospect of further computerization of the automobile. It rates high-market merit because it provides the most (if not the only) cost-effective way of meeting future regulatory fuel consumption and emission standards. Given this overall evaluation, White predicts that the innovation should be commercially successful.

In contrast, White takes a pessimistic view of the prospects for a U.S. SST. (He views Concorde as a technological success, but a commercial failure.) The inability to eliminate sonic boom, coupled with higher fuel consumption, lowers its inventive merit. The impact of atmospheric pollution at the plane's cruising altitude (65,000 feet) remains unknown. Supersonic flight over the continental United States is not allowed at present. Therefore, either engineering enhancements which permit economic and reliable performance at subsonic as well as supersonic speeds must be developed or the large U.S. transcontinental market must be foregone, with consequent losses of sales and economies of scale. Although Concorde passenger response to flight comfort is reportedly favorable, whether the time savings outweigh the high prices to the majority of potential passengers is unclear. Given this evaluation, White says that a U.S. SST will never be commercially appealing.

The details of White's approach are clearly dependent upon the specific technology-market in question, but it is hoped that these illustrative examples are sufficiently explicit to enable the reader to apply the approach (at least mentally) to revolutionary and possibly normal innovations within his own organization. Two final comments on the approach are warranted, however. First, the distinction between the separate considerations—inventive merit, embodiment merit, and so forth—is to some extent arbitrary rather than clear-cut. For example, the uncertainty concerning the impact of atmospheric pollution generated by fleets of SST's flying at 65,000 feet is clearly a "demerit." However, it is somewhat arbitrary whether it should be labeled an inventive, embodiment, or operational "demerit." This arbitrariness is unimportant since the key feature of White's approach is that it offers a *systematic, comprehensive, and exhaustive* identification and evaluation of all the technological and commercial factors that should be considered in evaluating a revolutionary innovation. Second, although comprehensive, it is obvious that the approach is qualitative rather than quantitative. *This fact is positively advantageous.* The uncertainties of both technological and commercial outcomes of a revolutionary innovation are too great to render conventional fiscal analysis meaningful. What *is* required is rigorous, logical, and systematic qualitative analysis. This point is best summed up by quoting White directly.

> *In these criteria we have avoided terms such as "return on investment" and "return on assets managed." Our view is that these issues are overwhelmed whenever a new inventive concept can be placed in a beneficial embodiment which will enhance its value in a major latent market with lowered operational costs. If evolutions of an innovation must be based on assumptions of narrow differences in return on investment, they are quite possibly based on fallacy. What we proposed is a logic structure to identify a small class of innovations of greater promise whose success will transcend the cash value of any normal investment.*

We now consider normal innovations, that is, the second row cells of the technology-market matrix of figure 4.3.

9.3
Radical Normal Innovations

In chapter 2, it was suggested that the term *normal* innovation embodied a continuum from very minor cost-reducing incremental innovations to major or radical innovations which, though technologically substantive, fell short of inducing a Kuhnian paradigm-shift. The scope of impacts of such neorevolutionary radical innovations dictates that they should be evaluated by the White approach. Indeed, one example he cited—the application of microprocessor technology to the automobile—could be viewed as a radical (as opposed to revolutionary) innovation in the automobile industry. Therefore we need not consider these evaluations further because they are implicitly considered in the above approach.

9.4
Incremental Normal Innovations

The reader could apply White's approach (at least mentally) to incremental as well as revolutionary and radical innovations in his own organization because the approach provides a useful framework for reviewing any potential new product or process. When considering an incrementally innovative project, however, we should, by definition, have much more background knowledge of the relevant technology and expected market; so it should be possible to perform a more detailed evaluation. In chapter 2, I suggested that any innovation should be evaluated in the context of its contribution to the objective and goals of the business, expressed in the notion of "gap analysis" outlined in chapter 7 (section 7.4 and figure 7.2). We should therefore perform the evaluation in the context of its impact at corporate as well as functional levels and we consider each of these in turn.

The Corporate Context

Given that an incremental project is evaluated in the context of its contribution towards filling the gap between corporate goals and the future contributions which can be obtained from its current product lines, the evaluation should therefore provide "more likely," "pessimistic," and "optimistic" estimates of expected profits, revenues, costs, and corporate resource requirements (including a timetable for completion of successive phases of the innovation process).

The time estimates are important because the organization may wish to phase in an orderly sequence of innovations both to meet corporate goals (recall figure 7.2) and to balance the use of common organizational resources shared between several projects. Further, as implied earlier, the "timing" of the market launch of an innovation may critically affect its chances of market success.

The Market Context

I have already emphasized the importance of the continued consideration of market definition and development from the secondary development phase onwards of the innovation process. Ultimately the commercial success of an innovation is dependent upon its generation of *product sales* based upon the satisfaction of a perceived *user need.* An overall conclusion of the Sappho study on technological innovations was that "user need understood" was the most critical factor in determining their commercial outcome.[2] Despite the somewhat platitudinous nature of this observation, it is remarkable how frequently insufficient attention is given to this factor. The purpose of the project is to develop and produce an innovative product or products which will generate profitable sales; therefore, as Blowatt[3] suggests, one of the first questions which should be raised is: "Who is going to buy the product, and why?" Remarkably, the evidence suggests that innovations frequently fail, because of the failure to ask or to answer critically this deceptively naive and obvious question. Three examples will be cited to support this observation.

Blowatt cites a flying saucer story. Two Canadian engineers developed an authentic flying saucer powered by eight propellors driven by Wankel engines located around its inner circumference. It could fly at 200 miles per hour at 10,000 feet, carrying a 450-pound payload and requiring no airstrip, so that it could take off and land in literally one's own backyard. There were over 12,000 licensed pilots in Canada and the engineers, figuring that half would like their own flying saucer, estimated the Canadian market alone to be $60 million. Further analysis of its customer appeal unfortunately proved to be less sanguine, for three reasons. First, even with economics of mass production, the saucer could not be sold for under $10,000 and it was debatable whether these pilots would be willing or able to pay this price for this particular product. Second, government regulations required a two to five-year testing period before an airworthiness certificate would be granted. Third, because of its aeronautical characteristics, it appeared that a helicopter pilot's license would be required to fly it (and there were far fewer than 12,000 helicopter pilots in Canada), so that the initial market estimate was grossly optimistic. In consequence of these further discoveries, the project was abandoned and Blowatt reports that Sakowitz's

1977 *Catalogue of Unique Christmas Gifts* offered the prototype saucer for sale at a price of $600,000!

The second example, cited by Gerstenfeld, was the development of the polarized car headlights and occurred on the opposite side of the Atlantic, in West Germany.[4] A filter was developed which polarized the light beams from the headlights of a car, so that its driver could see ahead clearly, without dazzling oncoming drivers. It took four years to develop a technically sound product which failed in the market because changes in legislation were required to make the device legally acceptable, and the product although beneficial to oncoming motorists, who were the effective "users," offered much less direct benefits to its purchaser. Therefore, the motoring public showed little interest in the device and would presumably only purchase it if polarized headlights were made mandatory. The Polaroid Corporation, which is in the "polarized light" business, originally planned to develop and market the same product. They abandoned it in favor of entering the polarized light-based "instant photography" business.

The third example was quoted to the author by a technology manager in a small company. His company test developed a piece of laboratory equipment in the light of the needs of its end users—technologists who claimed that the equipment was technically superior to anything else available and that they would use it if it was made available. The company therefore manufactured and marketed the equipment, but again it proved a failure. What the company had failed to realize is that the purchasing decision for this item of equipment was made, not by laboratory personnel, but by purchasing agents. The product, as well as being technically superior, was significantly more expensive than alternative equipment offered by competitors. Purchasing agents, who were less sensitive to the technical merits of the competing products, therefore refused to sanction the purchase of a more expensive item of equipment.

All three of these products were developed to technical success, after substantial fiscal expenditure and sophisticated technical thinking. If the merest fraction of money and sophisticated thinking had been spent upon the identification of the mode of end use—the end users and the purchasers—the naive marketing mistakes would have been avoided. In the first two innovations, only a brief discussion with government agencies would have identified the regulations that constrained their mode of end use. In the first example, the engineers would have quickly recognized their target market of end users as helicopter, not fixed-wing, pilots! In the second and third innovations, cursory thought and investigation would have differentiated between end users and purchasers and would have discovered that the benefits accrued to the former, while the extra costs were paid by the latter!

These three management parables illustrate the vital importance of establishing a clear insight into the environmental niche or market that the innovation will fill. Blowatt provides a most useful survey of the factors to be considered and data sources which may be accessed in developing a market analysis and plan for a proposed new product. The analysis should identify

1. The market niche in terms of size, geographic distribution, and structure and the customers (individuals, companies, government agencies, and so forth) who should buy the product; the use needs of these customers and (for other than individuals) purchasing procedures in the user organizations; the advantages/disadvantages of the product *in the eyes of these users* as compared with competitive products; and the dynamic behavior of the market in terms of growth/decline, seasonality and fashion.

2. The impact of the economic and regulative environment on this market; how it might be affected by changes in the economic and political climate; and the environmental, health, and safety considerations and regulations that must be satisfied.

3. The nature of the competition—its strengths/weaknesses, the likely competitive reactions to the introduction of the product; and whether the product has adequate patent protection and does not violate competitors' patent rights.

Data required for such an analysis should typically be available from publicly accessible (see Blowatt for suggestions) and internal company sources. By definition, an incremental innovation considered here will usually be offered through an established product promotion, distribution, sales and servicing system, so the marketing function should be able to provide ample support for the market evaluation. Given strong downstream coupling, this support should be forthcoming, probably under the aegis of the person who enacts the role of the "market gate keeper" in the project team. (This role is discussed further in chapter 12.) Sometimes an incremental innovation may be targeted onto a new market. In this case a more substantial market analysis will be needed and a new market development strategy formulated.

The market analysis should establish estimates of annual sales revenue, based upon the following interrelated considerations, possibly disaggregated over a number of market segments:

a. Total sales estimate, including possibly license income from offshore markets.

b. Most likely, minimum and maximum unit product price.

c. Most likely, minimum and maximum performance level for the product.

d. Most likely, latest and earliest date of product launch and the expected growth and decline pattern over its life cycle.

e. Channels for promoting, distributing and selling the product. Any customer education and "after sales" servicing programs required.

f. The launch and overall strategy for marketing the product.

In making these analyses, it is vital to consider potential competitive developments or reactions (see **c** above). Competing companies may be developing new products to compete in the envisaged market and, insofar as possible, this contingency must be allowed for. Gerstenfeld quotes another example to illustrate this point as well as the need to monitor competitive developments regularly while the project continues.[5] Over two years, a company developed a method for imbedding aluminium oxide whiskers in a plastic base, to produce a reinforced plastic material of far greater tensile strength for little increased weight, compared with those currently available. At the beginning of the project, it had identified a ready market for such a product. Unfortunately, by the time development was completed and the innovative product was ready for launching the market had disappeared! A competing company had developed a similar but superior product at lower cost using boron carbide instead of aluminium oxide. The moral is try and keep "tabs" on the competition and be prepared to terminate a project or reorient its direction in the light of competitive developments.

Another marketing consideration was illustrated earlier in the Polaroid Corporation's decision to enter the "instant photography" rather than the "potential headlights" market. Sometimes during the process of technical and marketing development, a company recognizes that the original target market envisaged for the innovation will prove unacceptable and a new market is identified. One such company is ABCO Ltd. of Nova Scotia, Canada. This company pioneered the development of fiberglass-reinforced plastic piping systems, which it originally intended to market in the already familiar pulp and paper industry. The new technology proved to be unacceptable in that market, but the company found a new market for it in the utilities industry. This departure into a new-technology-market niche, which is described in Martin and Rossen,[6] proved to be very successful to the company.

The Technical Development Context

The development phases of the innovation process are primarily concerned with reducing technical uncertainty and establishing the technical feasibility of the product at varying performance levels. In consequence, the estimates of technical outcomes, development duration times, and costs can clearly be expected to be subject to large errors. The level of uncertainty, and these consequent errors, may vary with both technology, project, and the technological strategy of the firm. For an aggressive innovator seeking to exercise technological leadership, working close to the state-of-the-art with sparsity of precedent, the level of uncertainty may be relatively large. For a firm following an applications engineering strategy it may be relatively small. Whatever the situation, estimates of the times and costs of achieving milestone events will be needed to complete the overall evaluation exercise. They will also be needed to monitor project progress and be useful as an input to overall R & D management and planning (see chapters 11 and 12).

The Manufacturing Context

Apart from the limits to accuracy imposed by the uncertainties of the development process (including pilot/prototype development), in general manufacturing parameters should be the least difficult to estimate. An incremental innovation is unlikely to require radical changes from current manufacturing methods so, when compared with marketing and development estimates, relatively accurate estimates of capital and operating cost versus output levels over the estimated sales range should be readily derivable, taking into account possible economies of scale and "learning curve" savings (see chapter 13) as experience is gained with product manufacture. The costs of distributing, selling, and providing an after-sales service for the product plus a spare parts inventory must also be estimated.

Regulative and Environmental Impact Context

For some industries, notably the pharmaceuticals and aviation industries, the development process includes extensive government regulated clinical trials or prototype testing procedures. Since the late 1960s, however, widespread concern for environmental pollution has led to the widespread enactment of environmental control legislation in most Western developed countries. Clearly, any thorough project evaluation procedure should include an appropriate assessment of the environmental impacts of the proposed innovation, and the effects of these impacts on estimates of the costs and benefits to be derived from the project. Wisemma describes a useful checklist approach for doing this.[7]

9.5
Evaluation as a Concomitant Process

In chapter 2 we viewed an invention or potential innovation as a technological mutation in the technological evolutionary process which should be evaluated by an extension of Popper's evolutionary conjecture-refutation methodology. This evaluation should identify the niche (or niches) which the innovation might competitively occupy in the evolving technological marketplace or technosphere. A cybernetic schematic framework (figure 2.5) was also suggested for viewing the evaluation process, which we extended further in defining the innovation management function in figure 4.2. The frameworks considered so far in this chapter are consistent with the Popperian approach, but at first sight appear to emphasize a static or instanteous evaluation followed by a "go/no-go" decision. This impression is misleading. As Popper indicates trial-and-error learning selection is really a trial-and-*elimination-of-error* process.[8] In our context, this "learning" or "elimination-of-error" process must be conducted at both the technological and *marketing (or commercial) levels*. The successive phases of Bright's treatment of the technological innovation process represent error-elimination or uncertainty reduction at the technological level, but successful innovation management requires that error-elimination or uncertainty reduction be simultaneously or concomitantly conducted at the marketing level (as reflected in figure 4.2). That is, the technico-commercial evaluation of proposed innovation should be a dynamic or "ongoing" process. The arguments and framework for conducting this process are outlined in this section.

Schmidt-Tiedemann, who is a member of the executive board of Philips GmbH and director of the Philips Research Laboratory, in Hamburg, West Germany, outlines one such concomitance model which is of interest.[9] It is an extension of the Bright type of linear and the cybernetic frameworks introduced in chapter 2, and that author claims its application increased the "acceptance rate" of R & D projects by downstream functions in one Philips laboratory to 53 percent. Cooper has also proposed a framework for the "ongoing" evaluation of industrial new products, based upon extensive studies of successful and unsuccessful innovations.[10] The framework suggested here is primarily based upon Cooper's work. Cooper, in Project Newprod, studied the outcomes of 195 innovations (102 success and 93 failures) introduced by firms in Ontario and Quebec, Canada.[11] From his analysis of the results from this project, plus those from about a dozen similar ones in the United States, Western Europe, Hungary, and Japan, he identified six lessons for innovation managers and project team members:

A Much Stronger Market Orientation is Needed.

This is essentially the point stressed earlier. He identifies the lack of market research as the most frequent cause of failure and suggests that market research results should be integrated with technological outcomes from early in the process and, most important, as inputs to the product design, engineering, and product development activities.

New Product Success is Largely Amenable to Management Action.

Most studies suggest that the actions of team members (especially the product or project champion, see chapter 12) and the impact of key (especially market research) activities determine innovation outcomes. That is, the design and implementation of the innovation process critically determines its success.

There is No Easy Explanation for What Makes a New Product a Success.

Most studies identified a large number of factors influencing success and, quoting Cooper, "success depends on doing many things well, while failure can result from a single error."[12] Innovation management requires the management of numerous multidisciplinary tasks into an uncertain venture, in which the failure of any one task may lead to overall failure. Thus an overall planning framework is required.

The Product Itself—a Unique Product with Real Customer Advantage—is Central to Success.

At first sight, this observation appears to be obvious, but more considered thought suggests otherwise. The product must be unique and superior *in the eyes of the customer*, which implies that clear understanding of the customers' needs, performance criteria, use practices, and so forth, is required before advanced product development begins. To obtain such understanding requires significant market analysis and research effort and the subsequent marriage of technological design and development to customer needs.

A Well-Conceived, Properly Executed Launch is Vital to Success.

A well-conceived marketing strategy and plan leading to a properly targeted and executed launch effort must be integrated into the overall innovation process.

*Internal Communication and Coordination
Between Internal Groups Greatly Fosters
Successful Innovation.*

Since successful technological innovation requires the multidisciplinary and multifunctional participation of scientists, engineers, marketing, and production staff, and so forth, the creation and maintenance of good internal communication within the project team and with other organizational groups is vital. This issue is discussed further in chapters 12 and 13.

Based upon his research analysis and these six "lessons," Cooper proposes an innovation process model which constitutes a seven-stage action guide for innovation managers. Although designed for industrial manufactured products, it is applicable in broad outline to any innovation, and we describe a somewhat modified version of it here. It is complementary to evaluation frameworks for revolutionary and normal innovations described earlier in the chapter, since they may be used in performing the details assessment required in the process. The overall process is illustrated in figure 9.3, in the context of corporations with a central as well as divisional R & D facilities. As the project progresses, it is viewed as moving from corporate to divisional R & D and then into production. The corporate-divisional transfer would clearly not apply to a company operating a single R & D facility. We consider each of the seven steps in turn.

i. *Product Idea.* The product idea may possibly emerge out of nondirected research, coupled with technology and marketing monitoring activities, that is, the matching of a technological possibility with a potential market niche. As was suggested at the beginning of the chapter, there should be an informal peer group screening of the idea at this early stage. This should identify whether the idea is of sufficient technological interest, feasibility, and challenge, congruent with the company's technological strategy and "doable" within the company's resource. This screening and the successive evaluations may be made in competition with other projects or product ideas, so may be subject to the simpler of the screening methods described in the next chapter. If the project survives this screening, it will pass into the *primary development phase* of the process.

ii. *Preliminary Assessment.* Primary development should clarify the technical feasibility of the idea and give a rough idea of the resources and money required to yield a commercial product, so a preliminary market assessment should be made. This can be a fairly "quick and dirty" desk marketing study eliciting information from the company sales force, readily accessible statistical data and research reports, industry experts, and perhaps a few potential custo-

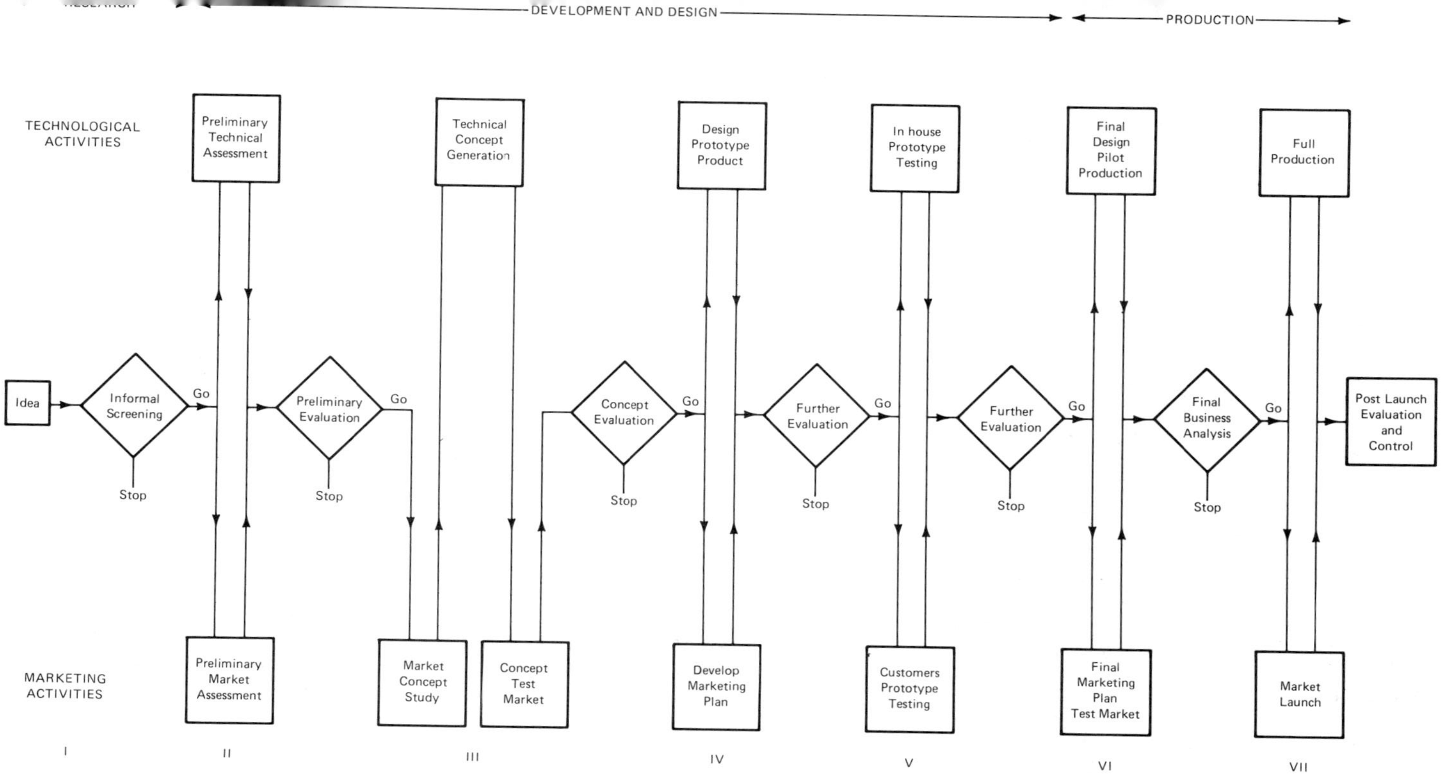

Figure 9.3 Cooper's Evaluation Process.

175

mers. If it survives a more rigorous screening process after this assessment, the project will continue to the *secondary development phase,* possibly being transferred from the corporate to the divisional R & D facility.

iii. *Product—Market Concept Definition.* Secondary development should further clarify and delineate the performance characteristics of the proposed product, thus faciliating the identification of its target market and how it will be positioned vis ā vis market segments and competitive products. Cooper comments that this concept definition stage is frequently omitted by industrial product firms, often with disastrous results. The concept definition should include several elements. First, it should identify a market niche for the product; that is, a customer segment dissatisfied with competitive products currently available or a niche where new technology can gain a competitive advantage. Second, it should identify what improved benefits or features are desired by these customers and the product concept and design (and its technical feasibility) required to satisfy them. Third, the concept should be informally market tested. That is, potential customers should be shown diagrams, models, or descriptions of the proposed product and their reactions noted. Product concept changes may be required in the light of these reactions, but they should enable an assessment of market acceptance and tentative sales forecast to be made. It will also initiate the market planning process and enable reasonable estimation of production costs to be made. Thus a reasonable financial and commercial analysis can be made by this stage.

To illustrate concept definition Cooper cites the example of a high-quality truck manufacturer who wished to enter the construction dump truck market. The initial product concept was to build a high-quality dump truck which appeared to be technically feasible and for which there existed a large growing market, but a detailed definition of the product concept and market positioning was lacking. A market study identified the likes, dislikes, and preferences of dump truck fleet operators. This showed that truck "downtime" due to breakdowns was a major and costly problem for operators. The revised product concept was defined as "a truck that can be repaired and back on the road overnight...no matter how serious the breakdown." To implement this concept, a modular vehicle was developed in which every major unreliable component could be removed and replaced quickly, using standard parts available from local suppliers. The minimum downtime product concept proved to be technically feasible at a competitive price.

The concept can now be subjected to a more detailed evaluation and, if continued, proceed to *tertiary design and development* leading to prototype or small-scale pilot production. By now such work will definitely be performed in the divisional R & D facility.

iv. *Tertiary Development, Design, and Market Planning.* Detailed technical design and development and market planning now
continue simultaneously. The market planning will determine the
pricing, selling, and advertising strategies, and the distribution and
after-sales servicing networks networks required. It may require
further buyer behavior studies. The technical output from this stage
will be prototype products which may be subject to customer testing.

v. *Prototype Trials.* One or more prototypes (depending upon
the product) will now be subject to detailed test trials both within the
company and with selected users. These trials conducted in a possibly more rugged and demanding user environment than obtained
in a development laboratory may identify technical and design
flaws before larger-scale production is begun. It may also identify
product modifications which should be incorporated to improve user
acceptance.

Cooper cites the example of a design fault in a dial-in-hand wall
telephone which was being developed by a major telephone manufacturer. Fifty prototypes were subjected to "in-house" durability
and reliability tests and a further fifty were installed in nontechnical
employees' houses for customer testing. This latter test identified a
potential disastrous design fault that had remained undetected in
laboratory testing: the receiver fell off the hook if a nearby door was
slammed in certain types of houses! This fault was eliminated by a
quite minor and inexpensive design change. After these trials, the
project is subjected to a further evaluation and, if continued, passes
onto stage VI.

vi. *Larger-Scale Pilot Production and Test Marketing.* The project incorporating any product design changes (identified above) will
probably now be transferred to the production facility and a *larger-
scale pilot production line* set up there. This stage may identify any
changes in manufacturing methods which may be required in scaling up to full-scale production. It will scale up output by at least one
order of magnitude and provide sufficient units to conduct a formal
test market, that is, to sell the product using the proposed marketing
plan to a limited sample of customers or in a limited marketing
areas. This should identify any remaining design faults which
should be eliminated or changes that should be made to increase
customer acceptance. Most important, it should also identify any
changes that may be required in the marketing plan.

When this stage is completed, and any desired product design,
manufacturing methods, and marketing plan changes have been
made, the final evaluation is performed. This will constitute a comprehensive business analysis and plan for the product, incorporating the considerations listed earlier in the chapter. If accepted the
innovation now proceeds to the final stage.

vii. *Full Production and Product Launch.* The final stage incorporates the adoption of the innovation by the operations management functions of the firm, the scale up to full production output, and the launching of the product through the implementation of the marketing plan in the full market area. Although further technical and manufacturing problems may need to be "debugged" during this final scaling-up process, if the previous stages have been thoroughly and effectively managed, barring any major unexpected changes in market circumstances, this launch should go ahead in a straightforward manner. Postlaunch monitoring will be conducted to identify customer acceptance, market share, sales, unit production costs, and so forth.

Cooper's process model provides a useful conceptual framework for managing the innovation process, which implicitly implements the Popperian conjecture-refutation process. Obviously the detailed stages may vary between differing product/process technologies, and it is presented in the context of an offensive-defensive innovator with a comprehensive technological base (as defined in chapter 3) here. However, a company following an applications engineering strategy from a truncated technological base could readily apply the latter stages of the process. Finally, it is of interest to note how the process reflects a transition from the divergent to convergent thinking discussed in the previous chapter. The early stages reflect a divergent exploration of alternative potential technology-market synergies while the latter stages converge on a specific focused product concept. Schmidt-Tiedemann, in describing his process model, argues that as an R & D project proceeds, the largely divergent search for new ideas or problem solutions is replaced by the convergent pursuit of a certain goal.[13] As he puts it, switching at the right time from divergent to convergent thinking is largely a matter of technical and commercial judgment or gut feeling. To converge too late means spending too much effort on losers. To converge too early may mean throwing away winners. This brings us to the topic of project selection which is discussed in the next chapter.

References

1. George R. White, "Management Criteria for Effective Innovation," *Technology Review* 80 (February 1978): 15-23.

2. *Success and Failure in Industrial Innovation* (London: Center for the Study of Industrial Innovation, 1972).

3. K. R. Blowatt, "Marketing for the Technical Entrepreneur,"in *The Technical Entrepreneur,* ed. Donald S. Scott and Ronald M. Blair (Ontario: Press Porcepic, 1979), pp. 135-83.

4. Arthur Gerstenfeld, *Innovation: A Study of Technological Policy* (Washington: University Press of America, 1977), chap. 5.

5. Arthur Gerstenfeld, *Effective Management of Research and Development* (Reading, Mass.: Addison-Wesley Publishing Co., 1970), pp. 29-30.

6. Michael J. C. Martin and Philip J. Rossen, "R & D Philosophy at ABCO," in *Four Cases on the Management of Technological Innovation and Entrepreneuriship,* Technological Innovation Studies Program (Ottawa: Department of Industry, Trade and Commerce, November, 1983).

7. Johan G. Wissema, "Putting Technology Assessment to Work," Research Management 24, no. 5 (September 1981): 11-17.

8. Sir Karl R. Popper, "The Rationality of Scientific Revolutions," in *Scientific Revolutions,* ed. Ian Hacking (Oxford: Oxford University Press, 1981), pp. 80-106.

9. K. J. Schimdt-Tiedemann, "A New Model of the Innovation Process," *Research Management* 25, no. 2 (March 1982): 18-21.

10. Robert G. Cooper, "A Process Model for Industrial New Product Development," in *IEEE Transactions on Engineering Management,* forthcoming.

11. Robert G. Cooper, *Project Newprod: What Makes a New Product a Winner?* (Montreal: Quebec Industrial Innovation Centre, 1980).

12. Cooper, "A Process Model for Industrial New Product Development," p. 11.

13. Schmidt-Tiedemann, "A New Model of the Innovation Process," p. 19.

10

Project Selection
Approaches

If you can look into the seeds of time, And say which grain will grow and which will not.

Macbeth, *Act I, Scene iii.*

10.1
Introduction

Given the fecundity of a good R & D function in generating potential innovations, it is virtually certain that a firm's R & D budget will be insufficient to support all the projects which are competing for funding. Therefore, any R & D-based firm will be required to operate some explicit or implicit procedure for R & D project ranking and selection. The purpose of this chapter is to examine some of the considerations and difficulties that are inherent in this issue and outline some approaches which appear to be of practical value. The chapter does not suggest any one methodology that can be applied slavishly to all situations. R & D project selection, like most areas of business decision-making, is best performed by informed management judgment at the appropriate level in the organization, using formal techniques as guides and disciplines to promote effective decision making. Therefore, we first focus on the appropriate managers who should participate in such a decision-making process.

10.2
Choice of Decision Takers

Whether R & D project selection should be made by general management depends upon the stage of the project in the innovation process, the resources required for its performance, and the *immediacy of the impact on the business of successful completion.* In chapter 7 I argued that the overall funding of long-term R & D should be set by corporate management and individual projects selected by peer group evaluation. I also argued there that both the overall funding and individual selection of projects from secondary development onwards should be determined by the innovation management function within the context of corporate gap analysis. Therefore, we may now discuss the formal techniques that may be used to evaluate and select such projects.

10.3
Project Selection Techniques

There is a large body of literature on this subject, and it would be beyond either the scope of this text or the interests and concerns of most of its readers to review this literature in great detail. We have already indicated the major considerations that apply to project selection, but to clarify further discussion, I will summarize these considerations.

First, R & D projects should be developing product concepts which will fill the future "gap" between revenue and profits from future sales of current products and the goals set by the corporate plan. Second, a good R & D facility is fairly prolific in generating projects with "new product" potential. Third, projects are typically interdependent rather than independent. That is, the knowledge and outputs obtained during the course of performing one R & D project may impact upon others; probably more important, projects compete for scarce personnel, laboratory space, and experimental equipment, as well as for finances, so that the extent that a project "consumes" these scarce resources must be considered. Fourth, projects are beset by technological, manufacturing, marketing, and environmental impact uncertainties, which dictate that the probability of *any one* project yielding a commercially successful outcome may not be high.

These four factors dictate that the intent of project selection should be to produce a balanced "project portfolio" which will deploy corporate resources to ensure the maximum probability of closing the future "gap" between corporate aspirations and current achievement.

Given these considerations, and the need to stimulate a creative R & D organizational climate (recall chapter 8), which may generate interpersonnel competition for R & D resources, most corporate and R & D managers would like to use "objective" methods of project evaluation and selection. In this way, they could hope to maximize the probability of achieving corporate goals and convince R & D staff of the equity of their project selection decisions, and thus seek to maintain organizational morale. Thus, over the last twenty-five years or so, massive efforts have been made to develop R & D project evaluation and selection approaches to satisfy this need. For example, as long ago as 1964, Baker and Pound found that there were then over eighty different formal models available in the literature to help management project selection decisions.[1] Unfortunately, few (if any) of these models were used in practice. Although there is some evidence that the use of formal selection models is increasing, in general more recent surveys have confirmed the findings of Baker and Pound. For example, one writer reports:

> *Indications are that most R & D laboratories either do not use such models at all...or have not adopted them for any length of time....Thus, the effectiveness of these models as aids in the R & D project selection decision process has not been fully clarified.*[2]

Since R & D project selection is a problem of allocating scarce resources in an economically and technologically uncertain situation, somewhat analogous to that facing the financial investor, it might appear that it could be handled using the quantitative analytical techniques which have been developed so successfully since World War II. Certainly, the numerous models quoted in the references cited above are rooted in the approaches of economic and financial analysis and operations research. Unfortunately, the consensus is that these approaches have proved to be of only limited practical value in this field of application, so we will first discuss the explanation of this observation.

By definition, the successful implementation of the technological innovation process requires the progressive reduction of ignorance and uncertainty (see chapter 11,) which remains incomplete at the end of the R & D phases of the process. This means that the numerical estimates required to be incorporated in any formal mathematical evaluation must be based upon subjective, and often unavoidably crude, judgments or "guestimates." The R & D staff closely involved with the project are often the only people capable of making these judgmental estimates. Since they may have a vested interest in the continuation of "their" project in face of competition from others, they may (either consciously or subconsciously) bias their estimates to favor its continuance. Models have been developed which incorporate uncertainties in model parameter and variable values, as well as their interactions, but these uncertainties dominate the subtleties and sophistication of model structures. Therefore, the final output from the evaluation process—the portfolio of projects ranked and selected for funding—may vary widely according to the values of the estimates chosen.

Despite these observations, it is important to distinguish between the use of the mathematical techniques of operations research (OR) and the OR approach itself. The latter is applicable. OR initially developed in World War II as a spin-off of the British development of radar, itself a revolutionary technological innovation, as a means of analyzing its operational application.[3] The OR approach is itself a formal application of the scientific approach, that is, the systematic, logical, and (as far as possible) empirical and quantitative analysis of the phenomena being studied. Quite frequently, when applying this approach to other management problem areas, the major benefits have been derived, not from the so-called optimal management decision making based upon the model developed, but from the improved and enriched management insights and interpersonal communications provided by the process of model development. That is, the benefits are derived from the management and organizational learning process itself, rather than its outcome. Similar results have been reported when developing R & D project evaluation and selection models. For example, Souder[4] makes the following comment upon the use of portfolio selection models:

> *The few successful applications of portfolio models reported in the literature have all stressed...decision process benefits, rather than the correctness of the model's outputs. In two studies in particular, the major benefits of the models were seen in terms of improved interdepartmental involvement and improved systematic decision making. The actual outputs from the models were never implemented.*

For this reason, and the critical importance of the subject to R & D management, it is useful to overview the approaches to project selection and evaluation which have been developed. The approaches are described in order of increasing mathematical sophistication and vary in applicability to the different stages of the R & D process, some being more applicable to basic research as opposed to advanced development. Some evaluate each project independently using subjective or numerical measures of merit, while others allow for the interactions between projects and the selection of project portfolios. The overview is mainly based upon Souder, Fusfield,[5] and Vertinsky and Schwartz.[6]

Intuitive Individual or Group Evaluations and Selections

An individual or committee selects projects, based on a number of subjectively evaluated criteria.

1. The "halo" effect—that is, the past track record or "halo" of each project team.

2. The "squeaky wheel method (A)"—project teams which complain the loudest get their projects funded.

3. The "squeaky wheel method (B)"—projects are selected so as to minimize the disruption from the selection of the previous period.

4. The "white charger technique"—this method is selection based upon direct individual influence or power of the project team, its power of advocacy, or possibly it being the last team to present its case to the selectors.

Intuitive judgment is a traditional approach, which in the absence of any alternatives was probably used also exclusively up to the 1950s. In 1965 Seiler[7] found that one-half of the R & D managers he surveyed used this method and Cetron reported 56 percent of a sample which he surveyed used individual subjective assessments.[8]

Checklists

The next stage in analytical sophistication from purely intuitive judgment, is a project evaluation using a checklist of factors, which determines the likelihood of an innovation's becoming commercially successful. Dean[9] studied thirty-four firms and identified thirty-two R & D, manufacturing, and marketing factors which were most frequently considered in evaluation projects. These considerations are listed in table 10.1,[10] which although incomplete, does incorporate many of the criteria that are relevant to project evaluation. Unfortunately, Dean found that the average company used only about a quarter of these considerations in project evaluation, so that they could hardly be said to be operating comprehensive checklists! However, R & D or corporate management should find little difficulty in developing a checklist similar to Dean's congruent with the technological strategy of the firm. Dessauer reports that at Xerox, R & D projects are evaluated by a corporate committee composed of R & D engineering, finance, and marketing functional representatives, using a comparable checklist.[11]

Project Profiles

The disadvantages of a simple checklist, even if it exhaustively lists all evaluation criteria, is that a project is checked against each criterion purely for its presence or absence without consideration for the varying degrees to which that criterion is satisfied. Bright[12] suggests extending the checklist to a profile in which the relevance of a project against each criterion is rated on a five-point Likert scale, as shown in figure 10.1. This literally produces a project profile by which the strengths and weaknesses of a project can be readily perceived.

1. Compatibility with company objectives;
2. Compatibility with other long-term plans;
3. Availability of scientific skills in R & D;
4. Critical technical problems likely to arise;
5. Balance of R & D program;
6. Interaction with other R & D projects;
7. Competitors' R & D programs;
8. Size of potential market;
9. Factors affecting expansion of the market;
10. Influence of government regulations and control;
11. Export potential;
12. Probable reaction of competitors;
13. Possibility of licensing and know-how agreements;
14. Possibility of R & D cooperation with consultants or other organization;
15. Effect on sales of other products;
16. Availability and price of materials needed;
17. Possibilities of 'spin-off' exploitation of innovation;
18. Availability of production skills and equipment;
19. Availability of marketing skills and experience;
20. Advertising requirements;
21. Technical sales and service provision;
22. Effects on company 'image';
23. Risks to health or life;
24. Probable development, production and marketing costs;
25. Possibility of patent protection;
26. Scale and timing of necessary investment;
27. Location of new or extended plant(s);
28. Attitude of key R & D personnel;
29. Attitude of principal executives;
30. Attitude of production and marketing departments;
31. Attitude of trade unions;
32. Overall effect on company growth.

Source: B. V. Dean, *Evaluating, Selecting and Controlling R&D Projects* (New York: AMACOM, a division of American Management Associations, 1968), p. 49. Reprinted with permission.

TABLE 10.1 DEAN'S CHECKLIST.

Merit Numbers

The scoring of each criterion on a Likert scale represents only a first step towards quantification of project features. Clearly some project features or criteria will be more important than others, and it would be desirable to reflect this fact in the evaluation scheme. If each criterion is given a weighing (perhaps derived from a relevance tree analyisis), then we may extend the above method to calculate an overall "merit number" for the project. Wells is reported to have

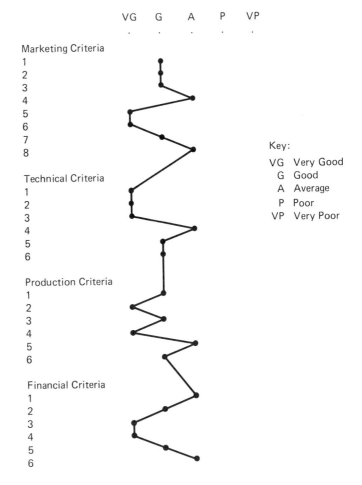

Figure 10.1 Project Profile.

incorporated an interactive weighing system and relevance trees in his evaluation model.[13]

If the criteria listed are comprehensive and include considerations of costs, profits, and probabilities of outcomes, a comprehensive ranking should be possible. The projects can then be selected for funding based upon the ranking subject to the constraint of the overall budget available.

Index Methods

R & D project appraisal is not markedly unlike public sector project appraisal, where cost-benefit analyses are frequently employed. Both expected costs and benefits are evaluated, separately summed, and then expressed as an overall index of total benefits divided by total costs or *benefit-cost ratio,* as a measure of return on financial investment. Projects can then be ranked in order of decreasing benefit cost ratios, and the projects can be funded ordinally until the budget available is exhausted.

The most popular model used for R & D project evaluation was developed by Ansoff and is shown in table 10.2. As Souder points out, index models have a seductive appeal because of their apparent simplicity, which is, in reality, a mirage. All the terms listed in table 10.2 must be estimated subjectively, so the realistic implementation of the method requires the application of consistent estimation procedures across the board to all projects. The problems of parameter estimation in R & D projects are discussed later in chapter 11. The method is likely to be most useful in selecting between alternative incremental innovations when the terms may be estimated with reasonable accuracy and consistency.

Risk Analysis Models

Risk analysis, or the computer simulation of expected future positive and negative cash flows generated by a capital investment project to derive an overall risk profile or statistical distribution of ROI, has been used extensively in other areas of project appraisal. The principles of the approach are well described by Hertz.[14] Furthermore, in

$$\text{Figure of Merit (profit)} = FM_p = \frac{(M_t + M_b) \times E \times P_s \times P_p \times S}{C_d \times J}$$

$$\text{Figure of Merit (risk)} = FM_r = \frac{C_{ar}}{FM_p}$$

Where M_t Technological Merit
M_b Business Merit
E Estimated total earnings over lifetime
P_s Probability of Success
P_p Probability of successful market penetration
S Strategic fit
C_d Total development cost
J Savings factor from shared resources
C_{ar} Total cost of applied research
F Total cost of extra facilities, staff etc.
M_p Figure of Merit (profit)

TABLE 10.2 ANSOFF'S FORMULA

principle anyway, it should be possible readily to extend index models into computer simulation programs which will generate risk profiles such as figure 10.2, which provide more effective measures than simple benefit-cost indexes. Unfortunately, the uncertainties of parameter estimation discussed earlier limit the credibility that might be attached to such profiles, so the approach is of limited applicability in the early phases of the innovation process. It is of more value in later phases of the process—perhaps from advanced development onwards—when most R & D uncertainties have been resolved. Equally, it offers a useful approach to evaluating a new product development with minor technological changes or new combinations of existing technological know-how, so that parameters may be more readily estimated. Bobis et al. describe a risk analysis type model in the Organic Chemicals Division of American Cyanamid Division to redistribute its research budget.[15] Another application of the approaches, this time at Xerox, is described by Zoppoth.[16]

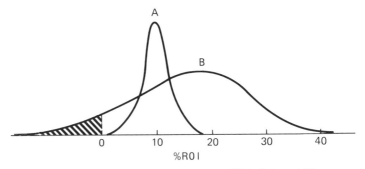

Project A offers a modest expected ROI of about 10%
Project B offers an higher expected ROI of about 20%,
but runs the risk of incurring a loss (shaded).

Figure 10.2 Risk Analysis.

Risk-Return Profiles

If, by the use of an index method, risk analysis, or other methods, an overall estimate of success/failure and expected return for each project can be derived, a risk-return profile may be constructed. This is illustrated in figure 10.3 for nine hypothetical projects A, B, C, D, E, F, G, H, and I. The axes of the graph are return (perhaps expected profitability) against risk (probability of failure), with boundaries marking the minimum acceptable return and maximum acceptable risk levels, respectively. Each project may be located on the graph through its return/risk coordinates. A offers an unacceptable low return, B offers an unacceptable high risk, and C is unacceptable in terms of both risk and return. The remaining projects offer acceptable

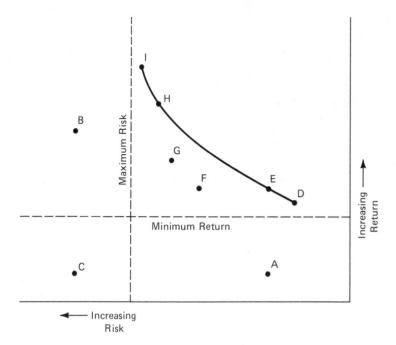

Figure 10.3 Risk-Return Profile.

risk-returns, with those further to the right offering the "best" risk-return combinations. A risk-return profile or frontier can be drawn through the "best" projects, and if sufficient funds are available these projects should be funded. Projects located to the left of this profile may be funded if sufficient money is available (provided also they lie within the acceptable risk-return boundaries). Subject to the unreliability of estimates discussed above, this approach provides a useful visual aid for ranking projects and perceiving the risk characteristics of a project portfolio. In the (admittedly artificial) example shown, projects D, E, H, and I lie on the frontier so, in principle, should be funded. However projects D and E offer low returns and low risks, while in contrast, projects H and I offer high returns and high risks, so that collectively D, E, H, and I constitute a questionably balanced portfolio. Management may therefore seek out one or two medium risk-return projects to obtain an improved portfolio balance.

Statistical Decision Analysis Models

Given the uncertain nature of outcomes in the R & D, some workers have formulated R & D project selection as a problem in statistical decision theory. Decision payoff matrices of projects against alternative outcomes are constructed, and projects which offer the highest expected value are selected.

A related decision analysis approach, particularly when used in conjunction with Bayesian subjective probability estimation, is decision tree analysis which has been particularly associated with new product development and is described in that context by MGee.[17] Because decision tree analysis is an approach to the analysis of sequential decisions under uncertainty, it can be used as a tool both to evaluate and manage R & D projects. The expected benefits from a project can be estimated as an input to other evaluation and selection methods. Once a project is funded, a decision tree model, possibly in conjunction with a PERT/CPM network model, may be maintained and updated by the project manager to monitor the progress of the project. In the event of disappointing project outcomes, it can provide a useful aid in deciding whether to continue or abort a project.

Statistical decision analysis offers valuable tools and insights for project evaluation, selection, and control to the R & D manager who has the statistical understanding to use it intelligently, particularly if used in conjunction with improved cost and outcome estimation procedures discussed in chapter 11. An interesting application of risk and decision analyses approaches, again at Xerox, is described by Hudson, Chambers, and Johnston.[18]

Mathematical Programming Techniques

The most sophisticated methods of project evaluation and selection are based upon mathematical programming models. These approaches may be based upon the selection of a portfolio of projects as divisible entities simultaneously (using linear programming), or sequentionally (using dynamic programming), or as indivisible entities (using integer programming). They all offer, as Thompson and Vincent[19] express it, the "lure of optimality," that is, they select projects and allocate resources to those selected so as to maximize the benefits or returns from the overall portfolio. Bell and Read[20] report that the Central Electricity Generating Board of the United Kingdom has used linear programming as an aid to project selection, while Lockett and Gear[21] describe an R & D portfolio selection model incorporating decision tree analysis and linear programming. Souder[22] describes a dynamic programming approach applied at Monsanto. Integer programming models have been applied in pharmaceuticals at Smith, Kline and French[23] and in the steel industry.[24] Another pharmaceutical company, Johnson and Johnson, has used a mixed integer model (where some projects are indivisible and the remainder divisible) for R & D portfolio selection.[25]

191
Selection
Committee
Procedure—
The Q Sort
Method

10.4
Selection Committee Procedure— The Q Sort Method

Whatever method of project evaluation is used, the ultimate selection decision is made by resource allocators who typically constitute a committee of senior R & D and/or corporate managers with differing functional backgrounds. Such managers may use one or more of the above methods of appraisal in their deliberations, but must ultimately achieve a group consensus in project portfolio selection.

The dangers of bias developing in committee or group decision making were discussed in chapter 5, and the Delphi method of expert opinion forecasting was described as a method of reducing this bias. Of course, similar biases may develop in committees responsible for project selection. Souder, therefore, has proposed a group project selection procedure which he calls the Q-Sort/Nominal Interacting (QS/NI) process, which bears some similarity to the Delphi method. This procedure is as follows:

Each member of the group or committee is given a deck of cards, one for each project to be considered, each outlining the characteristics of the project. He begins by privately and anonymously classifying each project into one of five categories ranging from a very high to a very low level of priority, using the sorting process illustrated in figure 10.4. The sorting criteria will be based upon an agreed set of project selection criteria (see figure 10.6).

Once this nominal period is completed, interaction begins. The results of the classifications of the whole group are tabulated on a tally sheet anonymously and displayed to the group on an overhead projector, as shown in figure 10.5. Agreements/disagreements can be identified, without violating individual anonymity, and discussion can now take place. Individuals may challenge others' classification and the latter may choose to defend their positions. Alternatively, they may choose to remain silent and preserve their anonymity, particularly if they wish to retain a minority position. During this interaction, however, the arguments in favor of the various classifications are likely to be canvassed, so individuals may wish to modify their choices, in the light of the diversity of opinion that has been expressed.

Group members are then allowed a second anonymous nominal period during which they modify their choices. The revised tally sheet is then projected, and a second interaction period begins. As in the Delphi method, nominating-interaction "rounds" may be repeated until, as nearly as possible, a consensus is reached. Souder reports that this consensus is usually reached after two or three rounds.

RESULT AT EACH STEP

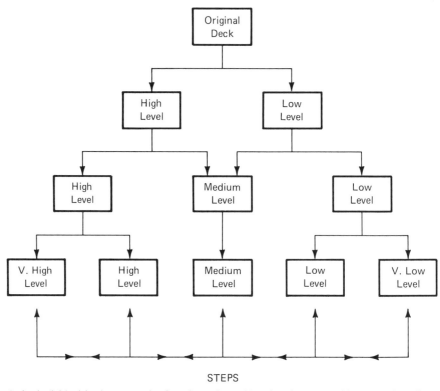

STEPS

1. An individual is given a stack of cards, each card bearing the name, title or number of one project. The individual is asked to perform the following sorting operations. A specified criterion (e.g. "priority") is the basis for sorting.

2. Divide the deck into two piles, one representing a high level of the specified criterion, the other a low level. (The piles need not be equal.)

3. Select cards from each pile to form a third pile representing the medium level of the criterion.

4. Select cards from the high level pile to yield another pile representing the very high level of the criterion; select cards from the low level pile to yield another pile representing the very low level of the criterion.

5. Finally, survey the selections and shift any cards that seem out of place until the classifications are satisfactory.

Source: W. E. Souder, "A System for Using R & D Project Evaluation Methods," *Research Management 20,* no. 5 (Sept. 1978), 29-37. Reprinted with permission of Industrial Research Institute.

Figure 10.4 Q-Sort Mechanics.

Souder further proposed a multistage application of the QS/NI process which would be applicable across the range of R & D projects performed in the offensive-defensive innovative firm, and which reflects the technological and budgetary considerations in the R & D resource allocation process. This is illustrated in figure 10.6. Projects are first classified as exploratory, applied, and developmental within the framework of corporate goals, which are themselves

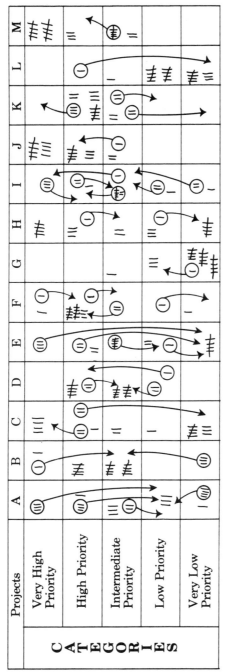

Tally chart for 20 subjects and 13 projects at the end of the following sequence of periods: Nominal—Interfacing—Nominal. Projects evidencing consensus at the end of the second nominal period = A, B, D, E, F, G, I, J, L, M. All projects evidenced consensus at the end of an additional subsequent interacting—nominal sequence. Consensus is measured by standard statistical tests and the subject's feelings.

Source: W. E. Souder, "A System for Using R & D Project Evaluation Methods," *Research Management 20,* no. 5 (Sept. 1978), 29-37. Reprinted with permission of Industrial Research Institute.

Figure 10.5 Example of Tally Sheet.

ranked by a QS/NI process. Projects in each classification may then be evaluated and selected by second-level QS/NI processes, which also employ appropriate criteria or models. Thus for exploratory projects, where information is meager and uncertainties are large, simple checklists or profile methods may be used. On the other hand, for applied and developmental projects, where more information is available and more resources are required, more sophisticated models may be incorporated as decision aids in the QS/NI processes. Souder's multistage QS/NI scheme appears to provide a framework for the intelligent use of project evaluation and selection models. More important, it provides a potential framework for integrating at both corporate and functional management levels the considerations of technological strategies, goals, forecasts, and R & D resource allocations we have discussed in earlier chapters.

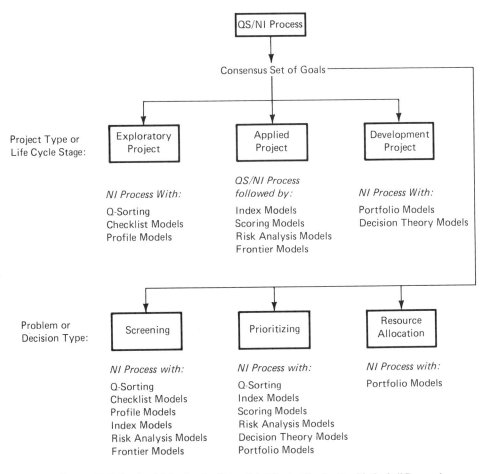

Source: W. E. Souder, "A System for Using R & D Project Evaluation Methods," *Research Management 20,* no. 5 (Sept. 1978), 29-37. Reprinted with permission of Industrial Research Institute.

Figure 10.6 Project Selection and Evaluation System.

References

1. N.P. Baker and W.H. Pound, "R&D Project Selection: Where We Stand,"*IEEE Transactions on Engineering Management* 11, no. 4 (December 1964):124-34.

2. W.E. Souder, "Analytical Effectiveness of Mathematical Models for R&D Projects Selecting,"*Management Science* 19, no.8 (April 1973): 907-23.

3. Sir Robert Watson-Watt, *Three Steps to Victory* (London: Oldhams Press, 1958).

4. William E. Souder, "A System for Using R&D Project Evaluation Methods," *Research Management* 21, no.5 (September 1978): 29-37.

5. Alan R. Fusfield, "A Review of the Major Types of Project Selection Techniques and Suggestions for New Approaches" (Pugh-Roberts Assoc., 1976).

6. Ilan Vertinsky and S.L. Schwartz, *Assessment of R&D Projects Evaluation and Selection Procedures,* Report no. 49, Technological Innovation Studies Program (Ottawa: Department of Industry, Trade and Commerce, December, 1977).

7. Robert E. Seiler, *Improving the Effectiveness of Research and Development* (New York: McGraw-Hill Book Co., 1965).

8. Marvin Cetron, *Innovation Group Newsletter,* no. II (Technology Communications Inc., New York, October 1970).

9. Burton V. Dean, *Evaluating, Selecting, and Controlling R&D Projects* (New York: American Management Associations, 1968).

10. Christopher Freeman, *The Economics of Industrial Innovation,* 2d.ed. (London: Francis Pinter, 1982), pp. 157-58.

11. John Dessauer, "Some Thoughts on the Allocation of Resources to Research and Development Activities," *Research Management* 10, no.2 (March 1967): 77-89.

12. James R. Bright, *Research, Development, and Technological Innovation* (Homewood, Ill., Richard D. Irwin, 1964), p. 424.

13. H.A. Wells, "Weapons Systems Planner's Guide," *IEEE Transactions on Engineering Management* 14, no.1 (March 1967): 14-16.

14. David B. Hertz, "Risk Analysis in Capital Investment," *Harvard Business Review* 42, no.1 (January-February 1964): 95-106; David B. Hertz, "Investment Policies that Pay Off," *Harvard Business Review* 49, no.1 (January-February, 1968): 96-108.

15. A.H. Bobis, T.F. Cooke, and J.H. Paden, "A Funds Allocation Method to Improve the Odds for Research Success," *Research Management* 14, no.2 (March 1971): 34-39.

16. R.C. Zoppoth, "The Use of Systems Analysis in New Product Development," *Long-Range Planning* 5, no.1 (March 1972): 23-36.

17. John F. MGee, "Decision Trees for Decision Making," *Harvard Business Review* 42, no.3 (July-August 1964): 126-38.

18. Ronald G. Hudson, John C. Chambers, and Robert G. Johnston, "New Product Planning Decisions Under Uncertainity," *Interfaces* 8, no.1, pt. 2 (November 1977): 82-96.

19. P.N. Thompson and David Vincent, "OR in Research and Development—A Critical Review," *R&D Management* 6, no. 3 (June 1976): 109-13.

20. D.C. Bell and A.W. Read, "The Application of a Research Project Selection Method," *R&D Management* 1, no.1 (October 1970): 35-42.

21. A.G. Lockett and A.E. Gear, "Representation and Analysis of Multistage Problems in R&D," *Management Science* 19, no.8 (April 1973): 947-60.

22. W.E. Souder, "A Scoring Model Methodology for Rating Management Science Models," *Management Science* 18, no.10 (June 1972): 526-43.

23. M.A. Cochran et al., "Investment Model for R&D Project Evaluation and Selection," *IEEE Transactions on Engineering Management* 18, no.3 (August 1971): 89-100.

24. C.J. Beattie, "Allocating Resources to Research in Practice," in *Applications of Mathematical Programming Techniques,* ed. E. M. L. Beale (New York: American Elsevier Publishing Co., 1970), pp. 281-92.

25. Dov Grossman and S.N. Gupta, "Dynamic Time-Staged Model for R&D Portfolio Planning—A Real World Case," *IEEE Transactions on Engineering Management* 21, no.4 (November 1974): 141-47.

11

Project Management and Control

The moral of the story is this: The inherent preferences of organizations are clarity, certainty and perfection. The inherent nature of human relationships involves ambiguity, uncertainty, and imperfection. How one honors, balances, and integrates the needs of both is the real trick of management.

A. J. Gambino and Morris Gartenberg, **Industrial R & D Management**

11.1
Introduction

In chapters 9 and 10, we examined some approaches and techniques for evaluating and selecting R & D projects, and here we will turn our attention to the "ongoing" management of a project once it has been "selected." Once again we shall focus upon analytical approaches and techniques and largely ignore behavioral aspects of the subject. Those aspects will be discussed in some detail in chapter 12.

11.2
Budgetary and Accounting Aspects

The process of overall budget setting , project evaluation, and selection discussed in chapters 7, 9, and 10 should be embedded in a comprehensive budgeting system in the R & D function which will be disaggregated down to individual project level. The overall subject of the "best" method of budgeting and accounting for R & D expenditures, for example, whether they should be treated as "investments" or "current expenses," is a specialist accountancy issue which is outside the scope of this text. The issues are discussed in some detail in other studies.[1] Winkofsky, Mason, and Sounder stress that the R & D budgeting and project selection process should be an information and communication process combining "bottom-up" (from "bench-level" scientist) and "top-down" (from corporate management) planning in iterative cycles.[2] The degree of formality versus informality of this process will probably be dependent upon the size of the R & D function. Whatever process is adopted, care should be taken to ensure that it is fully understood and accepted by R & D staff as well as management to minimize the risk of conflict developing between the two groups, arising from their differing professional orientations. This latter issue is discussed further in section 11.5 and in the next chapter. We will not consider budgeting and accounting further except to quote the words of Lin and Vasarhetyi.[3] They state that an integrated financial and technical budget should exist; frequent revisions and evaluations should be performed; and budget variances are not necessarily undesirable, but when larger than a prescribed amount, they should be examined, thus providing for a full examination of the project. These requirements are implicit in discussions of individual project management and control, which follow.

11.3
Project Progress as an Uncertainty Reduction Process

We may view the successive phases of the innovation process as a progressive reduction of technical, commercial, and marketing uncertainties, using a variation of the technology-market matrix suggested by Scott and Szonyi[4] (figures 2, 3, and 4) and shown in figure 11.1. In the early phases of the innovation process technical,

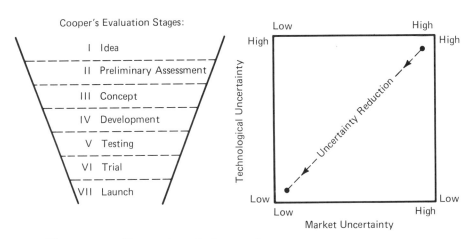

Figure 11.1 Uncertainty Reduction Process.

commercial, and marketing uncertainties will be high, corresponding to a location in the upper right-hand cell of the matrix in figure 11.1. As we move through the innovation evaluation described in chapter 9 uncertainty is reduced, so we approach the bottom left-hand cell that corresponds to a commercially successful innovation when these uncertainties have been eliminated.

Thus, the early state of an R & D project evaluation corresponds to the upper-right cell in this matrix, and only fairly crude subjective estimates of projective outcomes can be made. We therefore initially examine the sources and magnitudes of estimation errors in R & D project evaluations.

11.4
Estimation Procedures in Project Evaluations

The innovation process consists of the progressive reduction of considerable uncertainty (particularly if the innovation proposes significant technological novelty), so that early estimates of outcomes must and should be subject to considerable errors. This is especially true in the early research (which really is characterized by ignorance rather than uncertainty) and primary development phases of the innovation process. Here, detailed estimations are possibly of less consequence, if projects are subjected to peer group evaluations and selections and require a relatively small commitment of corporate resources. However, once a project moves into more advanced development, the increasing commitment of corporate resources coupled with the increasing competition between projects dictate that estimation procedures should facilitate realistic evaluations and comparisons between projects to be made. Often estimation errors are large compared with the differences in yields between competing projects, so quite small differences in project estimates that are well within the error ranges could produce marked differences in project selections. For example Meadows,[5] in one of the few reported studies of estimation errors, compared two projects and calculated that estimation errors of only 10 percent would have yielded a higher return on a project at present making a loss, as compared to one yielding a 230 percent ROI!

This being so, the emphasis in estimation procedures should be on the development of *consistent measures* of the most likely, pessimistic and optimistic values of paramenters between projects, rather than the pursuit of spurious accuracy. If consistent estimation methods are used for all projects, it should be possible to set up overall rankings with some confidence. The robustness of these rankings can then be subjected to some simple sensitivity analysis, to see how they are maintained in the face of the variations of outcomes that may obtain in practice.

Identifying the Project Elements

Later in this chapter (section 11.6) I shall cite network analytic techniques that are useful aids to project management. Both the application of these techniques and the requirements for consistency measures between projects dictate that a detailed identification of project elements is required. Obviously also, this identification is a necessary requirement for the implementation of evaluation, assessment, and selection procedures discussed in the previous two chapters. The following framework for identifying these elements is based upon Monteith.[6]

1. Define the project.

2. Divide the work into activities or *tasks* which may be independent or dependent upon the completion of predecessor tasks.

3. Decide upon all courses of action by which each task can be completed. This will identify the problems to be solved before the task can be completed.

4. Identify the alternative approaches that may be pursued, either in series or parallel, to solve these problems (note that the relevance tree techniques discussed in chapter 5 may be of use here).

5. Select the most promising approach for solving each of the problems, specify time and/or expenditure limits on these approaches and the alternative approaches which will be used should the first ones selected be unsuccessful. That is, clarify the strategy for implementing alternative approaches.

6. Allocate priority to courses of action which are likely to yield the best technical and cost outcomes.

7. Delay the performance of any independent tasks that will prove to be worthless, if another task or tasks proves to be infeasible.

The logical application of the above analysis should identify most likely, optimistic, and pessimistic estimates of the cost time and resource (labor, equipment, and so forth) requirements for a project and should include appropriate overheads and contingency cost allowances. If applied consistently to all projects, it should enable realistic comparisons to be made.

Although the above, essentially commonsensical, framework exudes the "warm glow of motherhood" it would be utopian to expect unbiased estimates to be made in practice. There are two major sources for such biases: unconscious differences produced by differences in individuals' personal subjective judgments of outcomes, and optimistic biases which individuals (consciously and unconsciously) introduce into an evaluation to increase the chances of a project receiving continued support.

Subjective Probability Estimates

Approaches for eliciting subjective probability estimates in management in general have been developed and studied quite extensively over the past two decades. Useful reviews of these approaches are provided by Chesley.[7] Ebert[8] and Souder[9] consider such approaches in the context of the R & D project evaluation process, while Allen[10] suggests an alternative approach based upon credibility assessment. Rubinstein and Schroder discuss some of the behavioral factors that may affect such estimates.[11] Within the R & D context, most studies focus on subjective estimates of the overall probability of technical success and, like the sophisticated OR project selection models described in the preceding chapter, have yet to be regularly applied in practice. Therefore, we will concentrate our attention on empirical studies of estimation errors in R & D projects.

Optimistic Biases and Empirical Studies of Estimation Errors

Given that R & D projects are competing for funding, it would be unrealistic to expect no optimistic biasing of estimates by staff seeking support for "their" projects. As Freeman expresses it, *"the social context of project 'estimation' is a process of political advocacy and clash of interest groups rather than a sober assessment of measurable probabilities."*[12] Realistically one must expect such optimistic biasing to be present, and the few empirical studies in errors support this expectation. Thomas[13] found this to be true in two scientific instruments firms he studied, while the *ex post* studies of estimation error distributions, discussed below, suggest that they are skewed in an optimistic direction. One interesting observation Thomas made was on the justification engineers offered for biasing their estimates. One company operated a profit-sharing scheme, so that it was a matter of self-interest for R & D staff to develop only the most profitable projects. Despite this fact, staff biased their estimates optimistically, not to gain support for projects, which they technically favored, but to emphasize to management the unreliability of estimation procedures. What tends to make matters worse is the way in which research managers tend to compound cost errors. As Twiss[14] points out, in a project that is commercially questionable but technically appealing to both R & D staff and management, the original estimator may be asked to reduce his cost estimations in two ways.

One way is to remove elements from the project that are "desirable" rather than "essential." Once the project has been funded, these elements may be reinstalled later in the process, in the light of technical outcomes, and treated as cost overruns. The second way is to reduce contingency allowances which by definition cannot be specified in detail. Because of the "sunk" costs incurred and the political and psychological trauma of terminating a project partially completed, the expenditure of these extra costs can be rationally justified at a later stage.

These considerations can undoubtedly significantly increase estimation errors and may occasionally lead to disastrous cost overruns. It is reasonable to conjecture that the collapse of Rolls-Royce, following the RB211 imbroglio, can be partially attributable to optimistic estimating.

Despite the importance of the topic, there have been relatively few studies reported on the relationship between *a priori* estimates and *a posteriori* actual outcomes in projects. Edwin Mansfield is a major contributor to the literature on the economics of technological innovation, and with coworkers he has conducted investigations that are relevant here. He reports a study of actual versus estimated costs, duration times, and first-year sales for about fifty projects in a U.S. ethical drug firm.[15] The means and standard deviations, in parentheses, for the ratios of actuals: estimates were 1.78 (0.74); 1.61 (1.15); and 1.29 (1.38), respectively. Eighty-eight percent of the projects incurred cost overruns and 68 percent time overruns, respectively. These results are congruent with those obtained in earlier studies by Marshall and Meckling[16] and Peck and Scherer[17] in analyzing military R & D. Across the Atlantic in the United Kingdom, Norris[18] studied 475 projects in the electrical utility, biological, chemical, and engineering process industries and found lower cost and higher time ratios. Separate mean ratios were calculated for smaller projects from each of the organizations involved, and for a group of larger projects. These mean ratios ranged from 0.97 to 1.51 for cost and 1.39 to 3.04 for duration time, respectively. Allen and November[19] included sales in their study in the U.K. chemicals industry and found comparable errors to Mansfield in sales estimates.

The most interesting analysis was performed by Norris, using his own results and some of these obtained by Meadows[20] and Thomas.[21] Four useful conclusions could be deduced from Norris's analyses.

First, duration ratios exceed cost ratios and since labor costs are the main costs in most projects, unexpectedly long durations must have been due to delays rather than unexpectedly high work contents. He suggests that optimistic duration estimates are not based on optimistic estimates of resource requirements, but rather upon the total amount of work that can be undertaken concurrently and the availability of access to shared resources (such as equipment) not under direct project management control. That is, there is a general tendency to accept more work than can be readily performed in the time allowed to ensure that resources (including labor) are not unemployed when uncontrollable external delays occur. (Thus the situation may be analogous to "job-shop scheduling" in manufacturing. To seek maximum machine and labor utilization, more jobs are scheduled than the shop can handle, leading to work-in-progress queues at successive stages in the manufacturing process, and job duration times longer than their intrinsic work contents warrant.)

Second, given that project progress is an uncertainty reduction process, it is intuitively reasonable to expect estimation errors to be lower for less as opposed to more innovative projects. In one organization, projects could be categorized as "routine" (where the work required appeared foreseeable from past experience) and "speculative" (where it did not). In another organization, projects could be categorized as "less basic" (where the aim was to solve a specified practical problem) and "more basic" (where the aim was to generate background knowledge and understanding). In neither organization was there any difference in the estimation errors between the two categories, nor were these lower for more as opposed to less recent projects. Thus there was no evidence of overall learning effects in these two research organizations. In contrast, in a third organization (in which projects could not be categorized as above), duration error ratios dropped from 2.88 in less recent, to 2.05 in more recent projects. Thus this third organization did appear to be learning to improve its estimates and/or to be improving its project planning and control procedures.

Third, again, given that it is an uncertainty reduction process, it is intuitively reasonable to expect estimation error ratios of both cost and duration to fall as a project progresses towards completion. Norris found that in these projects, errors remained constant *or increased* as they progressed. A possible explanation for these counter-intuitive results is that the "relative level of uncertainty or ignorance" remains constant or increases as a project approaches completion, but this explanation lacks plausibility. A more likely explanation is that the knowledge and experience gained as a project progressed *was not effectively used* to revise earlier estimates.

Fourth, Norris hypothesized that each organization would exhibit a particular degree of bias in its actual:estimate ratios which is represented by the mean value of this ratio found for each organization. Each "organizational mean value" will reflect the underlying uncertainty of the technology involved and the inaccuracy of the organization's estimation procedures. In each organization, the residual dispersion of actual:estimate ratios of individual projects about this mean is taken to reflect the residual technological uncertainties peculiar to individual projects. These latter dispersions or distributions of "debiased" ratios were subjected to logarithmic (base e) transformations. Norris found that most of these residual transformed distributions (which, by definition, will have a common mean of zero) had similar standard deviations lying in the narrow range 0.53 to 0.63. This result appears to support the argument of Dean, Mantell, and Roepcke that actual project costs follow a log-normal distribution.[22] Given the uncertain nature of the R & D process itself and the competitive nature of project selection procedures, it can be expected that the errors will be rather large and biased toward optimistic project evaluations. However, the first three conclusions suggest that biases could be reduced both initially and during the progress of a project by conscientiously applying the framework suggested earlier. The fourth conclusion suggests that each organization has *its own characteristic error generation mechanism and outcome variation.*

Thus if an organization builds up a historical record of its own actual:estimation ratios (which may follow a log-normal distribution), it should be possible to set statistical confidence limits on estimates. In fact, the situation is comparable to establishing statistical quality and cost controls in manufacturing processes. There the pattern of variation in the key parameters of the product generated by the underlying technological mechanisms can be identified and appropriate quality control limits set up, based upon this variation. Similarly, manufacturing operations monitor cost centers to ensure that costs remain within established mean and variance standards. The desirability of establishing somewhat similar procedures in the R & D function leads us to the subject of project control, commenting upon both its behavioral and analytical aspects.

11.5
Project Control—Behavioral Aspects

First of all, it must be stressed that we are not using the term *control* in a coercive sense. Our review of technological strategies in chapter 4 stressed that offensive/defensive innovators need to cultivate an "organic" research climate, as reflected in the pioneering studies of Burns and Stalker.[23] This viewpoint was reinforced in the discussion of creativity in chapter 8. Therefore, in the R & D setting, there is no question that the management style must be participative. Professional R & D staff and managers are peers, and there should be no polarization between the two, because of the adoption of "us and them" attitudes. Some R & D staff (particularly in the R-intensive functions of offensive-defensive innovators), pledge their primary allegiance to their disciplines and are totally uninterested in management. They are committed to the pursuit of technological achievement and excellence and are relatively indifferent to "mundane" management and commercial concerns. The Kuhnian educational process and the individual's identification with the values and mores of his profession does generate professional chauvinism in R & D laboratories as elsewhere (notably among university faculties!), and the insensitive and coercive application of otherwise worthy project control procedures would undoubtedly be doomed to failure!

Despite those considerations, it must also be recognized that commercially successful technological innovation in a competitive market usually requires the timely and cost-effective completion of the secondary development and onward phases in the innovation process. This requirement dictates that from the secondary development phase onwards, projects should be "managed" to ensure that they meet their cost and time performance targets insofar as possible. This requirement was also reflected in chapter 4, when it was argued that the successful implementation of an applications engineering strategy required the cultivation of a more "directive" development climate. As an R & D project progresses, consuming an increasing amount of scarce organization resources, management has an increasing responsibility to exercise discrete and behaviorally sensitive control. The work on the critical functions of project teams discussed in chapter 12 suggests that this responsibility may be best exercised by the person assuming the "project manager" role. Furthermore, Texas Instruments has achieved an outstanding record of innovation and entrepreneurship, within the framework of the quite bureaucratic Objectives, Strategies, and Tactics (OST) system (see chapter 16). Thus superior technological innovation and responsible R & D management control are not mutually incompatible requirements.

Szakonyi provides a useful discussion of a participative management approach to project control.[24] The stimulation of a participative and cooperative climate for developing project control systems need not prove to be too difficult. Seminars to discuss the problems of project evaluation, management, and control can be offered, and all R & D staff can be invited to attend. Alternative approaches to handling these problems can be examined and an approach suited to the local situation designed. In chapter 4, the need for innovation managers as well as operation managers was stressed, particularly if the company is to sustain a record of successful technological innovation. An important feature of the OST system in Texas Instruments is the way it encourages individuals to enact both roles and become "two-hat" persons as both operations and innovation managers. Because successful innovation requires good management, it requires good R & D professionals *and* good R & D managers. In chapter 12, we will discuss aspects of the personal development of R & D staff, and argue in favor of the dual-ladder system for developing and rewarding staff equitably, either as "researchers" or "managers." It will also be argued that R & D professionals should be encouraged to consider both the managerial and technical aspects of R & D and explore their interests in and aptitude for management. Those aspiring toward managerial careers (and the management ladder) who appear to possess the requisite administrative and personality skills can be given the opportunity to gain initial management experience, by assigning them both management and technical responsibilities in a project team. Then, as "two-hat" persons (project managers and scientist/engineer), they can gain initial management experience without surrendering their professional R & D identity. Later, in the light of this experience, they can select which ladder to climb. Such project team members can identify with both the managerial and technical aspects of project development. They may also probably supply the nucleus of team members who will transfer the project to manufacturing operations or establish a separate business venture, as discussed in later chapters.

11.6
Project Control—Analytical Aspects

The framework suggested earlier, plus a historical analysis of past error distributions along the lines described by Norris, should provide R & D management with a sound basis for exercising project control. Given the quality of record keeping in most organizations, the historical records are likely to be (at best) incomplete. Nevertheless, it should be possible to generate a data base fairly quickly, and then refine it over time, to establish confidence limits on estimate errors. This analysis should also identify sources of estimation errors and which (if any) can be eliminated or reduced.

In chapter 10 we discussed the contribution of OR techniques to project appraisal and control and concluded that, despite a massive output of research papers on the subject, there are limited actual applications of sophisticated mathematical modeling to that decision problem. When we turn to project management and control, however, the situation is rather different. Network analytic techniques have been used quite extensively as decision aids to R & D project managers, and it is probably true to say that they could be used more extensively than has been the case.

Gantt or bar charts may be used, and project management is particularly associated with PERT/CPM, one of the most popular and effective of the battery of logical and mathematical techniques that have been developed under the rubric of OR since World War II. PERT and CPM use essentially simple network diagrams to represent complex projects and identify the timetable or schedule and critical path that will minimize their completion time. CPM was developed by DuPont to schedule the maintenance of complex chemical plants, but *PERT was developed specifically to manage a complex R & D project.* It was originally developed to manage the Polaris missile development program and it is claimed that it reduced the completion time for that project by two years. Since the original introduction of these techniques in the late 1950s, the scope of application to R & D project management has expanded quite considerably.

In its original and simplest form, PERT represents the logical interrelationships between the component activities or tasks in a project in a network. Then, based upon estimates of their "most likely," "pessimistic," and "optimistic" durations, it identifies the schedule that minimizes the expected completion time for the project. Such a network should be prepared within the framework suggested earlier and should become an integral part of the project management. In fact, since the 1960s, many U.S. government agencies (and some, in other countries) have made the development of a PERT network mandatory for companies performing R & D contract work for them. Therefore, there is a considerable body of expertise on the use of PERT in R & D project management.

In this simplest form PERT may be used to monitor project progress to ensure that all tasks are being completed within schedule, or if the completions of tasks are delayed, the impact of these delays can be determined on the overall completion time of the project. As such, it is a valuable aid to project management, but it is possibly even more valuable in its extensions. In its earliest forms, it is most readily applicable to the tertiary design and development and prototype trials phases of the innovation process, but more recent extensions have extended its scope of application to earlier and later phases.[25]

The early phases are characterized by high levels of technical and commercial uncertainties and outcomes. The PERT network can handle uncertainties in activity duration times through the use of three time estimates, but cannot handle uncertainties in outcomes and the possibility that one or more of a number of alternative development paths may be followed. This is because it is a *deterministic* network in which no activity in the network can be started until all its predecessor activities have been completed and overall project completion is defined as the completion of the terminal activity or activities. To accommodate the possibility of alternative development paths, *probabilistic or stochastic* networks are required. These networks incorporate probabilistic nodes (or events) from which each of two of more mutually exclusive following branches (or activities) has a certain probability of being taken. The sum of these probabilities adds up to one, so one path is certain to be followed. The most widely used probabilistic network technique is Graphical Evaluation and Review Technique (GERT) which may incorporate both deterministic and probabilistic branching at nodes.[26] It can also incorporate recurring activities and cost information. More recently, another probabilistic network technique known as Venture Evaluation and Review Technique (VERT) has been developed which incorporates performance as well as time, cost, and probability measures.[27] It also incorporates the facility to calculate critical paths based upon time (as in PERT), cost, or performance considerations.

GERT and VERT networks may be initially drawn up in an earlier development phase of the innovation process and can be incorporated into repeated computer simulations to simulate to a sample of alternative project outcomes. GERT simulations may be updated and used as the project progresses to simulate the costs of alternative manufacturing approaches. VERT simulations may be updated and used as it progresses to simulate both expected costs and financial returns from the project. Thus a realistic and continually updated VERT network may be used as a tool in Cooper's ongoing evaluation process described in chapter 9. Its approach is similar to the risk analysis models cited in chapter 10.

Probabilistic network approaches offer considerable promise in project management but, as yet, have not enjoyed the widespread application of deterministic approaches. We will therefore turn to a discussion of the extensions of the deterministic approaches that have been applied quite widely over the past ten to twenty years. The original approach was labeled PERT/TIME because only the duration times of activities were used, but subsequently activity costs were incorporated into PERT/COST systems. Two pairs of duration times and costs may be estimated for some activities: first, the NORMAL times and costs. Second, the estimated duration times of some activities may be reduced by "crashing," that is, by deploying more (or more expensive) resources on their completion. For example, workshop staff may be paid overtime rates to accelerate the assembly of a prototype item of equipment, or faster and more expensive testing facilities may be used in an experiment. In such cases, CRASH times and costs are estimated which are less than the normal time and greater than the normal cost estimates, respectively.

This information may then be used to aid the financial control of the project. Project "progress" may be defined in terms of completed activities, so initially graphs of elapsed time versus progress and cost versus progress may be plotted, assuming that the project will be completed in normal time at normal cost. However, as the project progresses, delays may occur which, unless redressed, will delay the overall project completion time. Such delays may be eliminated or reduced by crashing selected uncompleted activities at an extra cost. By identifying the duration time-cost tradeoffs between activities, PERT/COST networks can be employed to identify alternative crash schedules by computer. The schedule chosen may depend upon subjective judgments of the overall tradeoff of the disadvantages of time delays versus increased costs in particular situations. The over-all outcome will be a time overrun or a cost overrun (or some combination of both), depending on the choice made. Some industries (such as aviation) may prefer cost overruns to time overruns, while others may prefer the latter.

Earlier we saw that time overruns are frequently caused by delays in gaining access to facilities shared with other users. Typically, certain laboratory resources (including labor and equipment) are shared by a number of projects, and such delays are to be expected if no attempt is made to schedule the overall assignment of these resources among projects. PERT/RAMPS or Resource and Multiple Project Scheduling techniques have been developed to handle such situations. The individual project PERT networks may be aggregated into an overall multiproject PERT network, and alternative activity-resource schedules evaluated. As when considering activity crashing, even quite small networks can generate quite large numbers of alternative activity-resource loading and sequencing possibilities, so PERT/RAMPS software packages have been developed to identify and evaluate alternative activities-resources loading schedules by computer, to determine the "best" overall allocation and scheduling of resources to projects.

Once a project progresses to the pilot production phase of the innovation process, there is often a requirement to manufacture perhaps a few hundred units of a complex assembled product. That is, one is moving from scheduling and controlling an individual project to scheduling and controlling an ongoing assembly operation. PERT/LOB, or Line of Balance, is particularly useful for scheduling such small-scale production. PERT/COST, PERT/RAMPS, and PERT/LOB are all described in texts on PERT methods.[28] Finally, if an innovation moves into full-scale production as a manufacturing assembly operation MRP, or Materials Requirements Planning, techniques, which are essentially extensions of PERT/LOB, may be used. MRP techniques are described in most production/operation management tests.[29]

Provided they are conscientiously developed and updated at a realistic level of detail—*by the project staff involved* in a participative management framework—networking techniques can provide a useful basis for planning, scheduling, and controlling activities in most phases of the innovation process. Note that they may readily incorporate the market planning aspects discussed in chapter 9. Such networks can be prepared and implemented by the staff involved as an integral part of the financial planning and budgeting process discussed earlier in the chapter. PERT-based methods are conceptually straightforward and, while GERT and VERT methods are more sophisticated mathematically, most graduate level scientists and engineers should have little difficulty in understanding and applying them.

References

1. A. J. Gambino and Morris Gartenberg, *Industrial R&D Management* (New York: National Association of Accountants, 1978); H. Bierman, Jr., and R. E. Dukes, "Accounting for Research and Development Costs," *Journal of Accountancy* 139, no. 4 (April 1975): 48-55; W. T. Lin and M. A. Vasarhetyi, "Accounting and Financial Control for R&D Expenditure," in *Management of Research and Innovation*, eds. Burton V. Dean and J. L. Goldhar (TIMS Studies in the Management Sciences, 15, Amsterdam: North Holland Publishing Co., 1980): 99-123.

2. E. P. Winkofsky, R. M. Mason, and W. E. Souder, "R&D Budgeting and Project Selection: A Review of Practices and Models," in Dean and Goldhar, eds. *Management of Research and Innovation*, pp. 183-197.

3. Lin and Vasarhetyi, "Accounting and Financial Control for R&D Expenditure," p. 211.

4. D. S. Scott and A. J. Szonyi, "Evaluating a New Venture," in *The Technical Entrepreneur*, ed. Donald S. Scott and Ronald M. Blair (Ontario: Press Porcepic, 1979), pp. 77-78.

5. D. L. Meadows, "Estimate, Accuracy, and Project Selection Models in Industrial Research," *Industrial Management Review* 9, no. 3 (Spring 1968): 105-21.

6. G. Stuart Monteith, *R&D Administration* (London: Iliffe Books, 1969), pp. 69-70.

7. G. R. Chesley, "Elicitation of Subjective Probabilities: A Review," *Accounting Review* 50, no. 2 (April 1975): 325-37; G. R. Chesley, "Subjective Probability Elicitation Techniques: A Performance Comparison," *Journal of Accounting Research* 16, no. 2 (Autumn 1978): 225-41.

8. R. J. Ebert, "Methodology for Improving Subjective R&D Estimates," *IEEE Transaction on Engineering Management* 17, no. 3 (August 1970): 108-16

9. W. E. Souder, "The Validity of Subjective Probability of Success Forecasts by R&D Project Managers," *IEEE Transactions on Engineering Management* 16, no. 1 (February 1969): 35-49; W. E. Souder, "The Quality of Subjective Probabilities of Technical Success in R&D," *R&D Management* 6, no. 1 (October 1975): 15-22.

10. D. H. Allen, "Credibility and the Assessment of R&D Projects," *Long-Range Planning* 5, no. 2 (June 1972): 53-64.

11. A. H. Rubenstein and H. H. Schroder, "Managerial Difference in Assessing Probabilities of Technical Success for R&D Projects," *Management Science* 24, no. 2 (October 1977): 137-48

12. Christopher Freeman, *The Economics of Innovation*, 2d ed. (London: Francis Pinter, 1982), p.151.

13. Howard Thomas, *Econometric and Decision Analysis: Studies in R&D in The Electronics Industry* (Ph.D. diss., University of Edinburgh, 1970).

14. Brian C. Twiss, *Managing Technological Innovation*, 2d. ed. (London: Longman Group, 1980), p.130.

15. Edwin Mansfield et al., *Research and Innovation in the Modern Corporation* (New York: W. W. Norton, 1971).

16. A. W. Marshall and W. H. Meckling, "Predictability of the Costs, Times, and Success of Development," in *The Rate and Direction of Inventive Activity* (Princeton: Princeton University Press, 1962).

17. M. J. Peck and F. M. Scherer, *The Weapons Acquisition Process: An Economic Analysis* (Cambridge: Harvard University Press, 1962).

18. K. P. Norris, "The Accuracy of Project Cost and Duration Estimates in Industrial R&D," *R&D Management* 2, no. 1 (October 1971): 25-36.

19. D. H. Allen and P. J. November, "A Practical Study of the Accuracy of Forecasts in Novel Projects," *Chemical Engineer* no. 229 (June 1969): 252-62.

20. Meadows, "Estimate, Accuracy, and Project Selection Models."

21. Howard Thomas, "Some Evidence on the Accuracy of Forecasts in R&D Projects," *R&D Management* 1, no. 2 (February 1971): 55-69.

22. Burton V. Dean, Samuel J. Mantell, Jr., and Lewis A. Roepcke, "Research Project Cost Distributions and Budget Forecasting," *IEEE Transactions on Engineering Management* 16, no. 4 (November 1969): 176-89.

23. Tom Burns and C. M. Stalker, *The Management of Innovation* (London: Tavistock Publications, 1961).

24. Robert Szakonyi, "Part I. How to Successfully Keep R&D Projects on Track: An Overview."

25. Burton V. Dean and A. K. Chaudhurak, "Project Scheduling—A Critical Review," in Dean and Goldhar, eds., *Management of Research and Innovation,* pp.215-33; L. A. Digman and Gary I. Dean, "A Framework for Evaluating Network Planning and Control Techniques," *Research Management* 24, no. 1 (January 1981): 10-17.

26. Lawrence J. Moore and Edward R. Clayton, *GERT Modeling and Simulation: Fundamentals and Applications* (New York: Petrocelli/Charter, 1976).

27. Gerald L. Moeller and Lester A. Digman, "VERT: A Technique to Assess Risks," *Proceedings of 10th Annual Conference, American Institute of Decision Sciences* (October 1978): 292.

28. Joseph J. Moder and Cecil P. Phillips, *Project Management with CPM and PERT,* 2d ed. (New York: Van Nostrand Reinhold Co., 1970); Peter P. Schoderbek and Lester A. Digman, "Third Generation, PERT/LOB," *Harvard Business Review* 45, no. 5 (September-October 1967): 100-110.

29. Richard B. Chase and Nicholas J. Acquilano, *Production and Operations Management: A Life Cycle Approach,* 3d ed. (Homewood, Ill: Richard D. Irwin, 1981), chaps. 16 and 17.

12

Project Needs and the Personal Needs of R & D Staff

There was, it appeared, a mysterious rite of initiation through which, in one way or another, almost every member of the team passed. The term that the old hands used for this rite...was "signing up." By signing up for the project you agreed to do whatever was necessary for success...From a manager's point of view, the practical virtues of the ritual were manifold. Labor was no longer coerced. Labor volunteered.

Tracy Kidder, "Building A Team" in **The Soul of a New Machine**

12.1
Introduction

In chapter 8 we surveyed ways of improving and maintaining the R & D inventive capacity of the organization, but stressed the critical importance of other skills in the R & D, as well as downstream phases, if innovation is to succeed. In this chapter we will explore these skill requirements in some detail and in the context of the personal developments needs of R & D staff. In chapter 13 we shall examine alternative organizational frameworks for successfully carrying innovation beyond the R & D phase.

12.2
Conflict between Individual and Organizational Values in R & D

It is generally accepted that the dissonance between individual and organizational values and goals is perhaps greater in the R & D function than anywhere else in the organization except on the production shop floor. Furthermore, this dissonance is likely to be particularly marked if the company follows an offensive/defensive strategy, pursuing nondirected research in a professionally prestigious central laboratory which is both geographically and culturally distant from its operating divisions. Such a company may have a continued requirement to recruit creative higher-degree engineering and science graduates from colleges and universities to maintain its proximity to the state of the art in research. The personal development of such graduates (who, for their parts, will be attracted by the prestige of the laboratory) is a matter requiring the sincere and concerned attentions of both individuals and the organization. In general, engineering graduates readily accept the value systems of business, but this generalization does not hold for many higher-degree science graduates.

The specialist educational process of a higher-degree graduate in science is essentially a Kuhnian apprenticeship in "puzzle-solving" skills, which makes him *au fait* with the currently dominant paradigm (or paradigms) in his field and provides the graduate with his initial research training. By virtue of his personality and education, his internal commitment is to his discipline and the goal of achieving significant professional recognition and esteem by making significant research contributions in his field. The graduate is thus interested in "publication" rather than "innovation." Indeed, his acceptance of a research post in industry may have been dictated by the absence of a suitably well-paid position in academia or government, rather than a belief in the values and goals of business. The graduate may even view the company, if not with disdain, then perhaps at least with indifference as the philistine patron of his scientific (as opposed to artistic)"genius."

The existence of this attitude among government (as opposed to industrial) R & D staff was illustrated in a study of the problems faced by small companies attempting to innovate inventions made in government R & D laboratories.[1] The investigators interviewed the president or vice-president of about fifty companies who had taken government R & D inventions and sought to develop them to commercially viable products, and asked them to comment on the problems they faced in transferring the "invention" from the government's to their own laboratories. Among these interviewees, there was a widespread feeling that some government "bench-level" R & D staff were concerned only with the generation of new knowledge and were disinterested in its commercial exploitation and the values and pressures of business. Two comments made which illustrate this feeling follow:

> *The objective of government and university research is publish or perish while in industry it is publish and perish. There is a major gap here in objectives between the two.*

and

*There is a stigma on profit. Government scientists' assist-
ance could be valuable but unfortunately there is a boy scout
concept of purity in Government R&D.*

The commitment of research staff (at least early in their careers)
to their discipline and their development as researchers, rather than
to the corporate goals, must be accepted by R & D management. It
must be recognized that research staff identify more closely with the
informal network and status system of their professional peers
rather than that of the organization. This, of course, is in contrast to
most managers who identify with the formal and informal networks
and status system of the organization. These days it is accepted that
the implicitly psychological employment contract does not dictate
that the employee should be totally committed to the values and
goals of the employer. One role of a manager is to reconcile the goals
of the organization and his subordinates, so far as possible, in the
best interests of both. That is, he should adopt a [9,9] orientation in
the terminology of Blake and Mouton's managerial grid.[2] The
adoption of this [9,9] orientation is probably more important in R &
D management than in any other functional area. It implies that
staff should be encouraged to participate in the formulation of R & D
goals and then seek the congruence between their personal and these
goals. Needless to say, such participation is implicit in approaches
to R & D budget setting and project selection discussed in chapter 7
onwards. By so doing, the problems created by differences of outlook
between R & D staff and managers will be ameliorated and hopefully
the former will become more sympathetic to the total innovation
process and business culture. Furthermore, as I shall argue in the
remainder of this chapter, a systematic examination of R & D staff
development needs and innovation needs does suggest that the two
are broadly congruent.

12.3
Development Alternatives for R & D Staff

A technological graduate whose first appointment is as a researcher
in an industrial R & D function, broadly speaking, has three alternative personal development paths available to him within the
organization. Sooner or later, either consciously or unconsciously,
he probably chooses to follow one of these.

One alternative is research. Individuals can remain as bench-level
researchers seeking self-actualization through continued research
output, gaining increasing insight into their fields and earning
increasing professional esteem. Following this path implies seeking

the minimum management and administrative responsibility, congruent with the initiation and supervision of perhaps an expanding portfolio or research projects. Such a role may imply modest formal status within the management structure since it may be effectively executed at perhaps a section leader level. Such individuals may enjoy considerable esteem among their professional peer group both within and outside the organization. Thus individuals may achieve maximum *personal* self-actualization by adopting this role, particularly as it may offer the maximum scope for personal autonomy, something creative people value highly. One possible disadvantage of the role is that it may not offer what we might label maximum *family* self-actualization. First of all, salary distributions may be generally lower for researchers than managers in most companies, so individuals may forgo substantial long-term income by selecting this path. Therefore, they may have to forgo the opportunity to offer their families a higher material quality of life—perhaps a higher-quality private education for children, higher-quality family health care, and so forth. Second, although they may enjoy the high esteem of their peers, this may not extend to society at large. Apart from the high-status professions, society at large tends to evaluate a family's status through its manifest material standard of living or the company hierarchical status enjoyed by the provider. Therefore, unless the individuals are really outstanding researchers, both they and their families may forgo some social esteem through the decision to remain a "researcher," rather than become a "manager." This is a factor that can affect a person's choice of career development.

The second alternative is research management. As an alternative to the above option, a person may choose to remain in the R & D function, but pursue increasing managerial responsibilities. Particularly if the choice is made to move from the research to the development phases of the innovation process, a person may be able to build up his management experience by taking increasing responsibility in administering increasingly complex projects. He may thereby progress to senior R & D management responsibility.

As stated in chapter 7, one role of senior R & D management is to mediate between goals and needs of corporate management (and other, functional areas) and those of the R & D function. To be effective in this role, a person must possess both management and R & D credibility. That is, he must enjoy the esteem and professional respect of both managers and R & D staff in the company, and given that he does, R & D management may offer a viable option.

The third alternative is other management positions. Although R & D management is an attractive personal development avenue, except possibly in high-growth situations, it is unlikely that the demand for R & D managers will exceed the supply, that is, R & D staff with the personal resources and motivation to develop their careers in that direction. Thus, after some project management experience, if they are to remain with the company, some R & D staff may wish to seek further growth opportunities through management positions elsewhere in the organization. Some high-technology companies do use the R & D function as a nursery for future managers.

The frameworks for developing R & D staff into innovation managers are discussed later. We will now briefly discuss the development needs of people who wish to remain "researchers," within the framework of a "dual-ladder" system.

12.4
The Dual-Ladder Structure

First of all, it is reasonable to assume that a creative individual will only wish to remain in research as long as he can maintain sufficient productivity to satisfy his internally set standards. This raises the issue of the aging process in creativity. There is a body of lay and, for that matter, management opinion which holds that creativity is a youthful quality and that an individual will have produced his best work before the age of thirty or forty. There is some but by no means conclusive evidence to support this belief. Outstanding mathematicians do have a habit of producing their best work before the age of thirty, but other outstanding scientists (as well as artists) have maintained highly creative outputs over a relatively extensive life span. Newton's massive contributions to physics and mathematics were spread over several decades while Edison (an inventor-entrepreneur) maintained his inventive output over a similar period. In art, Milton produced his masterpiece (*Paradise Lost*) when comparatively old. Similarly, at a more modest level of creativity, some researchers are able to maintain a satisfying level of output (to themselves and their employers), throughout a lengthy professional career. Although the larger industrial laboratories will wish to maintain a steady recruitment of young researchers from leading university research schools to maintain their proximity to the state of the art, a cadre of mature experienced researchers, *who have maintained their personal proximity to that state,* can provide research coaching and leadership that would otherwise be lacking. Therefore, the offensive/defensive innovative company has a definite long-term incentive for encouraging its most able researchers to make a permanent career in research. The issue then becomes one of providing such people with adequate remuneration and status. One method of dealing with this issue has been for laboratories to create dual hierarchies—one managerial and one technical through which people can progress as "managers" or "researchers."

One traditional method of rewarding R & D staff for good technical performance is to "promote" them into management positions. This is a method whereby the firm may "lose" the services of a good researcher and "gain" the services of a poor manager—clearly an undesirable outcome. In order to develop a satisfying career plan for R & D staff which motivates and rewards their good performances as technologists, rather than managers, a number of major companies employ dual-ladder systems of personnel recognition and promotion in their R & D functions. Although "successes" and "failures" have been reported with such systems, detailed examinations of individual corporate experiences suggest that the dual-ladder system provides an intrinsically worthy and effective means of R & D staff development and only fails when it is misapplied or corrupted in some way or other. Two U.S. oil companies—Mobil and Amoco—report* having used such systems since the 1940s and, despite having had some periods of frustration with them, both companies speak highly of the value of dual-ladder systems and are pledged to their continued use.[3]

As its name implies, the central feature of the dual-ladder system is to create the two parallel hierarchical sets of positions or titles in the organization—one for "researchers" and one for "managers." Most organizations that use it appear to try exactly to match "rungs" in the research and management ladders and, so far as is possible, to provide similar fringe benefits and status connotations as well as a common salary scale. Military officers of identical rank may perform widely different duties or tasks, dependent upon the specialist branch of the service to which they belong, but will enjoy a broadly identical status which is primarily dependent upon their identical rank. Similarly, researchers and managers occupying equivalent rungs in the dual ladder enjoy broadly similar rewards and outward recognition in the organization—success which readily communicates itself to others in the organization, the person's family and friends. Needless to say, the detailed formats of dual ladder systems vary from company to company, but table 12.1 illustrates the concept by showing the dual-ladder structure for two large companies in different high-technology industries—Westinghouse Research Laboratories in the United States and ICI Ltd in the United Kingdom.[4] In both firms there are four matching rungs, though ICI does have a terminal position which is purely administrative. Table 12.2 shows the criteria that Westinghouse reported using in evaluating R & D staff for promotion in their technical ladder. The consensus of companies which have *successfully* instituted a dual-ladder system is that its success is dependent upon using a high professional standard of undoubted integrity in its application, that is, ensure that the system recognizes and rewards professional excellence.

Research Management 20, no. 4 (July 1977) contains a series of short articles on individual corporate experiences with the dual-ladder system, which provides a useful overview of the topic.

ICI

Scientific Ladder		Administrative Ladder
	1st rung	
Senior Research Officers		Some Section Leaders
	2nd rung	
Senior Scientist etc.		Section Heads
		Section Managers
	3rd rung	
ICI Research Associate		Group or Assistant
		Research Managers
	4th rung	
Senior ICI Research Associate		Associate Research Managers
		Research Managers etc.
	Terminal	
		Research Director

Westinghouse

Management	Individual contributor
Director	Director
Department Manager	Consultant
Section Manager	Advisory scientist or
	advisory engineer
Supervisor	Fellow scientist or
	fellow engineer

Source: S. L. Meisel, "The Dual Ladder—The Rungs and Promotion Criteria," *Research Management 20*, no. 4 (July 1977), 24-26. Reprinted with permission of Industrial Research Institute.

TABLE 12.1 TWO LADDERS.

Companies that have *unsuccessfully* instituted such a system, appear to have failed because (possibly for quite legitimate reasons) they have been unable to establish or maintain the required standard of professional excellence. These considerations suggest the following "do's" and "don't's" for installing dual-ladder system.

Do ensure that the ladder structure is integrated into the overall human resource planning procedures in the corporation. In particular, ensure that the system is incorporated into the job evaluation and career planning procedures used in the R & D function and elsewhere in the corporation. The critical functions approach (see next section) to the analysis of R & D project needs provides a framework for job evaluation and career planning that could be readily linked with a dual-ladder system.

Do ensure that common technological and managerial rungs have at least roughly equivalent salaries, responsibilities, statuses, and fringe benefits. Such equivalence would be reflected in personal locations and titles in organizational charts, committee member-

Advisory Scientist or Engineer

1. Mastery of scientific and technical field Guide: Ph.D. or equivalent in research accomplishment
2. Accomplishments as demonstrated by
   a. Papers (external and internal), reports, memos, and other publications —number and quality
   b. Discovery or invention leading to company benefit
   c. Recognition by peers
   d. Strong consulting activity
   e. Invited participation in professional societies
3. Demonstrated ability to plan independently and follow through on significant programs
4. Demonstrated ability to
   a. Exert influence and exercise judgment in matters that affect the laboratories or company (mandatory)
   b. Assume responsibility for special tasks
   c. Guide work of others

The supervision of others is neither a necessary nor a sufficient condition for the position of Advisory Scientist or Engineer.

Fellow Scientist or Engineer

The Fellow Scientist and Engineer classification is based on same criteria as for the Advisory Scientist and Engineer except that, while high, they are applied less stringently.

Source: S. L. Meisel, "The Dual Ladder—The Rungs and Promotion Criteria," *Research Management 20*, no. 4 (July 1977), 24-26. Reprinted with permission of Industrial Research Institute.

TABLE 12.2 PROMOTION CRITERIA.

ships, reporting procedures, responsibilities (including project selection), laboratory and office space, and so forth.

Do ensure that promotions up the technological ladder are based upon reasonably defined criteria (as illustrated in table 12.2) and professionally agreed upon and recognized standards of excellence appropriate to the remuneration and status offered by the position. The establishment and maintenance of these standards appear fundamental to the acceptance of the system by the R & D staff, and for its continued success. Promotion to the more senior positions on the technological ladder should be made at corporate level by a committee consisting of senior and distinguished technologists (not all of whom need necessarily belong to the corporation).

The composition and membership of such a committee must obviously be dependent upon the local situation, but the success of the system will be largely dependent upon the integrity and credibility of its decisions. Given the "seniority" of the more senior positions and the caliber of suitable candidates for promotion to this level, decisions ideally need to be made by one central (that is, corporate level) committee with an "arms-length" relationship to the candidates. If possible, a candidate's local supervisors and superiors should *not* sit on this committee, but rather act as advocates on his behalf. Obviously, the strength of their advocacy would be dependent upon the extent to which these superiors thought the candidate merited promotion, but such an arrangement protects the integrity of the procedure from corruption by local pressure or political patronage. Promotion to the less senior rungs may have to be made at a more local level in a large organization, if for no other reason than the relatively large numbers of R & D staff involved, but promotion criteria and procedures should be designed to maintain the integrity of the system.

Whatever detailed criteria for promotion are used, a mandatory criterion should be an appropriate past and future level of contribution to corporate goals (see 4a. table 12.2). At first sight it might appear that this mandatory criterion is in conflict with the arguments presented earlier. However, we can expect candidates for the more senior rungs to be distinguished technologists in their fields who consequently enjoy continued job mobility. In these circumstances, it is unlikely that such individuals would have stayed with the corporation unless they could reconcile their personal research goals with those of their employers. Conflicts between senior R & D staff and R & D and corporate management can and do arise, but they are more typically concerned with which new innovations or products should be supported in order to achieve corporate goals, rather than the goals themselves. Furthermore, persons seeking to promote the adoption of "their" innovation by the firm are likely to be "entrepreneurs" or "project champions" and "managers" rather than being solely concerned with a research career. In other words, their appropriate career path may be the management rungs on the ladder.

The pitfalls to be avoided mostly concern the protection of the integrity of the rank structure.

Never promote staff to a rank that their R & D "track record" does not justify; that is, ensure that promotion recognizes legitimate R & D performance and is not used in some "honorary" capacity to reward non-R & D contributions.

Never ever use R & D rungs as "dumping grounds" for failed managers. If a manager proves ineffective in a line R & D management position, there is a humane and pragmatic temptation to move him sideways into the corresponding technological rung. If the dual system is working well otherwise, it might be argued that by moving him to that rung, he will retain his salary and status, but occupy a position in which little harm can be done. *Nothing could be further from the truth.* The use of the R & D rungs in this way immediately discredits the dual system in the eyes of R & D staff and rapidly destroys its credibility. Smith and Szabo of Union Carbide also indicate that the credibility of their company's system was compromised when some managers were appointed to technological posi-tions (often one or two rungs lower) to protect their job security during a period of corporate cutbacks.[5] Although the desirability of the retention of such personnel was not disputed in this case, the use of the ladder in this way eroded its credibility and viability.

It appears certain that the maintenance of a credible system requires a continual commitment on management's part. As Dr. Meisel (vice president of research for Mobil R & D Corp. at the time he presented his paper) put it, "the dual-ladder system is no perpet-ual motion machine."[6] It needs to be subject to changes in the light of changing disciplinary needs and corporate objectives and staff cutbacks or recruitment increases due to economic fluctuations. Typically, there may be several technological ladders required for different scientific and engineering specialties, and staffing re-quirements in each specialty may change markedly over a few years, due to shifts in the state of the art or corporate goals. Because a specialist researcher's professional life may last forty years, this can create difficult human resource management problems, as any con-temporary university president will confirm! As far as possible, promotion criteria and standards need to be maintained equitably across different disciplines and (if a multinational corporation is attempting to use the system in its R & D laboratories in different countries) across national geographic boundaries.

These considerations require a significant ongoing administra-tive effort which must be initiated and maintained without undue bureaucracy. Furthermore, the existence of a dual-ladder signifi-cantly impacts upon management as well as R & D staff. R & D managers may find that they have researchers reporting to them who are their seniors in the system. This means that their managerial authority is less well defined, and the researchers may be able readily to bypass them in the formal and informal communications networks of the organization. Managers may also envy the relative freedom from corporate disciplines enjoyed by researchers; indeed, overall, members of the technologist ladder may exist as a kind of counterculture to management in the organization. According to the proponents of the dual systems, these inevitable tensions can be made productive rather than counterproductive. As Moore and Davies of ICI express it: "Like the effect of grit in oysters, pearls have been produced but at some discomfort to the host—and at some discomfort also to the Scientific Ladder people."[7] Finally, perhaps the best overall argument for using a ladder system is summed up by Dr. Meisel who quotes the words of one of Mobil's young research chemists: "I don't think I would have come here if Mobil didn't have a professional ladder. But most other places have a dual ladder too."[8] In other words, corporate R & D laboratories may not find it easy to attract good R & D staff if they have a dual-ladder system, *but they will find it even more difficult to do so if they don't!*

12.5
Staffing Needs of the R & D Function

Earlier we argued that R & D managers should adopt [9,9] orientation in terms of Blake and Mouton's managerial grid. Readers familiar with that grid recall it postulates two orthogonal dimensions of concern—"concern for production" and "concern for people," with "production" and "people" really being used as synonyms for "organizational needs" and "individual needs." In section 12.2 we explored the individual needs of R & D staff and in this section we explore the other dimension of that grid—the organizational needs of the R & D function. In section 12.5 we shall outline one approach to "marrying" these two sets of needs, that is, in establishing an R & D function which has a [9,9] orientation.

Blake and Mouton's choice of the designation "concern for production" as a synonym for organizational needs is quite apposite in the present discussion. Readers conversant with the history of management thought will recall that the development of a discipline of "production management" as a systematic body of knowledge could be said to have started with the pioneering work of F. W. Taylor. He argued that efficient production imposed a requirement of task specialization and the obligation to match the right person to the right task in production as a line-team process. Although Taylor's ideas were carried too far in subsequent developments in mass production methods, leading to the more recent counterdevelopments of job enlargement and enrichment concepts, the broad concept of task specialization remains valid.

Furthermore, "production-line thinking," rooted historically in his ideas, has been applied effectively quite recently to service and manufacturing industries. For example, in chapter 4 we referred to the MacDonald's "technological hamburger" as an example of the innovative use of production line methods in the fast-food industry. Since the authority and effectiveness of task specialization have been recognized in other functional management areas, it is reasonable to suppose that they are also valid in R & D and innovation management. However, despite this comparison with Taylor's classic work on "Scientific Management," it must be stressed that the "management of science" in a business (or indeed any other) setting cannot be based upon purely rational considerations. The potential for conflict between individual and organizational needs in R & D plus the competition for support between R & D projects, precludes this. Thus, management at both the R & D organizational and separate project levels is best viewed in an organizational politics model framework as suggested by Dill and Pearson.[9] Much useful work on communication networks and role needs in R & D has been performed since the late 1960s.[10] We shall not examine such work here, but rather focus on one approach which reflects both "rational" and "political" considerations and can be readily developed by expanding the considerations implicit in the innovation chain equation analogy introduced in chapter 2. It is also consistent with Cooper's ongoing evaluation process described in chapter 9.

The nature of task specialization in R & D management has been developed by Pugh-Roberts Assoc. Inc., a Boston-based management consultancy company, and described in Fusfield and Rhoades, Roberts, and Fusfield.[11] These authors studied the R & D functions in a number of organizations to determine the tasks required to expedite the innovation process. They identified six distinct tasks (idea generation, technological gate keeping, market gate keeping, project championing, project managing, and coaching) which must be performed effectively if innovation is to be successful. Fusfield describes these six tasks as the *critical functions* in R & D management. They relate directly to the terms or ingredients in the innovation-chain reaction, as shown in figure 12.1. Although any one person may be able to perform perhaps up to three of these tasks effectively, no one person possesses either the aptitude or desire to perform them all effectively. Therefore, to ensure that an innovation is potentially successful, it is imperative that R & D management selects a project team that is *collectively capable* of performing *all* these tasks. This implies that the project team is built upon task and role specialization (rather like a baseball or cricket team), in which one member of the team is recognized as having prime responsibility for the effective performance of each of the tasks. The activities and personal aptitudes required by these critical functions, based upon Fusfield's descriptions, are now discussed.

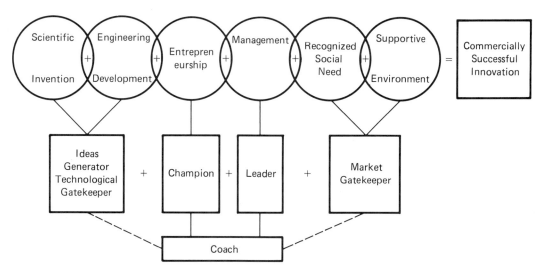

Figure 12.1 Critical Function and the Chain Equation.

Idea Generating

An idea generator is an inventor or creative individual. Therefore, he is likely to exhibit the characteristics we discussed in chapter 8. The idea generator will possess specialist expertise, be good at problem solving, and at generating or testing the feasibility of new ideas. He will enjoy conceptualization, working with abstractions, and seeing new and different ways of viewing things. He may be an individual contributor to the team and prefer to work alone.

Technological Gate Keeping

The role of an information gate keeper in R & D was suggested by Allen as a person who provided a human interface between the R & D staff and the outside environment, particularly the external R & D community.[12] This person thus provided an information channel whereby R & D staff could keep abreast with the state of the art both in terms of new technological ideas and new commercial exploitations of the technology. Fusfield has separated these two aspects of the role into two critical functions—technological and market gate keeping. The technological gate keeper possesses a high level of technical competence and esteem and is very widely read and known. He keeps informed of related developments outside the function, through reading a wide range of journals, attending conferences, and maintaining a wide network of outside professional contacts. He is personable and enjoys helping people and so serves as an information resource for colleagues in the organization. The technological gate keeper can provide invaluable support for the idea generator, since he can expand the latter's repertoire of ideas, concepts, and facts which can be incorporated into the latter's conscious and nonconscious creative mentation. He can also provide informal coordination between project personnel.

Marketing Gate Keeping

In the framework of the technology-market matrix, the technological gate keeper may be viewed as providing the information resource for generating new technological realizations and capabilities. Similarly, the market gate keeper may be viewed as providing the information resource for identifying potentially exploitable market opportunities. His personality characteristics are similar to the technological counterpart, but this gate keeper's expertise and interest will be in technological marketing. Again, he will possess technical competence and esteem, but in the area of product application rather than realization. Again he will read journals, attend conferences, and maintain a network of professional contacts concerned with the markets in which the company exploits its generic technological skills. The marketing gate keeper will discuss their problems and problems and needs with customer's researchers and design engineers thereby monitoring the market and identifying with the technological gate keepers of customer companies. He can play an important role in ensuring that the innovation is meeting an identified market (avoiding the dangers of technology-push), which may change during project development, and help coordinate the required marketing effort.

Project Championing

Most case histories of innovations stress the importance of the project champion role. Technological innovation, which is a synonym for technological change, requires not only that the uncertainties and hazards associated with a novel technical ensemble should be resolved and surmounted, but also that internal and external resistances to change must be overcome. Unless there is at least one person, often in a position of some influence, who is totally committed to the innovation and has the dogged determination to overcome all the barriers, it is unlikely to succeed. Expressed more succinctly this means that: "Behind every successful technological innovation there growls a bull-dog!" The project champion is the individual who provides for the continuance of the project despite resistance and set-backs, until it succeeds. The champion has wide technological interests which are applications oriented, but without a strong propensity to contribute to the basic knowledge in his field. He is energetic and determined and aggressive in championing his own "career" and willing to take risks. He is skilled in selling ideas to others and securing organizational resources. He is the typical technological entrepreneur whose personality characteristics are discussed further in chapter 14.

Project Managing and Leading

Although the champion or entrepreneur provides the drive or momentum for the project, he may be unsuited to the task of "managing" the project, that is, be incapable of the attention to detail required to coordinate the various facets of the project and ensure that they are completed on time. This requires a person with management and administrative interests, who is recognized as the focus for questions, information, and decision making. This person provides the team with leadership and motivation and is sensitive to accommodating others' needs. He is interested in all management functions, knows how to use the corporate structure to get things done, and seeks to match the project goals with organizational needs. He plans, organizes, and coordinates the project, ensuring that administrative requirements are met and that the project moves forward smoothly.

The first five critical functions can be directly related to the consituents of the innovation-chain reaction as shown in figure 12.1, but the final function requires rather more explanation. Not infrequently an innovative project is conceived and promoted by an informal coalition of "young Turks." In that case, although the project team may incorporate all the above critical functions, it may still be ineffective because its youthfulness undermines its organizational credibility. This is particularly so if the project championship and leadership roles are performed by relative "juniors" in the organization hierachy. The project may consequently be aborted through inadequate corporate support. In some cases, this may lead to the resignation of the project team from the firm and the formation of a "spin-off" company dedicated to the commercial exploitation of the proposed innovation, possibly funded by private and venture capital. This spin-off phenomenon is, of course, well documented in the innovation literature and is discussed in chapter 14. If the project looks viable, its abortion is clearly undesirable from the company's viewpoint, whether or not it is resurrected elsewhere. To avoid such outcomes, Fusfield postulates a project coaching role.

A project coach acts as a "Godfather" (without the perjorative criminal associations!) or patron and advisor to the project team. He is often a more senior person in the company whose patronage invests the project with credibility and confidence. A coach is experienced in developing new ideas and provides objective guidance and information or coaching to the less experienced team members helping them to develop their talents. He provides access to the power base of the organization, helping the team obtain organizational resources and buffering it from unnecessary constraints. He thus ensures that the project does not fail through lack of company support.

Fusfield's critical functions provide a useful typology of innovative project team needs, and so we will now discuss how they may be used to develop a [9,9] orientation in R & D management.

12.6
Matching Individual and Organizational Needs in R & D

In section 12.2 it was suggested that there are three possible personal development paths that young R & D staff might follow. Although these may be viewed as the extremes of a triangular continuum, they do suggest the critical function roles that would appeal to staff following each of the three paths, as is shown in

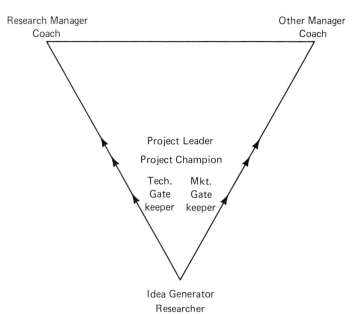

Figure 12.2 Development Paths.

figure 12.2. This figure is essentially self-explanatory, but some brief comments may be made. Clearly the "researcher" is going to be the idea generator (because in the present context, the terms are synonymous), and he may also play a technological gate-keeping role. Anyone might become a project champion, since it implies emotional commitment and determination, qualities that any researcher or manager may possess. However, "inventor-entrepreneur" (applied to those such as Edison) is essentially a synonym for "idea generator-project champion." Championing his own invention through the successive innovation phases can provide a valuable and chastening experience for the "researcher," since he may thereby experience for the first time and in an intensely personal way, the real difficulties of the innovation process. The "research manager" is likely to develop through "project-leading" roles and, once he gains the experience and seniority, is well equipped to enact the "coaching" role. The "other manager" may also develop through "project leader" and/or "market gate-keeper" roles and again, later he may be very well placed in the company to enact the "coaching" role.

To match individual staff and organizational development needs, Fusfield recommends the performance of a *critical function assessment*. This assessment identifies the personnel required in each critical function, based upon the R & D goals and strategy and the portfolio of projects; the existing strengths in each function; and the factors supporting and inhibiting the performance of each function. This information is obtained through questionnaires, discussions, and workshops with R & D staff. Individuals complete questionnaires to identify their critical function strengths and those required in their current R & D roles. From the assessment, mismatches between ideal requirements and the strengths of staff can be identified to highlight weaknesses and suggest areas for improvement.

Fusfield quotes the results of such a survey of a unit with ninety R & D staff. This unit appeared to have an unbalanced allocation of skills to needs and suffered from a shortfall of project leaders and market gate keepers. It appeared to be an organization that will be good at generating ideas or inventions, but relatively weak at carrying them through to innovations or perceiving their true market relevance. Despite a declared market-pull orientation, the distribution of critical strengths was biased towards technology-push.

It was obvious that senior management was obligated to remedy the weakness exposed, and the following actions were taken:

1. Every individual discussed his critical function score with management, to reach agreement as to the individual's skills as a basis for future career planning. The jobs of staff who were mismatched to their innovation roles were modified accordingly.

2. Training in project-leading skills was provided which involved the participation of personnel from outside units, to enhance the recognition of project leadership in the total organization. This exercise clarified the project-leading role and ensured that upper management was sensitive and supportive towards it.

3. Recruiting practices were changed to ensure that applicants were evaluated in terms of their critical functions (as well as technical) strengths and interests, and employment offers were based upon critical function as well as disciplinary needs. Interviewers were chosen to reflect a balance of critical function to ensure no unconscious bias in recruiting. Historically, it appeared that idea generators had unconsciously favored and attracted more idea generators.

4. All future jobs were defined in terms of the critical functions as well as technical factors. The management-by-objectives or MBO procedure being used was expanded to incorporate critical functions, and innovation teams were selected to ensure a balanced distribution of these functions.

The critical functions treatment provides a pragmatic typology for developing [9,9] innovation management, and one other facet needs to be discussed. Both the discretionary scopes and the measuring and reward structures vary across the critical functions, but it should be recognized that they must honestly adhere to the performance criteria of the critical function if the approach is to succeed. If an idea generator perceives that the company awards most of the credit for a successful idea to perhaps the project champion, this person is either going to be his own project champion (even though perhaps ill suited for it), or he may well suppress the idea rather than let someone else take the credit for it. Similarly, people will not undertake gate-keeping roles that produce intangible outputs, unless the organization recognizes the worth of these activities. Therefore, it is imperative that a person's rewards and recognition are based upon the criteria set by the critical function role(s) he is enacting.

12.7
Matrix Structure

The technical demands of R & D (particularly in an offensive-defensive innovator) impose conflicting requirements on the laboratory organizational structure as well as its staff. The fundamental and applied research phases of innovation are primarily discipline oriented. Furthermore, young scientists and (to a lesser extent) engineers entering an industrial research laboratory straight from the university identify with their respective disciplines. These two factors dictate that a strong research capability should be based upon research groups in the appropriate disciplines (chemistry, physics, and so forth). Even when innovations have progressed into development projects, they place varying demands upon expensive laboratory services and facilities such as engineering, machine shop, computing services, so that it makes economic sense to manage these centrally to provide a common service to all project "users." To take an extreme example, projects in an oceanographic R & D establishment may require the performance of seagoing trials. Oceanographic research vessels are expensive items of "laboratory equipment," and it is obviously economically absurd to lease or purchase a ship for *one* voyage for *one* project. In reality, any one voyage of such a vessel would be designed to accommodate the needs of a number of projects, including perhaps oceanographic prospecting devices, diving gear, and so forth. Such considerations dictate a laboratory management structure based upon disciplines and laboratory service functions, as shown schematically in figure 12.3.

In contrast to the arguments, however, as an innovation progresses from research through to development, it becomes more "multifunctional" in character and places an increased demand on a widening range of laboratory and other company services—

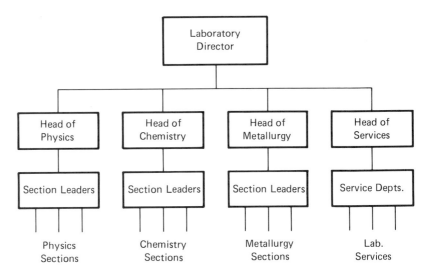

Figure 12.3 Discipline Structure.

demands that must be made in the face of competition from other projects. Furthermore, if a project team is cohesive, as it should be, team members will increasingly identify with the project, as opposed to their disciplinary or service function background. This cohesion and identification will be weakened if each member still sees his disciplinary or service function divisional head as his only "boss"— who will largely determine his future career prospects. Effective project management must therefore "cut across" the structure shown in figure 12.4, with team members identifying an allegiance to the project and its leader, as much as their divisional head. Many laboratories have accommodated this requirement by adopting a matrix organizational structure as shown in figure 12.5. The team membership is seconded from a number of divisions, depending upon the technical requirements of the project and the personnel available, and reports to a project leader. Given that project teams may have a transient existence and/or include part-time members, individual members will usually retain their divisional affiliations. They thus become "two-boss" persons, often having reporting responsibilities to divisional heads *and* project leaders. Matrix structures therefore violate the classical management dictum of a single line of command, and they have been criticized for the conflicting reporting requirements they impose. However, despite doubts concerning the efficiency of matrix structures, many companies claim to have adopted them to manage projects successfully, both in R & D laboratories and in manufacturing operations. They are discussed further in the next chapter when we will consider the transfer of projects from R & D to operations.

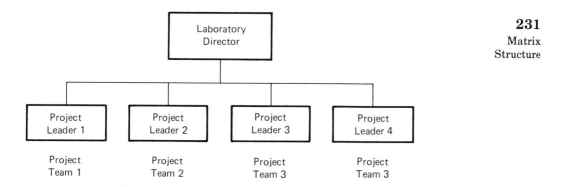

Figure 12.4 Project Structure.

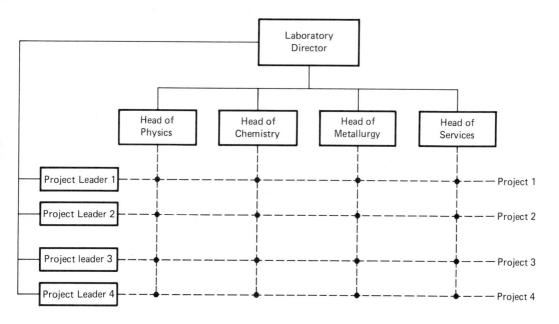

Figure 12.5 Matrix Structure (R & D).

References

1. M. J. C. Martin, J. H. Scheibelhut, and Robert Clements, *Transfer of Technology from Government Laboratories to Industry,* Technological Innovation Studies Program Report No. 53 (Ottawa: Department of Industry, Trade and Commerce, November 1978).

2. R. R. Blake and J. S. Mouton, *The Managerial Grid* (Houston, Tex.: Gulf Publishing Co., 1964).

3. S. L. Meisel, "The Dual Ladder—The Rungs and Promotion Criteria," Research Management 20, no.4 (July 1977): 24-6; E. W. Cantrell et al.,

"The Dual Ladder—Success and Failures," *Research Management* 20, no.4 (July 1977): 30-33.

4. E. X. Hallenberg, "The Dual Advancement Ladder Provides Unique Recognition for the Scientist," *Research Management* 13, no.3 (May 1970); 221-27; D. C. Moore and D. S. Davies, "The Dual Ladder—Establishing and Operating It," *Research Management* 22, no.4 (July 1977): 14-19.

5. J. J. Smith and T. T. Szabo, "The Dual Ladder—Importance of Flexibility, Job Content, and Individual Temperament," *Research Management* 20, no.4 (July 1977): 20-23.

6. Meisel, "The Dual Ladder."

7. Moore and Davies, "The Dual Ladder."

8. Meisel, "The Dual Ladder."

9. D. D. Dill and A. W. Pearson, "Managing the Effectiveness Project Managers: Implications of a Political Model of Influence" (Paper presented at the EURO V-TIMS XXV International Meeting, Lausanne, Switzerland: July 1982).

10. W. A. Fischer, "Scientific and Technical Information and the Performance of R & D Groups," in *Management of Research and Innovation,* ed. B. V. Dean and J. L. Goldhar, TIMS Studies in the Management Sciences, 15 (Amsterdam: North Holland Publishing Co., 1980); W. A. Jernakowicz, "Organizational Structures in the R & D Sphere," *R & D Management* 8, (1978): 107-13; Ralph Katz and M. L. Tushman, "An Investigation into the Managerial Roles and Paths of Gatekeepers and Project Supervisors in a Major R & D Facility," *R & D Management* 11, no.3 (July 1981): 103-10.

11. A. R. Fusfield, Unpublished Conference Presentation; R. G. Rhoades E. B. Roberts, and A. R. Fusfield, "A Correlation of R & D Laboratory Performance with Critical Functions Analysis," *R & D Management* 9, no.1 (October 1978): 13-17.

12. T. J. Allen, "Roles in Technical Communication Networks," in *Communication among Scientists and Engineers,* ed. C. E. Nelsen and D. K. Pollock (Lexington, Mass.: D.C. Heath/Lexington Books, 1970), pp. 191-208.

Part 4

The Operations Setting

In chapter 13 we will discuss barriers to the transfer of an innovation from an R & D to a production facility, the organizational frameworks for facilitating such transfers, and the role of the "learning curve" in innovation.

13

Transferring the Project from R & D to Production

It must be considered that there is nothing more difficult to carry out, nor more doubtful of success, nor dangerous to handle, than to initiate a new order of things. For the reformer has enemies in all those who profit by the old order, and only lukewarm defenders in all those who would profit by the new order, this lukewarmness arising partly from fear of their adversaries, who have the laws in their favour, and partly from the incredulity of mankind, who do not truly believe in anything new until they have had actual experience of it.

Machiavelli

13.1
Introduction

In this chapter we will examine the issues and problems faced in the transition from the R & D to the "production" phase in the innovation process. In a large offensive/defensive innovative company, this typically involves the transfer of the inventive "know-how" from a centralized R & D function to an operating division or production facility which is organizationally and (probably) geographically separate. The problems endemic to this transfer process lead naturally to the consideration of alternative organizational formats for the development of technological entrepreneurship and new ventures—an issue that will be discussed in the succeeding chapters.

If a successful offensive/defensive innovator has established strong downstream coupling (chapter 4), using the project evaluation process described in chapter 9, and a project team staff based upon critical functions criteria (chapter 12), this transfer could be comparatively straightforward. The transfer is likely to be made at stages V and VI in Cooper's framework, but the transfer from the central R & D facility may occur at an earlier stage if the organization has a divisional engineering development facility closely associated with production. It may also occur in two steps, the first from a central R & D laboratory to a divisional engineering develop-

ment facility and the second from that facility to production. Whichever situation obtains, the above considerations should have ensured that project staff will have already established informal
communications links with operations people. Nevertheless, it is
important to recognize that the differences in values and outlooks
of the individuals involved can create barriers to the transfer process and we begin the chapter by discussing these.

13.2
Specific Problems of Technology Transfer

Technology transfer has been researched and discussed extensively
in the literature during the last twenty years.[1] Much of the research
and discussion has focused upon international (particularly from
developed to developing countries) and interorganizational (particularly from government and university R & D institutions to private
industry) rather than intraorganizational transfers, and it is only
the latter that are of specific interest here. Steele[2] suggests that the
barriers to such transfers are as follows:

Technical Barriers

The essential purpose of the exercise is to transfer the technical
know-how of the project to production and to replicate its performance results in a scaled-up production rather than R & D environment, so we discuss technical barriers initially.

First of all, what will already be obvious to most experienced
scientists, engineers, and technology managers in industry is that
technical know-how cannot be transferred purely "on paper." It is
virtually impossible to document exhaustive, detailed, unambig-
uous, and error-free specifications for a project. Much of the experience and insight built up by solving the problems and overcoming
the "bugs" endemic to successful project progression can never be
meaningfully documented on paper. Part of this learning experience
may be incorporated in revised specifications and instructions,
but inevitably duplication of learning must occur, which may be
minimized if at least some members of the R & D project team are
personally involved in the transfer process. The "operations" project team then has immediate access to the R & D team's experience
and knowledge whenever bugs arise. In fact it may be argued that
it is virtually impossible to transfer technology effectively without
"people transfer" (at least temporarily), through intraorganizational secondments. If the secondment takes place from R & D to
operations it can play a useful role in the personal development of
R & D staff, as discussed in the previous chapter.

"People transfer" is imperative for other reasons too. A phenomenon well known to innovation managers is the "not-invented-here" or "NIH syndrome." This is when inventions or innovations introduced into organizational units (including R & D laboratories) from outside sources quite frequently fail through lack of internal support. There are at least two fairly obvious psychological explanations for this observation. First, an externally conceived invention is "threatening" since it may be viewed as reflecting adversely on the professional competence of the internal staff. Second, as indicated in chapter 12, a "project champion" or "entrepreneur" plays an indispensable role in the project team. It is the individual enacting this role who maintains the momentum and support for the project through the vicissitudes of corporate resource allocations and politics. An R & D project introduced into operations through the transfer of paper documentation only is unlikely to find the sponsorship of a project champion in operations. Therefore, there is a considerable likelihood that it will be terminated prematurely through benign, if not malign, neglect.

Even with the transfer of some members of the project team, other technical barriers need to be overcome. The process of scaling up product output to the pilot/prototype level and beyond, within acceptable cost limits, almost inevitably involves technical design and performance changes and the modification of the original inventive concept. This process can frequently generate "pride of ownership" disputes between R & D staff and production engineers in operations. R & D staff "inventors" are sometimes reluctant to accept the technical compromises required for the large-scale manufacture of "their" product at a price which the market will accept. As one technology manager expressed it to this author in the context of similar transfers, "The scientist who developed the concept didn't give us much assistance. It was his baby. He didn't want any interference."[3] A related technical barrier is the attitudinal difference between R & D staff and operations managers towards "workability." To the former, an invention or concept "works" if it can be produced on a laboratory scale of manufacture probably with a technician in permanent attendance to repair or modify it. To the latter, an invention only "works" when it can be manufactured at the required full-scale output and cost level and is sufficiently robust to operate continually in its perhaps rugged "end-use" market. Steele quotes a general manager's definition of "working" as "something that works...24 hours a day, 7 days a week, 365 days a year!" There is a considerable difference between these two concepts of "workability."

First, the process of scaling up production will probably require considerations of mass production methods, tool-up cost (frequently a major cost element), learning effects, and the lower standards of assembly operators performances and skills, quality and asurance, the economics of shared components and modules with other related products, short- and long-term reliability, and after-sales service systems, all within acceptable cost constraints. Second, R & D staff sometimes lack insights into the end-use environment for a product. They may fail to recognize that the final user, who may be an unskilled worker, may not have the time, ability, or inclination to take much care in reading results from a piece of equipment. Furthermore, this environment may expose the equipment to considerable abuse. The final design of a product must accommodate these considerations. Such an accommodation must often degrade the technical performance of the invention, to the chagrin of its inventor. Some R & D personnel suffer from a professional hubris that equates scientific competence with manufacturing and commercial competence. The following comments made to the author by technology managers illustrate these points:

> *Scientists don't have to worry about marketing a product. They often don't realize that the guy who will have to operate a gadget won't have a Ph.D. They don't know the manufacturing problems of trying to fit a million wires into a saleable package.*

> *The scientist had devised a perfect instrument which was difficult, if not impossible, for a layman to use. I took twenty minutes to take a reading, which is alright for a scientist but not for a business that considers time a real cost. The lab model needs considerable re-designing for quicker readings and must be re-designed to manufacture at a lower price. One of the big problems is that the scientist has no market orientation. He is strictly tied up in his scientific toys.*

> *Also, when they (scientists) do become involved, they are reluctant to be a follower and accept the leadership and direction of the commercial entrepreneur. It seems that if they think themselves scientifically and technically competent then they believe themselves to be commercially competent.*

> *I think the scientist and manufacturer should work together to build a commercial product that takes into consideration the scientific principle, design engineering, quality that the customer wants and at an acceptable price.*

Nontechnical Attitudinal Barriers

The above barriers are all to some degree technical impedimenta to technology transfer, but there are cultural and value differences between research-oriented and operations-oriented staff which impede transfer.

First of all, there is a significant difference in attitudes towards "time." As Steele shrewdly points out, researchers view an event or outcome as the independent variable and time as the dependent variable. That is, a researcher may be able to plan the study of a particular research problem, but he cannot guarantee that the expenditure of a given amount of time will generate the event or outcome of a successful problem solution, discovery, or invention. In contrast, the manager views time as the independent variable and an event or outcome as the dependent variable. The manager may plan and budget for a given interval of time and then identify and guarantee the sequence of events or outcomes that may be completed in that interval. Furthermore, because a manager typically has to produce results against unremitting time deadlines, working from the annual budgeting and reporting cycle downwards, he is "time-oriented" and seeks to "digitalize" his time, that is, break it down into discrete units within which he completes specific preplanned tasks. On the other hand, a researcher produces results through discoveries or events that are unpredictable in time, therefore the researcher is "event-oriented" and treats time as a continuum that cannot be divided between different tasks. Because of these orientation differences, researchers almost invariably appear to be poor time performers to managers.

The progression of a project from the R & D phases to the operations phases of the innovation process reflects a progressive reduction in technological uncertainty. Project management and control can thus be increasingly dictated by time-oriented rather than event-oriented consideration. This shift in emphasis is reinforced by the typical dominance of a time orientation in an operations facility. Failure to recognize the shift can erect a barrier to the transfer process. Ideally, projects should not be transferred to operations until technological uncertainties have been resolved to the extent that any further technological "bugs" can be solved in a time-oriented, rather than event-oriented, management mode. Ideally, also, R & D staff remaining associated with the project should recognize this shift and change their own orientations accordingly.

Closely related to the event versus time orientations differences is the knowledge versus action orientations of researchers as opposed to managers. Research scientists place high emphasis on the importance of obtaining a theoretical understanding of the mechanism underlying the invention. They argue that the more comprehensive this understanding is, the less is the likelihood of technical "bugs" occurring later, and if they do occur, the more likely they are to be overcome. They are therefore prone to delaying the progress of the project, while a satisfying theoretical explanation for unexpected results is derived, or what may be worse, becomes readily sidetracked into peripheral investigations which follow from these unexpected results. While this attitude may be desirable in a pure or undirected researcher, it has less appeal when a project has moved into the later phases of the innovation process. In contrast to the researcher, the manager is action oriented. His job requires choosing between alternative courses of action, in a timely manner, with the best information that is available. If all the information pertinent to the decision is available, all well and good, but if not he will choose anyway, since some action must be taken.

Once a project has moved to an operations facility, it is likely and, on the whole legitimate, that a more action-oriented approach should be adopted. The technology of the project should be largely "proven" by then, and the increased costs and time delays associated with too exhaustive investigations may critically affect the chances of commercial success of the innovation. However, *it should be stressed* that each situation must be treated on its own merits. For example, in the aircraft and pharmaceutical industries, the risks to the consumer from the premature introduction of a product whose safety is subject to doubt may create situations where the knowledge orientation of the researcher takes a moral precedence over the action orientation of the manager.

Third, in private industry, innovations must ultimately be judged by commercial profit/ROI criteria. As a project moves into the operations phases of the innovation process, both decision makers and decisions are increasingly likely to be made against "bottom line" and product launch considerations. To most operations managers, this criterion appears morally legitimate (and if it does not, they would appear to be in the wrong job!). As we discussed in the previous chapter, some researchers view the profit motive with disinterest, arguing that their role as scientists is to generate knowledge for its own sake or for the good of society. These two viewpoints are not mutually antithetical since, following no less an authority than Peter Drucker, we may argue that today's profits are the costs that will finance tomorrow's innovations. This is an argument that is certainly congruent with our overall approach to corporate technological strategy formulation and implementation. However, some researchers may be either unaware of or disbelieve such arguments and, in any case, are unlikely to value profit as a criterion of success. Therefore, they are likely to measure internal rewards from their contributions to an innovation in terms of technical rather than commercial success.

All these differences in attitude and values may inhibit communication between the various professional groups involved in technology transfer and tend to reduce the chances of a commercially successful outcome for the innovation. On the other hand, a recognition that such factors may inhibit transfers does enable management to design more effective transfer mechanisms and, in line with the discussions in chapter 12, to plan the personal development of staff and staff the critical functions of project teams to promote technology transfer as smoothly as circumstances allow.

13.3
Management Framework for Transfer

The discussion so far has argued that effective technology transfer cannot occur without "people transfer" and has identified some of the barriers to this transfer. We now consider means for facilitating both technology and people transfer and reducing the barriers to the transfer process. The discussion is conducted in the context of an incremental innovation, which can be fairly readily pursued within the framework of the company's existing organization structure. Alternative frameworks incorporating the technological venture approach are discussed in chapter 16. Gerstenfeld outlines several alternative frameworks, including operations participation in R & D, R & D's participation in operations, and mixed transfer teams from both functions.[4] The appositeness of each is dependent upon a company's unique circumstances, organizational structure, and climate, so no specific framework can be recommended as best for all. However, because the mixed transfer team framework appears to be generally preferable, it will be outlined here. For exposition purposes, we discuss it in the context of the transfer of a project from a central R & D facility to a divisional engineering development facility at step IV of the process described in chapter 9.

By the time that this step is reached, the requirements of the earlier steps in the evaluation process should have ensured the informal participation of divisional personnel. The R & D project team (which should already incorporate the critical functions requirements) will provide the nucleus of the transfer team and can be expanded to include divisional personnel subject to three requirements. First, it is not dominated by people from either central R & D or divisional functions, so that the differences in attitudes and approaches that create the barriers to the transfer process may be discussed openly, and a cohesive team spirit developed. Second, all management functional interests are represented at an appropriate level of expertise and authority, and the critical functions requirements are still fulfilled if some R & D personnel withdraw from the project. Third, team membership includes people who are senior enough to give the project organizational credibility.

To ensure that it enjoys organizational credibility, Gerstenfeld suggests that the project team should report to a corporate coordinator who will be its sponsor. He will be a corporate manager, enjoy seniority and a diverse breadth of experience in the organization, will be skilled at securing the organizational resources and the access to senior and middle management, and will possess the advocacy skills required to expedite the progress of the project. Thus the coordinator "orchestrates" the transfer of the project from central R & D to the division and enacts the critical function coaching role. Quite possibly the original project coach will continue in that role as transfer coordinator. At one of steps IV through VI the critical functions team requirements should also be expanded.[5] A *production engineer* will be added to advise on the limitations and possibilities of production processes, the design "makeability" of the product, and to oversee and iron out bugs in its pilot to full production runs. A *resource controller* will also be needed to ensure that functional human as well as material resources are made available when required, and to monitor costs and participate in pricing procedures.

The transfer team approach, working under the sponsorship of a coordinator/coach senior manager appears, in principle, to be the best one, but it will likely vary between organizations, and within organizations for different projects and at different times. The approach adopted will also be dependent upon the human and financial resources available since the coordinator-mixed-team approach is an expensive consumer of both. However, whichever approach to transfer is adopted, to be effective it requires the staffing of teams that span the critical function needs and the technical and commercial requirements of the projects. It also strengthens the basis for any future venture team (see chapter 16).

13.4
Matrix Structures

So far in this chapter we have focused on the barriers to and formats for transferring innovations from R & D to divisional development or manufacturing operations, without regard to how the project team (howsoever it is compromised) is incorporated into the operations management structure. We will now consider this issue.

In chapter 12, it was argued that for offensive/defensive innovators, the research end of the R & D function, particularly non-directed basic research activities, are best performed in an organic organizational climate, within a management structure based upon the disciplines appropriate to the technological base of the company. Once the invention moves into the development phases of the innovation process, it was argued that the establishment of a project team based upon the technological needs of the project must cut across functional disciplines in a matrix structure (recall figure 12.5). The same considerations apply when the project is moved from corporate R & D to divisional development and then production operations. The project team and its requirement now cut across management functions and the requirements of ongoing product lines, and again a matrix structure would appear to be desirable. As stated in chapter 12, the efficiency of the matrix structure is disputed. So as well as discussing matrix structures in operations, here we will continue the discussion begun earlier in section 12.5, and will consider the benefits of matrix structures in general.

The benefits of a matrix structure are perhaps best judged in the context of a compromise between two alternative extremes. One possibility is the pure project organization, where operations are organized as an aggregate of largely autonomous projects, typified by the construction industry. At least in theory, it might be possible to manage the operations of any high-technology company as an aggregate of projects in various stages of the innovation cycle. In this framework, the composition of a project team would change with the varying critical function requirements of the successive phases of the innovation process. This framework is most nearly approached in industries involved in the assembly of large, complex high-technology products such as air transports and ships.

Most innovations, however, are incremental and more typically introduced into production operations in mature industries which have reached the cost-minimizing stage of their industry life cycles. Part of this cost-minimizing rationalization process involves the establishment of functional management structure in which potential economies of scale are exploited by manufacturing or processing a limited range of closely related products in mature established divisions. Even in the aircraft and ship-building industries, a project management structure is only partially realized because much of the subassembly work may be performed in functionally managed organizations. Thus realistically it can be expected that most incrementally innovative projects will be transferred from R & D to an operations facility that is primarily functionally managed. The disadvantage of this latter structure, in the context of innovation management, is that it creates an environment which is probably resistant to technological change. Operations staff will identify their loyalties, career aspirations, and goals with the dominant operations functions and respective departmental heads. For an innovative project to thrive in the operations facility, it will require able team members from that facility. If the management, leadership, and reward structure is dominated by a functional orientation, such staff will not wish to join an innovative project team, since they will perceive it as penalizing their career prospects. If they do accept such assignments, they may do so reluctantly and on a part-time basis, sharing their time between the project and an ongoing functional responsibility. In a functionally dominant organizational climate, they will probably assign higher priorities to their functional responsibilities and a relatively low commitment to the project. The project team will thereby fail to secure its legitimate share of organizational resources and suffer a decline in morale, cohesiveness, and entrepreneurial drive. Therefore, there are sound arguments for introducing a new product into operations through a matrix structure, with project team members reporting to both project and functional managers (see figure 13.1). Once the new product has achieved market acceptance, its production may be transferred to a conventional functional operations management structure.

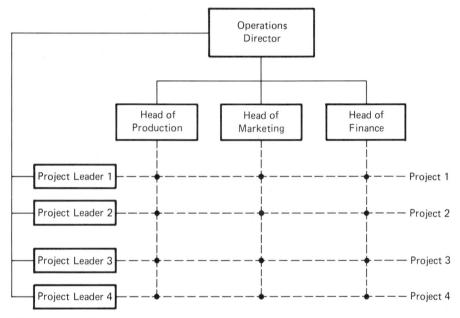

Figure 13.1 Matrix Structure (Operations).

The benefits of a matrix structure can be summarized quoting the words of Kolodny,[6] who has made a specialist study of the subject:

> *Matrix organization is designed to have the best of both project and functional worlds. It overcomes the weakness of poor task responsiveness on the part of the functional organization by channelling the knowledge of the specialists through project and program teams. It overcomes the weakness of limited long term specialist competence development on the part of the project organization by keeping specialists closely connected to functional homes.*

Matrix management structure in operations is similar to that in R & D (compare figures 12.5 and 13.1) except that the vertical columns of the matrix are now management functions, rather than subject disciplines. The main disadvantage is that the individual team member is a "two-boss" person, since he has reporting responsibilities to both the project manager and his functional manager. This person may therefore suffer from the role uncertainty and conflict of loyalty that this creates. Equally the project and functional managers may experience discomforts with the ambiguities created for them, particularly if they are competing for an individual team member's time, as well as other organizational resources. Such potential conflicts can be avoided or resolved, provided that top management understands its leadership behavior in a matrix organization.

Lawrence, Kolodny, and Davis point out that a matrix structure is really a diamond structure (figure 13.2), with both project and functional managers reporting to a common superior.[7] Successful matrix leadership behavior requires this person to initiate dual control, evaluation, and reward systems which balance project and functional needs and create a matrix culture which fosters open conflict resolution. Given the undoubted difficulties of implementing an effective matrix organization, it is pertinent to comment briefly on the research which has evaluated the approach. Kolodny provides most valuable reviews in the references cited earlier and two other investigations are specifically revealing. Marquis studied thirty-seven projects in the aerospace industry in which he identified four organizational formats.[8] They are (1) functional, (2) matrix with large functional areas and small project teams, (3) matrix with small functional areas and large project teams, and (4) project. He found that matrix structure (2) was the most effective format for achieving

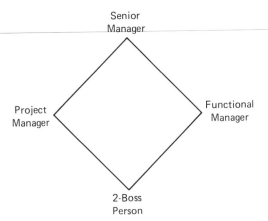

Figure 13.2 Matrix Diamond.

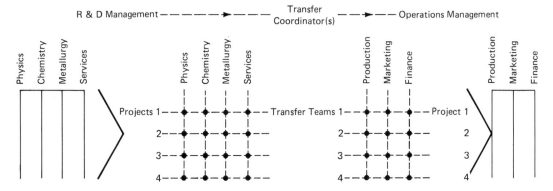

Figure 13.3 Overall Transfer Process.

technical success. Jermakowicz studied 100 projects spread across seven organizational formats with functional and project formats at each extreme and the matrix format in the middle.[9] He used two measures of effectiveness for evaluating these formats—"number of new products introduced in the organizational system" and "originality of new products." He found that while the highest level of originality occurred in a project format, the highest implementation rate was achieved in a matrix structure. Given that the implementation of a profitable innovation is the ultimate measure of success, it could be argued that the two studies yield broadly identical conclusions.

The overall evidence suggests that often the best framework for transferring projects from R & D to operations is via a mixed transfer team, possibly under the sponsorship of a senior coordinator, from an R & D management matrix structure to an operations management matrix structure. When the project becomes a fully-fledged new product, it may be transferred to a functionally structured production facility. Given that the original invention began in a discipline-structured research facility, the overall sequence of organizational structures in the innovation process is functional—matrix-functional, as illustrated in figure 13.3. This sequence has been achieved through intuitive trial-and-error organizational learning by many offensive/defensive innovators.

13.5
Learning Curves and the Technological Progress Function

Once a fully-fledged new product is transferred to functionally managed mainstream operations, its production and marketing may be integrated into operations. Once it becomes an operations (as opposed to innovation) management responsibilty it could be said to be no longer the concern of this text. Nevertheless, there is one important property of technological innovations in full-scale production setting which remains to be considered. That is the impact of learning effects on unit production costs and pricing of an innovative product. In chapter 2 we viewed technology as progressing though the cumulative trial-and-error elimination impacts of successive innovations. It is important to recognize that the trial-and-error elimination learning occurs not only in the inventive phases, but also throughout the innovation process, including the production and diffusion phases when an innovative new product has been adopted by mainstream operations. In production this property is reflected in the learning curve, which was first observed in the aircraft industry and subsequently in numerous other manufacturing and process operations.

Such learning patterns are documented by Hirschmann.[10] All readers will know from personal experience that the time required to perform a repetitive task decreases with repetition. That is, it can be performed more quickly the second time than the first, the third than the second, and so on. What is less well known is that the rate of improvement is consistent and appears to continue indefinitely. For example, if the improvement rate is 80 percent the hours required to produce one unit will be reduced by 20 percent each time output is doubled. That is, if it takes 100 hours to produce the first unit, it will take 100 x 0.80 = 80 hours produce the second, 80 x 0.80 = 64 hours to produce the fourth, and so on as shown in table 13.1. It is expressed mathematically in the learning function shown in table 13.2, where the parameter "b" is the improvement rate (b = 0.32 above). It follows from the form of this function that when we plot hours per unit against cumulative output, we obtain the curve shown in Figure 13.4 (a), and when we plot the logarithms of these measures we obtain the straight line shown in Figure 13.4 (b). The slope of this line measures the improvement rate "b."

The ubiquity of this production learning effect has been amply demonstrated in numerous industries including those cited by Hirschmann. The value of the improvement rate "b" varies between different production technologies and is particularly dependent

Total number of units produced	1	2	4	8	16	32
Man-hours required for last unit	100	80	64	51.2	40.96	32.77

TABLE 13.1   LEARNING EFFECT.

$$y = ax^{-b}$$

Where y is man-hours required to produce x-unit
x is the number of units produced
a is man-hours required to produce first-unit
(a = 100 hours in Table 13.1)
b is the learning or improvement parameter
(b = 0.32 in Table 13.1)

TABLE 13.2   LEARNING / IMPROVEMENT FUNCTION.

upon the relative proportions of manual assembly versus machine-paced work requirement. It typically lies between 80 percent and 90 percent. However, what is more important to recognize is that such learning effects are not confined to unit production, but also influence a broad range of other costs. The term *experience curve* is sometimes used to describe these broader learning effects that may apply to the total costs of successive versions of a product which incorporate a succession of incremental innovations. For example, Abernathy and Wayne show that the price of the dominant design, Model T Ford followed a 85 percent slope experience curve over the years 1909-23.[11]

The consistency of experiential learning effects may be exploited by companies in their pricing strategies when launching innovations. Companies in the aircraft and electronics industries (such as Boeing and Texas Instruments) may launch an innovative new product at a price below current production costs anticipating that the levels of future sales will ensure that costs fall below price from learning effects as output increases. This strategy of pricing ahead "down" the learning curve enables them to reap the benefits of large market share as well as learning effects. It must also be stressed that such exploitation must be based upon prudent management judgment. Firstly, if continued learning improvements are to be maintained, the organization must continually seek ways of improving production methods and products. That is, it must maintain a continued positive climate for improvement from all its members. Secondly, it is obvious that pricing down the learning curve can be financially disastrous if anticipated sales and therefore outputs are not realized. Thirdly, the continued exploitation of the learning effects from a successful product range should not be allowed to constrain technological flexibility. An excessive dependence on its benefits may lead to undue rigidity and the inability to respond to the future opportunities and threats of technological change. The limitations of the learning curve are discussed in the article by Abernathy and Wayne cited earlier.

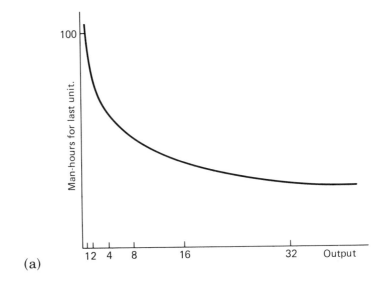

(a)

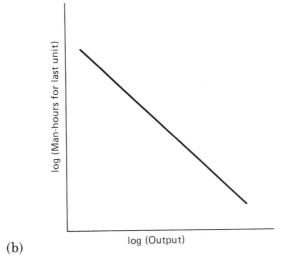

(b)

Figure 13.4 Learning Curve.

13.6
New Venture Operations

So far in this chapter we have concentrated upon the transfer of
technology from R & D to the mainstream operations in a company.
Because revolutionary innovations are based upon new technological paradigms, it is often technologically infeasible to transfer them
to existing operations, as they require the establishment of new
businessess and new industries to exploit them commercially. More-

over, as suggested in chapter 9, some radical innovations, while falling short of inducing a Kuhnian paradigm-shift, are sufficiently novel to require the introduction of radical new production technology. Also, even if an innovation posseses more modest technological novelty, it may have to struggle for continued survival in face of the opposition from operations personnel who believe (rationally or otherwise) that they have a vested interest in maintaining the technological *status quo*, particularly if the company suffers from the technological rigidity indicated above.

The causes of technological conservatism among operations personnel are discussed in chapter 16, but if an otherwise promising innovation remains unexploited by mainstream operations, the project team will be frustrated and may set up their own new company to do so (see next chapter). If this happens, the innovation will be "lost" to the present company. To avoid this outcome, an alternative approach to transferring technology to mainstream operations is to treat a normal, radical, or incremental innovation *as if it were* a revolutionary innovation and set up a separate business venture to manufacture and market it. Since "entrepreneurship" is a vital element in the innovation chain equation, it is hardly surprising that some companies have adopted a separate venture approach to exploiting innovations commercially. This separate ventures approach will be discussed in some detail in chapter 16, after we have examined another aspect of technological entrepreneurship—the spin-off phenomenon and the process for establishing new technology businesses, in chapters 14 and 15.

References

1. W.H.Gruber and D.G. Marquis, eds., *Factors in the Transfer of Technology* (Cambridge: MIT Press, 1969). Harold F. Davidson, Marvin J. Cetron, and Joel D. Goldhar, eds., *Technology Transfer* (Leiden, Netherlands: Noordhoff International Publishing, 1974).

2. Lowell W. Steele, *Innovation in Big Business* (New York: American Elsevier, 1975).

3. M.J.C. Martin, J.H. Scheibelhut, and R. Clements, *Transfer of Technology from Government Laboratories in Industry*, Technological Innovation Studies Program, report No. 53 (Ottawa, Department of Industry Trade and Commerce, November 1978).

4. Arthur Gerstenfeld, *Effective Management of Research Research and Development* (Reading, Mass.: Addison-Wesley, 1970), chap. 4.

5. E.B. Roberts, "Generating Effective Corporate Innovation," *Technology Review* 80, no.1 (October/November 1977): 27-33.

6. Harvey F. Kolodny, "Matrix Organisation Designs and New Product Success," *Research Management* 23, no.5 (September 1980): 29-33. Harvey F. Kolodny, "The Evolution to a Matrix Organization," *Academy of Management Review* 4, no.4 (October 1974): 543-53.

7. Paul R. Lawrence, Harvey F. Kolodny, and Stanley M. Davis, "The Human Side of the Matrix," in *Readings in the Management of Innovation,* Michael L. Tushman and William L. Moore, eds. (Marshfield, Mass.: Pitman Publishing Inc., 1982), pp. 504-19.

8. D.C. Marquis, "A Project Team + PERT = Success, Or Does It?" *Innovations* no.7 (May 1969).

9. Wladyslaw Jermakowicz, "Organizational Structures in the R&D Sphere," *R&D Management,* 8 (1978): 107-13.

10. Winfred B. Hirschmann, "Profit From the Learning Curve," *Harvard Business Review* 42, no.1 (January-February 1964): 125-39.

11. William J. Abernathy and Kenneth Wayne, "Limits of the Learning Curve," *Harvard Business Review* 52, no.4 (September-October 1974): 109-19, also reprinted in Tushman and Moore, eds., *Readings in the Management of Innovation,* pp. 109-21.

Part 5

The Entrepreneurial Setting

Part 5 considers the entrepreneurial aspects of the techno-logical innovation—entrepreneurship process. It begins by considering the key role played by small spin-off com-panies in pioneering innovation and the personal charac-teristics of technological entrepreneurs. Some readers may be contemplating launching their own technological ven-tures, so chapter 15 extensively discusses such start-up operations. In contrast, chapter 16 discusses frameworks for stimulating technological entrepreneurship in larger companies. The book concludes with some final comments on the management of technological innovation and entrepreneurship in a company setting.

14

Small Is Beautiful: Spin-Offs, Innovation, and Entrepreneurship in Small Businesses

The successful entrepreneur is driven not so much by greed as by the desire to create. He has more in common with the artist than with the business manager.
Paul Johnson, *"George Gilder Praises Capitalism"* **Wall Street Journal**

14.1
Introduction

The establishment and growth of new high-technology industries in the nineteenth and early twentieth centuries was particularly associated with the pioneering efforts of inventor-entrepreneurs. These were individuals who were both creative scientists and/or engineers *and creative businessmen.* Their technological and commercial vision, drive, and judgment created what sometimes ultimately became major high-technology corporations, often built around revolutionary innovations. Examples of such innovation giants can be found in most Western countries. They include Nobel (Swedish and dynamite), Linde (German and liquid air distillation), Brins (French and industrial oxygen), Solvay (Belgium and ammonia soda), Perkin (English and aniline dyes), Bell (Scots-Canadian and the telephone), Edison (American and electric light), and Marconi (Irish-Italian and radio) to name but a few. In chapter 1 we surveyed the experiences of the last of them. As the twentieth century progressed, however, with the increasing institutionalization and professionalism of R & D in universities, government, and industrial laboratories (often founded by such individuals), the role of these individual inventor-entrepreneurs appears to have declined, although Land (polaroid camera) and Carlson (xerography) are recent examples of the breed.

In chapter 2, we hypothesized that creative young technological entrepreneurs pledge adherence to revolutionary innovations just as creative young research scientists pledge adherence to new paradigms following Kuhnian scientific revolutions. During the Abernathy-Utterback fluid stage following a revolutionary innovation, new companies are often founded by such individuals to seek to exploit the new technology. Such new companies are usually founded or spun off by individuals from larger companies, government laboratories, and universities. This spin-off phenomenon is notably identified with specific industries (solid-state electronics and computers) and localities ("Silicon-Valley" in California and "Route 128" in Massachusetts). It might be conjectured that this phenomenon characterizes the establishment and growth of new high-technology industries at the present time as did the individual inventor-entrepreneur in the past. The success of a new high-technology company is partially dependent upon the entrepreneurial aptitudes of its founders, and entrepreneurship may be viewed as the commercial expression of the creative process described, in the R & D context, in chapter 8. Moreover, such successful small companies usually generate economic growth and employment opportunities in a region. Therefore, in this chapter, we briefly comment on the role of small-business technological entrepreneurship in economic growth, then examine the spin-off phenomenon—the personal characteristics of technological entrepreneurs and new venture initiation.

14.2
The Economic Role of Small High-Technology Business

As Gee and Tyler indicate, innovative companies have much better records for generating economic growth and employment than others, and an industrialized nation's economic ability to grow and maintain international competitiveness is largely determined by technological innovation.[1] This observation, and the contribution that smaller companies can make to innovation, is further supported when seen in combination with the results of other studies. Mansfield has found that the social return (that is, the benefits to society) of an innovation is typically double the private return (that is, the ROI to the company which introduced the innovation).[2] This finding is sadly reflected in EMI's experience with the CAT scanner (described in chapter 1).

Furthermore Kamien and Schwartz[3] note that smaller companies play a vital role in the process of technical change, while Shapero points out that the 1,000 largest U. S. firms are outperformed by smaller firms.[4] These contentions are supported by the

	Compounded Annual Growth Rates 1969–74			Absolute Increases in Employment
	Sales	*Tax*	*Employment*	*1969–74*
Mature Companies	11.4%	7.8%	0.6%	25,558
Innovative Companies	13.2%	8.5%	4.3%	106,598
Young "Tech." Companies	42.5%	34.1%	40.7%	34,369

Mature Companies: Bethlehem Steel, Du Pont, GE, GF, International Paper, P&G.
Innovative Companies: Polaroid, IBM, 3M, Xerox, Texas Instruments
Young "Tech." Companies: Data General, National Semiconductor, Compugraphic, Digital
Equipment, Marion Labs.

TABLE 14.1

data presented in table 14.1 from a comparative study of large mature, large innovative, and "young" high-technology companies made between 1969 and 1974.[5] What can be seen is that the young companies have far more impressive sales, employment, and tax compounded growth rates than large innovative and mature companies. Perhaps more impressively, between 1969-74, they created more absolute employment opportunities (34,369 versus 25,558 employees) than their mature counterparts from an employment base which was less than 1 percent of the latter's (7,579 as opposed to 786,793 employees in 1969)!

The superior growth records of these small companies suggest that they may demonstrate superior innovative records, and this suggestion is supported by British studies. Gibbons and Watkins, discussing the competitive ability of small firms, argue that large firms find it more difficult to react to technological change, so that small firms benefit from their faster "reaction time."[6] Freeman found that small firms made significant innovative contributions in many industries.[7] The Bolton Report also concluded that small firms make an important contribution to innovation in the United Kingdom.[8]

At the first sight it might be argued that the unemployment created by frequent small business failures must be set against the growth records of successful companies. This argument is of only limited validity, however, for two reasons. First, the mortality rate of emergent high-technology business (20 to 30 percent) is much lower than all small business (over 80 percent). Secondly, just as Mansfield demonstrates that the social benefits of innovation outweigh their private benefits, Shapero argues that a business failure generates social as well as individual learning. He also cites Henry Ford as an example of entrepreneurs who experienced several failures (in

254

Ford's case, two) before achieving business success. Thus techno-logical entrepreneurship, like technological evolution, is a trial-and-error learning-by-doing process. This technological and entrepre-neurial learning process is reflected in the spin-off phenomenon, which we will now discuss.

14.3
The Spin-Off Phenomenon

Earlier we have referred to *incubator organizations,* that is, institu-tions that appear particularly prone to employing individuals who later establish (that is, spin off) successful new companies. Govern-ment R & D facilities, universities, as well as private companies, have been credited with the propensity to spin off such companies. Stanford University can claim to have "hatched" some very suc-cessful high-technology companies after World War II. William Hewlett and David Packard and Russel Varian graduated from Stanford and later established Hewlett-Packard and Varian Asso-ciates, while in 1972, the Stanford Industrial Park contained sixty-five companies employing 16,000 people. The evidence suggests, however, that private companies are more effective incubators than the other institutions, and we will focus on these. There are two factors—one transient and the other permanent—that have stimu-lated the spin-off phenomenon. First, the Apollo space program launched by President Kennedy in 1960 and second, the personal frustrations experienced by individuals in the larger companies, both coupled with the business opportunities created by microelec-tronics technology. We will examine each of these factors, in turn, before studying the motivations of individuals who have spun off.

The influence of the Apollo program on the creation of new companies is described by Osborne.[9] This program generated the need for numerous industry-based, limited period R & D contracts funded by the U. S. government. Government policy allowed com-panies fulfilling such contracts to charge profits based upon a fixed percentage of total salaries paid. This policy encouraged some com-panies to hire more professional staff and pay them higher salaries than the R & D needs and labor market rates warranted—that is, to "pad" their payrolls to increase their profits. While such contracts lasted, the arrangement was fine for R & D staff. They were well paid and being underemployed, they could learn new subjects and devel-op new ideas on "the government's time." But they also knew that they enjoyed no job security. Once a government R & D contract ended, the staff could expect to be laid off. Therefore, they changed jobs frequently, hopping from one R & D contract to the next one in another company, becoming professional gypsies. If such footloose

256

Small Is
Beautiful:
Spin-Offs,
Innovation, and
Entrepreneurship
in Small
Business

individuals develop an idea or invention that looks commercially promising, it is less than likely that they will seek to exploit it with their current employers. They are more likely to set up a company to exploit it themselves. The burgeoning expansion of the solid-state electronics and computer industries in the 1960s and 1970s created many business opportunities of this kind, which Osborne compares with the California gold rush.

We have already seen that the comparative fecundity of scientists and engineers to generate potentially profitable inventions, coupled with the rising costs of a project as successive phases of innovation are completed, usually dictates that all projects cannot be supported to completion. Furthermore, projects sometimes fail to survive once they have been transferred to the manufacturing setting through lack of enthusiastic support from operations and/or general management. Whenever a project is aborted (or subjected to malign neglect) for whatever reason, it is hardly surprising that members of the project team often experience frustration and dissatisfaction with the treatment of "their" project. One reaction to this frustration and dissatisfaction is for such individuals to sever their association with their employers and set up a new company to manufacture and market the product they are developing. This is the second factor that has led to the "spin-off" phenomenon and the establishment of networks of small high-technology companies in specific localities. The Santa Clara Valley in California and the Greater Boston area in Massachusetts are most notable of these, but others exist in the United States and elsewhere; "Silicon Glen" for example, in Central Scotland and "Silicon-Valley-North" in Ottawa, Canada. Danilov, in describing the spin-off phenomenon, lists the following examples of spun-off companies—Control Data Corporation, Digital Equipment Corporation, Teledyne Inc., Aerojet General Corporation, High Voltage Engineering, and Itek Corporation.[10] Such companies have subsequently spun off second and third generation offspring—Spectra Physics from Varian Associates and Coherent Radiation Labortories from Spectra Physics. More notably perhaps, we have the Rand Corporation which spun off from Douglas Aircraft and was the "parent" of System Development Corporation and the Hudson Institute.

Probably the most famous of these incubator organizations, however, is Fairchild Semiconductors and the cascade of its spin-offs or progeny is shown in the "Begat Tree," figure 14.1 based on Draheim.[11] This figure shows that between 1957 and 1970, Fairchild begat numerous "fair-children, grand-children," and so forth. Although Fairchild's fecundity is exceptional, being related to the rapid growth opportunities manifest in the "California gold rush" climate of the solid-state electronics industry and the "go-go" venture investment climate of the period (partially financed from the high profits from the Apollo program cited earlier), some companies

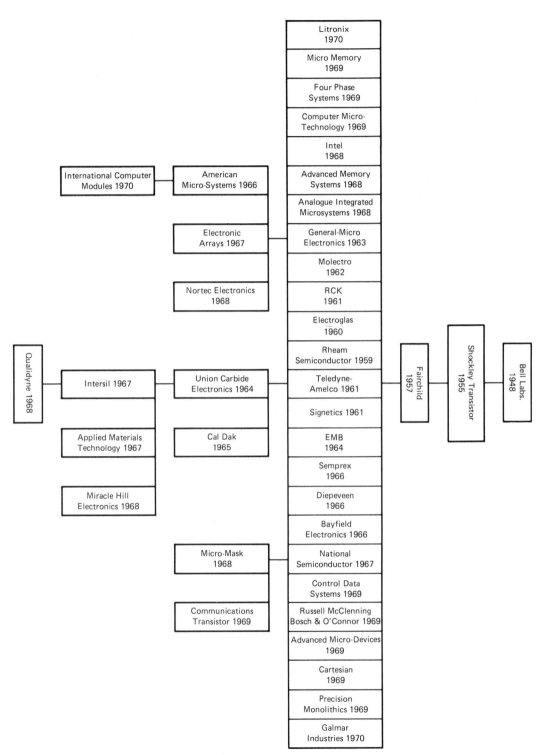

Figure 14.1 The "Fairchildren," 1970.

257

258

Small Is
Beautiful:
Spin-Offs,
Innovation, and
Entrepreneurship
in Small
Business

clearly act as better incubators than others. Cooper[12] conducted a study to identify factors that influence a company's propensity to spin off New Technology Business Functions or NTBF's.

Cooper studied about 250 NTBF's that were founded in the San Francisco area in the period 1960-69, from incubator parents whose total employment exceeded 77,000. The results of his study can be summarized as below.

Founder's Expertise

He found that the majority of the new firms set up seek to exploit their founder's expertise in the technology-market matrix of their parents. Of Cooper's new firms, 85.5 percent sought to exploit the same technologies and/or markets as their parents. Obviously this makes prudent commercial sense, since fledgling venture founders are likely to be more successful in technologies and markets that they already know well. Astute participants in a high-technology industry in the fluid growth stage of its life cycle will be able to identify profitable market niches or, as Cooper puts it "pockets of opportunity."

Location Choice

For several reasons NTBF's tend to locate close to their parents. First, the founders may wish to exploit an intimate network of fairly local professional and social contracts, both to finance and develop their product(s) and to exploit the target market. Second, there may be a period of part-time activity or even "moonlighting" prior to the founder's final break with the parents (see chapter 15) during which they develop their venture plan and possibly secure financial support. Alternatively, they may build up local reputations as consultants, which they may wish to exploit by performing consulting work during the early phases of NTBF development to generate supportive income. Third, apart from the fact that the founders may have to take a second mortgage on the family home as collateral for start-up capital, they may be unwilling to disrupt the lives and education of their families. This observation supports the evidence cited later; namely, that such venture founders have an average age of thirty-three years when founding their first NTBF and usually have supportive, stable marriages.

In two British studies, Dickinson and Watkins[13] and Watkins[14] reported broadly similar conclusions. In the former study, only two of a sample of twenty-one spin-off companies reported that the new entity was founded outside the founding proprietors' residential areas.

Other observations of Cooper suggest that industries in the fluid rather than mature stage of their technology life cycle are more likely to provide spin-off opportunities. At the mature stage, companies are often comparatively large with a high investment in capital equipment exploiting economies of scale and efficiency to minimize costs with innovations focused on process rather than product improvements. In these circumstances, it is more difficult to identify opportunities to found an NTBF that will attract the "front-end" capital required or to produce a product at a competitive cost.

Size of Incubator Firm

At first sight it might be expected that larger firms should show greater spin-off propensities than smaller ones. Larger organizations are more likely to suffer from organizational senescence and a bureaucratic climate that frustrates younger staff. In fact Cooper found that the spin-off rate, expressed as the inverse of the number of employees per spin-off was ten times greater for companies employing less than 500 persons, as compared with those employing more than 500. Note that the firms were classified as under 500 employees, over 500 employees, and subsidiaries under 500 employees, and that the spin-off rate of the latter was also eight times that for the larger firms. This result suggests that size rather than ownership affects spin-off propensity. The explanation for the higher spin-off rates in the smaller firms possibly lies in the positive attributes of their organizational climates as much as the negative attributes of those of the larger ones. First, individuals with the longer-term aspirations of setting up their own small businesses are more likely to be attracted to a position in a smaller firm. There they will probably be given broader responsibilities, in both range and substance, at an earlier age, than they would in a larger company. Second, they will be gaining direct experience of the problems of operating and managing small businesses, having closer contacts with different functional managers, as well as familiarity with the markets in which they operate. Third, they will directly observe a small businessman in action, namely, their own employer; and this, together with the examples of colleagues who have spun off business, may provide a positive "demonstration effect."

Cooper suggests other mechanisms that may increase the propensities of small business environments to spin off new ventures. He suggests that some of the small business subsidiaries studied may have been taken over quite recently, so that individuals spin off because they are unwilling to adjust to the loss of management

260

Small Is
Beautiful:
Spin-Offs,
Innovation, and
Entrepreneurship
in Small
Business

independence associated with the take over. Furthermore, some of these may have had substantial stakes in the equity of the takeover company and have achieved generous capital gains from the take-over deal. They are then ready to finance an independent new venture. One behavioral trait of some technological entrepreneurs, which supports this contention, is the tendency to establish success-fully, and then sell at profit, a succession of new ventures. Cooper also found that in the larger firms studied, the spin-off rate appears greater from departments which constituted the "small businesses" of the firm. Apart from the reasons cited above, he suggests that such departments may be not as well managed as the larger divisions of these firms and have relatively weak "bargaining power" in securing corporate resources for new product develop-ment. This conjecture is congruent with the reported reasons individuals offer for "going it alone" (see below). In contrast, Wat-kins did not find a small business bias in spin-off propensities in the United Kingdom.[15] Such a difference may perhaps be attributed to the absence of a "Silicon-Valley/Route 128" phenomenon and the lower number (in relative and absolute terms) of small high-technology businesses in the United Kingdom. Watkins also quotes the experience of one person who left Plessey to join a small com-pany in order to obtain some small business experience prior to founding his own NTBF.[16]

Motivation Factors

The personal frustration with employers obviously may be a signifi-cant factor in inducing an individual to found an NTBF, and this observation is born out by Cooper's results. He subjected thirty individuals to in-depth interviews to elicit their reasons for leaving their incubator-organizations. Seventy percent reported themselves to be highly frustrated in their previous jobs; 30 percent said they quit without specific alternative employment; and 40 percent said they would have left anyway. Cooper reports the major areas of concern among these people were poor selection and development of managers, and poor investment in products and technologies. Other writers suggest the importance of precipitating events or ma-jor displacements, such as bankruptcies, dismissals or extensive layoffs, and major company policy changes in stimulating spin-offs. Watkins quotes the example of a policy change in Plessey leading to twelve spin-offs in the United Kingdom. Such a displacement, to-gether with a sympathetic economic climate, may well trigger an individual with entrepreneurial leanings to found a NTBF, so it is pertinent to our discussion to examine the personal characteristics of the technological entrepreneur.

Characteristics of Technological Entrepreneurs

Before discussing their characteristics, it is useful to establish a working definition of the noun *entrepreneur,* since it is a word that has considerably broadened its meaning over the last 100 years. The 1897 edition of the *Oxford English Dictionary* defined an entrepreneur as: "the manager or director of a public musical institution: one who "sets up" entertainment specially musical performances." But the 1975 edition of *Webster* defines one as someone who undertakes to: "organize, manage and assume the risk of business." This latter definition is a realistic semantic approximation of its present general usage. Presumably its etymology is the French verb *entreprendre* (literally between-taker), meaning to undertake or assume a responsibility or task. Clearly an entrepreneur undertakes or assumes business responsibilities and tasks, and in France, it is often used to describe a small businessman or trader, particularly a building contractor. Its German equivalent—*unternehmer*—does literally mean undertaker, and it is perhaps unfortunate that this word has acquired a funereal colloquial usage in English. The entrepreneur typically assumes responsibility for *creating* and developing new living businesses rather than burying old dead ones, and it is this quality of creativeness that is important in technological entrepreneurship! In the context of technological evolution, this quality converts inventions or "technological mutations" into innovations that occupy useful social as well as technological niches.

As well as experiencing changing meanings, the noun *entepreneur* suffers from an unfortunate pejorative association. Many English-speaking peoples (particularly outside the United States) equate the entrepreneur with a "get-rich-quick fast-buck artist," "huckster," or "spivish" figure; and, for this reason, some larger companies use the term "project champion" to describe the role of the individual whose drive and tenacity will sustain an innovation to fruition. This misconception is both unfortunate and unfair as well because the evidence suggests that, if anything, many technical entrepreneurs have somewhat higher ethical standards than the average person.

As with creativity, different behavioral schools of thought have offered different explanations for an entrepreneur's behavior. However it appears that, at present anyway, there is no agreement from the results of the various studies on the psychological profiles of entrepreneurs, although the differences in part may be attributable to differing biases in the samples studied. Some have studied techno-

262

Small Is
Beautiful:
Spin-Offs,
Innovation, and
Entrepreneurship
in Small
Business

logical entrepreneurs (that is, people who have set up high-technology business) as opposed to all entrepreneurs (that is, people who have set up any kind of business, including small traders, Mom and Pop storekeepers, etc.), while others have studied business students or have included managers in their samples. Consequently, it is difficult to draw firm overall conclusions from these numerous studies. A useful recent review of such work by Brockhouse and Gasse[17] is provided and some overall generalizations can be made.

McClelland, in *The Achieving Society*,[18] has been a seminal contributor to the field and he summarizes his ideas in "Business Drive and National Achievements."[19] He argues that a society's economic development is stimulated and maintained by its entrepreneurs who are motivated by an high need for achievement (n-Ach). He describes a high need for achievement as:

> *the desire to do something better, faster, more efficiently, with less effort. It is not a generalized desire to succeed, nor is it related to doing well at all sorts of enterprise ...Rather it is peculiarly associated with moderate risk-taking because any task which allows one to choose the level of difficulty at which he works also permits him to figure out how to be more efficient at it, how to get the benefit (utility) for the least cost. And business, cross-culturally, is the specific activity which most encourages or demands using the calculus of cost-benefit.*

He characterizes an individual with a high n-Ach as someone:

1. Who likes situations in which he takes personal responsibility for finding solutions to problems.

2. Who has a tendency to set himself challenging but realistic achievement goals and to take "calculated risks."

3. Who also wants concrete feedback as to how well he is doing.

He observes that a society which offers equal opportunities to people regardless of origin is likely to show an high n-Ach. Hagan also argues that such development occurs when creative problem-solving ability is directed towards industrial innovation.[20] In fact both argue that economic development is stimulated by individuals with personal attributes and goals (whether they be labeled high in n-Ach or creative problem solvers), which set them ahead of the prevailing norms. Such individuals may be somewhat alienated from the prevailing social values, but have curious enquiring natures, rich stores of ideas, and confidence in their own evaluations of problem situations. They provide much of the driving force of economic development and may be archetypes for the many inventor-entrepreneurs, some of whom were listed at the beginning of this chapter. Thus,

they may also represent the archetype for the technological entrepreneur I am discussing here. Both the scientist and the technological entrepreneur have a strong need to reach high self-set goals.
Some individuals may set themselves goals that embrace R & D inventiveness and entrepreneurial aspirations, and may thus become
inventor-entrepreneurs. Associated with a high n-Ach is a desire for
feedback on how well they are doing, with an associated anxiety that
such feedback will reveal failure to achieve the desired goals. This
need for feedback is also consistent with the entrepreneurs' attitude
towards risk. To the layperson, an entrepreneur may be viewed as a
gambler or irrational risk taker. However, as Roberts and Wainer
indicate, the technical entrepreneur sees himself as a moderate risk
taker who would not take the gamble of founding a new business
venture, until he had estimated that it had a reasonable choice of
success.[21]

An alternative and possibly complementary explanation of
entrepreneurial behavior is offered by Rotter's "locus-of-control"
theory.[22] According to this theory, a person perceives the outcome of
events to be controlled by chance, powerful other persons (or factors)
or himself.

One who believes that he exercises personal control over outcomes is described as believing in an "internal" locus-of-control.
Clearly a person is unlikely to set up in business unless he believes in
an internal locus-of-control. Such "internal" individuals are more
striving and competent than their "external" counterparts (that is,
those who believe that chance or powerful others control outcomes)
and extract more information from ambiguous situations or environments. Rotter conjectures that a high n-Ach should be related to a
belief in an internal locus-of-control and this conjecture is supported
by other studies.[23] In another study, Borland asked over 300 business students to indicate their future expectancy of starting a business as well as determining their n-Ach scores and locus-of-control
beliefs.[24] She found that students with high n-Ach scores *and* internal locus-of-control expressed high expectancies, and for students
with low n-Ach scores, an increasing internal locus-of-control correlated with an increased expectancy of starting a business. Her results suggest that an internal locus-of-control is better than a high
n-Ach in measuring entrepreneurial aspirations. Brockhouse also
found that the owners of businesses that had survived for three
years held stronger internal locus-of-control beliefs than those businesses that had failed in the same period.[25] In contrast, Hall, Basley,
and Udell failed to find a correlation between locus-of-control scores
and entrepreneurial activity among business school alumni.[26]

Both high n-Ach and internal locus-of-control scores appear intuitively to be reasonable measures of entrepreneurial aptitudes
and, apart from the differing sample frames chosen, conflicting
results may be attributable to inadequate psychometric techniques
rather than invalid concepts. However, it is useful to look for more

264

Small Is
Beautiful:
Spin-Offs,
Innovation, and
Entrepreneurship
in Small
Business

pragmatic measures of entrepreneurial aptitudes. Probably the best summary of these aptitudes was proposed by Williamson.[27] He does not suggest that *all* successful entrepreneurs will possess *all* these personality characteristics, but rather that

> *the probabilities of entrepreneurial success may be expected to be proportional to the degree to which the aspiring entrepreneur possesses those characteristics which appear common to individuals who have started and successfully operated a new business.*

Williamson suggests ten characteristics:

Good physical health because establishing a new business requires the physical stamina to sustain long hours of hard work.

Superior conceptual and problem-solving abilities because the entrepreneur must learn from his mistakes and resolve complex technical and commercial problem situations quickly.

Broad generalist thinking because the above problem-solving skills must be based upon the ability to maintain continued overviews of situations and to relate and integrate diverse technical and commercial factors into the overall thrust of the venture.

High self-confidence and tolerance of anxiety so that the inevitable setbacks and adversities can be overcome and "defeats" turned into "victories."

Strong drive since persistence and tenacity plus a strong sense of urgency will ensure that "things get done" and that the venture maintains its momentum.

A basic need to control and direct because the successful entrepreneur wants to maintain overall control of the situation and rejects higher authority and externally imposed bureaucratic structures. He wishes to establish and maintain control mechanisms and fully accept responsibility and accountability for decisions made.

Willingness to take moderate risks based upon a rational analysis of alternative actions and the calculation of consequent risks before decision making

Very realistic viewpoint because he perceives and accepts situations as they are, seeks to monitor them continually and solve problems pragmatically. Although both cautious and suspicious at times, he is honest and dependable and expects these virtues in others.

Moderate interpersonal skills because the technical entrepreneur wants to "run his own show." This person is uninterested in delegation and a participative approach to decision making. While the NTBF remains small, he does not need the interpersonal skills of the professional manager, although if he does possess them, they will be personal assets as the firm grows larger.

Sufficient emotional stability because the successful establishment of a NTBF imposes physical, emotional, and time pressures which in turn impose a mandatory rugged ability to "keep one's cool" in continual crisis situations.

When we turn to the social and domestic (that is, what might be called biographical as opposed to psychological) backgrounds of technological entrepreneurs, we find that studies yield remarkably consistent results. First, McCelland and others have amply demonstrated that entrepreneurs frequently belong to ethnic or sectarian refugee minorities in the society, of which the Jews (in Western societies) and Chinese (in both Southeast Asian and Western societies) are obvious examples. The experience of minority status may predispose people towards entrepreneurial initiatives regardless of their ethnic or sectarian backgrounds. Litvak and Maule, in studies of technological entrepreneurs in predominantly Catholic Quebec, found a lower proportion of Catholics and a higher proportion of first-generation Canadian immigrants in their sample, than chance would indicate. In contrast, Catholic entrepreneurial minorities are to be found in the United States. Second, as Shapero and Sokol indicate, numerous studies in different countries show that a self-employed parent strongly influences an individual to set up in business.[28] In all of the studies, these authors cited that 50 to 89 percent of the entrepreneurs involved had at least one self-employed parent. This percentage is much higher than the corresponding population-wide values (12 percent in the United States). Such children initiate business ventures regardless of their parent's success or failure in self-employment.

Having reviewed some of the personal characteristics of entrepreneurs, we will turn to some specific examples.

14.5
Illustrations of Technological Entrepreneurs

We have already cited Cooper's and Watkin's detailed studies of NTBF's in the United States and United Kingdom, respectively. Others were performed by Shapero in the United States, France, Italy, India, and South Africa[29] and Litvak and Maule in Canada. Shapero's work provides a partial basis for the new venture initiation framework we will discuss in the next section. Probably the best summary of such studies is provided by Litvak and Maule, within the context of their study of small high-technology entrepreneurs in Canada.[30,31]

The authors mailed a questionnaire to 343 selected entrepreneurs throughout Canada, from which 112 usable responses were obtained. They found that seventy-three (65 percent) of these respondents were Canadian born and thirty-nine (35 percent) were non-

266

Small Is
Beautiful:
Spin-Offs,
Innovation, and
Entrepreneurship
in Small
Business

Canadian born. In 1971, the populationwide proportion of heads of household who were non-Canadian born was 23 percent, so much lower. The relatively high proportion of nonnative respondents tends to support the "stranger hypothesis." That is, that immigrants in a society may be more predisposed towards entrepreneurial initiatives because of their feelings of being "outsiders" in their present society and because of a need to justify their traumatic act of emigration from their native land. This hypothesis is supported, at least impressionistically, by the number of successful entrepreneurs who have been immigrants from other countries in the United States and United Kingdom as well as Canada. However, it should also be pointed out that the proportion of immigrants among scientists and engineers in Canada is high, due to the sustained growth of the Canadian economy (until the 1970s) which has created a continued demand for technologists in excess of the native supply available, and the continued "brain-drain" of Canadian scientists and engineers to the United States. It seems likely that all the above factors were contributing to produce a relatively large proportion of nonnative respondents.

Although the mean age of the respondents was about forty-seven years at the time of their response, their mean age at the time that they incorporated their first firm was thirty-three years. This agrees with the U.S. results of Roberts and Wainer and Cooper[32] who state:

> *The firm is started by two founders, both of whom are in the middle thirties. One usually can be described as the driving force. He conceives the idea and enlists the other founder. They come from the same established organization, which is where they got to know each other. Either both are in engineering development, or one is in engineering and the other is a product manager, or in marketing. Often they have achieved significant prior success, with titles such as Section Head, or Director of Engineering, being common.*

The Canadian respondents included a relatively large proportion with Protestant and Jewish backgrounds. The proportions with Protestant, Jewish, and Catholic backgrounds in the respondents, together with the populationwide proportions in parentheses, were as follows: 56 percent (43.6 percent); 10 percent (1.3 percent); and 27 percent (47.3 percent), respectively. The corresponding breakdown by religious affiliations in the United States (from 18) were 57 percent, 13 percent, and 19 percent. About one-half (clearly much higher than populationwide proportion) the respondents had fathers who were self-employed, reflecting a "paternal" as against "colleague" demonstration effect (as discussed earlier). This proportion corresponded to those reported in the United States and Britain. Most respondents held two (although a few up to ten) jobs before incorporating their first firm, as compared with an equi-modal distribution of one, two, and three jobs in the British study.[33] One interesting observation was that the proportions of Canadian respondents holding managerial positions increased from 4.6 percent in their first job to 38.7 percent in their last job before incorporation. Hardly surprisingly, 54 percent held university degrees (mainly in science or engineering) or technical diplomas, which is probably at least ten times the populationwide statistic for the corresponding age group.

Their responses to questions asking their reasons for starting a business supported the conclusions cited earlier in this chapter. Seventy-four percent cited "challenge" and 50 percent cited "being one's own boss" and "freedom to explore new ideas" as prime reasons for starting up on their own. These results were broadly consistent with the British study which listed "desire for independence" and "desire for increased job satisfaction" as the most frequent reasons cited, followed by "a release of creative urges" and "financial motivation" significantly less frequently. When asked for the prime reason for incorporating a new firm, 51.8 percent cited the development of an existing product for a new market or new product for an existing or new market, 20.5 percent the acquisition of partners, and 19.7 percent the acquisition of financial support. Of the respondents who cited new markets or product as their reason, 83.7 percent believed that their previous employers would have refused them permission to develop their ideas. Fifty-six percent of the respondents founded their first company with part-timers, a figure which corresponds closely with U.S. and British findings. Most first companies were financed through personal savings, bank loans, and loans from friends and relations (in that order) with venture capitalists providing little funding. Fifty-two percent of the respondents received federal or provincial government grants. Thirty-eight percent of respondents (compared with 33 percent in Britain and only 6 to 15 percent in the United States) cited financing as a major problem. The next most frequently cited problems were selling and managing personnel. Seventy-eight percent of the respondents formed more than one company. The mean number of firms formed per respondent was 3.25 with 2.87 still in operation at the time of the study. These results were also consistent with U.S. findings and suggest that such individuals learn to become better technological entrepreneurs through trial and error. About one-third of the sample expanded by miniconglomerate and a second third by horizontal integration.

14.6
Entrepreneurial Venture Initiation

Liles[34] and Shapero[35] provide frameworks for viewing the entrepreneurial venture initiation process, based upon their own ex-

268

Small Is
Beautiful:
Spin-Offs,
Innovation, and
Entrepreneurship
in Small
Business

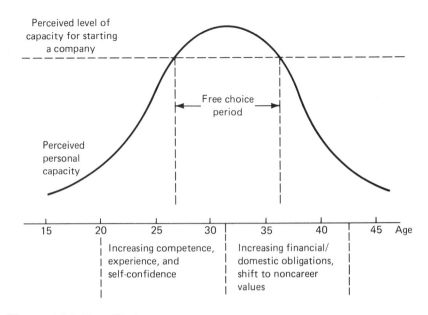

Figure 14.2 Free Choice Period.

tensive studies of entrepreneurship. The work we have cited in the previous two sections indicates that certain individuals have predispositions or readiness to initiate NTBF's. However, this *readiness to act* of prospective technological entrepreneurs is initially restrained by their need to acquire professional education and business experience and then by increasing family responsibilities. Such an individual typically enters the work force in the early or middle twenties and gains progressive experience and responsibility as shown in figure 14.2 leveling off as he approaches his fortieth birthday. During this time, he* also typically acquires a wife, children, increasing financial commitments (including a heavily mortgaged house), and possibly a reorientation of goals from career to familial and other concerns (shown in figure 14.2). Therefore, during the thirties, the individual experiences a *free-choice period* when he still retains a readiness to act and is, as yet, unrestrained by his noncareer concerns.

---

*To date almost all technological entrepreneurs are male. This fact does not imply that women lack entrepreneurial aptitudes, as they are outstandingly successful in other entrepreneurial roles in real estate, small businesses, and so forth. It does suggest that the sexist biases in Western educational systems and culture make it very difficult for women to acquire the requisite combination of education and experience to become a *technological* entrepreneur. This situation should change in the future because there is no reason why technological entrepreneurship should be an exclusively male prerogative.

Some people never experience this free-choice period because the other commitments become too large before they have the experience and confidence to set up a company. Many others pass through such a period but do not set up a company because they fail to experience a *precipitating event* which induces them to do so. It is a minority that experience precipitating events during their free-choice periods—who become technological entrepreneurs. We therefore now focus on the characteristics of such precipitating events.

The interactions of a readiness to act, precipating events, and so forth, are illustrated in figure 14.3. If a significant proportion of these factors is present in an individual's personal situation, it is reasonable to conjecture that there is a substantial chance that he will launch a new venture. An individual's readiness to act is determined by his psychological makeup, social background, and the experience of a demonstrative model in a parent and/or a colleague/ mentor in an incubator organization. If the individual experiences a precipitating event (job dissatisfaction, dismissal, or layoff) he will be predisposed to launching his own company. This predisposition

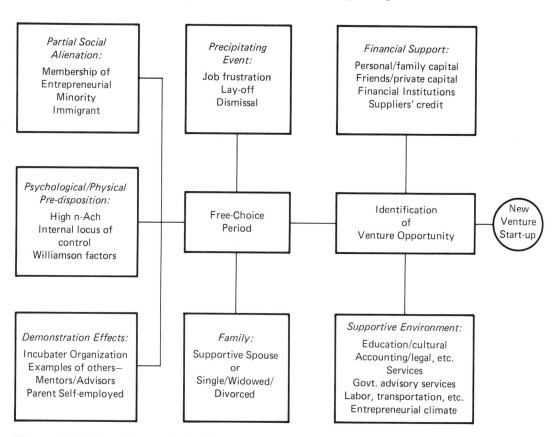

Figure 14.3 New Venture Initiation.

270

Small Is
Beautiful:
Spin-Offs,
Innovation, and
Entrepreneurship
in Small
Business

will be reinforced if the spouse is supportive (a crucial factor, see next chapter) or conversely if the person is single, divorced, or widowed without significant family commitments. Clearly the person should only launch a new company if he perceives a suitable opportunity that he is confident of exploiting (probably with one or more partners) and he can assemble the financial and other required resources from a supportive environment. The launch and establishment of such a new venture is a challenging process which we discuss, in detail, in the next chapter.

References

1. Edwin A. Gee and Chaplin Tyler, *Managing Innovation* (New York: John Wiley & Sons, 1976).

2. Edwin Mansfield, "Returns from Industrial Innovation," in *Technological Innovation: Government/Industry Co-operation*, ed. Arthur Gerstenfeld (New York: John Wiley & Sons, 1979), pp. 18-19.

3. Morton I. Kamien and Nancy L. Schwartz, "Market Structure and Innovation," in *Progress in Assessing Technological Innovation-1974*, ed. H. R. Clauser (Westport, Conn.: Technomic Publishing Co., 1975), pp. 32-35.

4. Albert Shapero, "Entrepreneurship and Economic Development," in *The Entrepreneurial Manager in the Small Business,* ed. William Naumes (Reading, Mass.: Addison-Wesley, 1978), pp. 183-202.

5. *Emerging Innovative Companies: An Endangered Species* (Chicago: National Venture Capital Association, 1976).

6. Michael Gibbons and D. S. Watkins, "Innovation and the Small Firm," *R&D Management* 1, no.1 (October 1970): 10-13.

7. Christopher Freeman, "The Role of Small Firms in Innovation in the UK since 1945," *Bolton Committee of Enquiry on Small Firms,* Research Report no.6 (London: HMSO Cmnd 4811, 1971).

8. Ibid.

9. Adam Osborne, *Running Wild: The Next Industrial Revolution* (Berkeley Calif.: Osborne McGraw-Hill Book Co., 1979).

10. V. J. Danilov, "The Spin-Off Phenonenon," *Industrial Research* 11, no.5 (May 1969): 54.

11. Kirk P. Draheim, "Factors Influencing the Formation of Technical Companies"; and Richard P. Howell, "Comparative Profile—Entrepreneurs Versus the Hired Executive: San Francisco Peninsula Semiconductor Industry," both in *Technical Entrepreneurship: A Symposium,* ed. Arnold C. Cooper and John L. Komives (Milwaukee, Wis.: Center for Venture Management, 1972).

12. Arnold C. Cooper, "Incubator Organizations and Technical Entrepreneurship," in Cooper and Komives, eds., *Technical Entrepreneurship: A Symposium.*

13. P. J. Dickinson and D. S. Watkins, "Initial Report on Some Financing Characteristics of Small Technology-Based Companies and Their Relation to Location," R&D Research Unit Internal Report (Manchester; England: Manchester Business School, 1971).

14. D. S. Watkins, ed., *Founding Your Own Business, Conference Proceedings* (Manchester; England: R&D Research Unit, Manchester Business School, 1972).

15. D. S. Watkins, "The Role of the Small Firm in Innovation Since 1945: Extension and Critique," R&D Research Unit Internal Report (Manchester: Manchester Business School, 1972).

16. D. S. Watkins, "Technical Entrepreneurship: A cis-Atlantic View," *R&D Management* 3, no.2 (February 1973): 65-70.

17. Robert H. Brockhouse, Sr.: "The Psychology of the Entrepreneur"; and Yvon Gasse, "Elaborations on the Psychology of the Entrepreneur," in *Encyclopedia of Entrepreneurship,* ed. Calvin A. Kent, Donald L. Sexton, and Karl H. Vesper (Englewood Cliffs, N.J.: Prentice-Hall, 1982), pp. 39-71.

18. David C. McClelland, *The Achieving Society* (New York: Irvington Publishers, Halstead Press, 1976).

19. David C. McClelland "Business Drive and National Achievements," *Harvard Business Review,* 40, no.4 (July-August, 1962): 99-112. Quotes from "New Introductions" and pp. 104-105.

20. Everett E. Hagan, *On the Theory of Social Change: How Economic Growth Begins* (Homewood, Ill.: Dorsey Press, 1962).

21. E. B. Roberts and H. A. Wainer, "Some Characteristics of Technical Entrepreneurs," *IEEE Transactions on Engineering Management* 18, no.3 (August 1971): 100-109.

22. J. B. Rotter, "Generalized Expectancies for Internal Versus External Control of Reinforcement," *Psychological Monographs: General and Applied* 80, no.1 (1966): 1-28.

23. P. E. McGhee and V. C. Crandall, "Beliefs in Internal-External Control Reinforcement and Academic Performance," *Child Development* 39, no.1 (March 1968): 91-102. P. Gurin et al: "Internal-External Control in the Motivational Dynamics of Negro Youth," *Journal of Social Issues* 25, no.3 (Summer 1969): 29-53. R. C. Loo, "Internal-External Control and Competent and Innovative Behaviour Among Negro College Students," *Journal of Personality and Social Psychology* 14, no.3 (March 1970): 263-70.

24. Candace Borland, "Locus of Control, Need for Achievement and Entrepreneurship" (Doctoral diss., University of Texas, 1974).

25. R. H. Brockhouse, "Psychological and Environmental Factors Which Distinguish the Successful from the Unsuccessful Entrepreneur: A Longitudinal Study," *Proceedings, Academy of Management Meeting,* August 1980.

26. D. L. Hall, J. J. Basley, and G. G. Udell, "Renewing the Hunt for the Heffalump: Identifying Potential Entrepreneurs by Personality Characteristics," *Journal of Small Business* 18, no.1 (January 1980): 11-18.

27. Byron Williamson, address to seminar, Life Planning Center, SMU School of Business, Vail, Colo: March 1974.

28. Albert Shapero and Lisa Sokol, "The Social Dimensions of Entrepreneurship," in Kent, Sexton, and Vesper, eds., *Encyclopedia of Entrepreneurship,* pp. 72-90.

29. Albert Shapero, *An Action Program for Entrepreneurship,* (Austin, Tex.: Multi-Disciplinary Research Press, 1971).

30. I. A. Litvak and C. J. Maule, "Some Characteristics of Successful Technical Entrepreneurs in Canada," *IEEE Transactions on Engineering Management* 20, no.3 (August 1973): 62-8.

272

Small Is
Beautiful:
Spin-Offs,
Innovation, and
Entrepreneurship
in Small
Business

31. I. A. Litvak and C. J. Maule, *Policies and Programmes for the Promotion of Technological Entrepreneurship in the U. S. and U. K.: Perspectives for Canada,* Technological Innovation Studies Program, Report no.27, (Ottawa; Department of Industry, Trade and Commerce, May 1975).

32. A. C. Cooper, "The Palo Alto Experience," *Industrial Research* 12, no.5 (May 1970): 58.

33. Watkins, *Founding Your Own Business.*

34. Patrick R. Liles, "Who are the Entrepreneurs?" in Naumes, ed., *The Entrepreneurial Manager in the Small Business,* pp. 10-21.

35. Shapero, "Entrepreneurship and Economic Development," pp. 31.

15

Creating the New Technological Venture

The day will come, as it must to 99% of all technical professionals, when the thought of developing your own ideas for your benefit crosses your mind. If you let this seed lodge and grow, then you begin to realize that the resulting fruit might require the establishment of your own business.
D. S. Scott, *"So You Want to Run Your Own Business," in*
The Technical Entrepreneur: Inventions, Innovations, and Business

15.1
Introduction

In chapters 8 and 14, we discussed aspects of individual creativity and entrepreneurship, respectively, suggesting that these qualities have much in common. In fact, the inventor-entrepreneurs of the nineteenth century could equally well be called *technologically creative-entrepreneurs,* that is, individuals who were twice-blessed by Providence with the flair and drive to generate both new technological ideas and the new companies and industries to exploit them. In chapter 14, we reviewed the relatively recent resurgence of creative-entrepreneurship through the spin-offs of new technological ventures from parent incubator organizations, a resurgence most visible in the microelectronics industry and "Silicon Valley." Although the enthusiasm for new venture creation waned somewhat in the less benign economic climate from the mid-1970's,[1] individuals are still launching new autonomous ventures, for the reasons discussed in chapter 14. In this chapter, therefore, we will discuss the problems and pitfalls of new venture creation.

We will develop the discussion in the context of autonomous venture creation for two reasons. First, the issues of sponsored ventureship in larger organizations are discussed in chapter 16. Second and more important, many R & D professionals and technology managers employed in larger organizations have an unfulfilled dream to launch their own NTBF's. As was stated at the end of the previous chapter, whether this dream is turned into a reality is dependent upon a combination of circumstances illustrated in figure 14.3. This chapter provides such individuals with an overview of the factors which they should consider before undertaking the undoubtedly hazardous step of launching their own businesses. There are numerous books on approaches to setting up and managing a small business, and we shall not attempt to review such a vast literature here. However, one of the best of these is by Timmons, Smollen, and Dingee and is strongly recommended reading for all aspiring new venture creators.[2] It is based upon new business creation programs implemented by the authors and their colleagues in the United States, Canada, Sweden, and Britain which, by 1979, had generated ventures valued at $70-80 million. Therefore, we shall draw upon their experiences as the major individual source material for this chapter. The works by Timmons and by Keirulf[3] also provide useful, brief reviews of the new venture creation literature.

15.2
Stages in the New Venture Creation Process

For the purposes of exposition, we may define three stages in the new venture creation process.

"Moonlighting" Stage

Autonomous ventures are spun off by individuals from incubator organizations which may be private firms, government, or industry agencies (such as the British research associations), or academic institutions. Unless a person is laid off or fired by his employer, he should not decide to launch an autonomous venture (even if he can get the money to do so!) without careful thought and planning. This gestation process will occur while the person is currently employed (hence the term *incubator organization*), when he is, in effect, "moonlighting." Timmons and associates suggest that 200-300 hours of "moonlighting" work is needed to produce a viable new venture proposal. This time estimate could be too conservative. Many, if not most, autonomous new ventures require external financing from government, banks, venture capitalists, or other sources. Such professional investors will assess the attractiveness of a proposed new venture as an investment opportunity, based upon the technical and commercial merits of the proposed innovation, the strengths and personalities of the individuals founding the proposed venture, and the business plan submitted to them. Professional investors are, in general, quite properly, knowledgeable, experienced, and astute evaluators of new venture proposals. In contrast, scientists and engineers seeking to exploit their own inventions can be commercially and managerially naive. Such individuals can find that the development of a venture proposal and business, which will survive the critical scrutiny of investment professionals, to be a time-consuming, traumatic, *but ultimately rewarding learning experience.* Such a critical scrutiny may well identify errors in thinking which might well have proved costly, or even disastrous, if they had been carried over into the launched venture. The detailed thinking to be undertaken in the moonlighting stage will be discussed later.

Postlaunch or Entrepreneurial Stage

A "successful" output from the moonlighting stage could be a venture proposal and business plan that secures external financial support. The development of this proposal and plan will identify many of the problems and pitfalls to be expected in initial operations. However, problems of plant construction and initial manufacturing operations, which may be unpredictable and uncertain *ex ante,* will have to be overcome. Furthermore the ultimate test of the marketability of the innovative product must await the market launch. In fact, despite good planning beforehand, the venture team, like generals commanding an army in battle, must be able to deal with a whole host of problems expeditiously as they arise. It may be recalled that Napoleon, when asked how he won battles replied: "On s'engage, et puis, on vois." Therefore, during the first few years of operation, the venture team can expect to be in a virtually permanent state of "crisis management." Some pitfalls endemic to new ventures in their early life can be identified and are discussed later.

Consolidation and Growth to Maturity

Once the venture has survived the launching period and established itself as at least an initially viable commercial entity, its founders will look to consolidation and probably, expansion. This stage presents both opportunities and threats. If the venture is generating a solid and growing sales revenue, procedures for manufacturing and marketing may have to be standardized, and the original venture team expanded to include functional management and specialist personnel, without creating undue red tape, or lowering organizational morale. One of the dangers faced by venture teams at this stage is the risk of *following success with failure.* It goes without saying that the venture teams we are discussing here will be both technologically and entrepreneurially oriented. If they enjoy commercial success with their first innovative project, they may become overconfident and rather than consolidating their position, may pursue more ambitious product ideas that are beyond their capabilities or market needs.

15.3
The Formation of the Venture Team

As we stated above, much of the planning required for the launching of a successful new technological venture is typically conducted in the "moonlighting" stage, before its founders sever their associations with their parent incubator organizations. Timmons and his associates contend that success depends above all else upon attention to three considerations:

> *(1) a capable lead entrepreneur, who has determined realistically his or her entrepreneurial strengths and shortcomings, and a balanced and compatible entrepreneurial team; (2) a feasible business idea; and (3) appropriate financing.*[4]

It is worth noting that these authors list entrepreneurial qualities before the feasibility of the business idea, that is, the quality of the entrepreneurial team is more important than the inventive idea, in determining the success of a venture. General George Doriot, who is regarded as the father of venture capital institutions (and who helped launch many high-technology companies, including Digital Equipment), is reputed to prefer a Grade A team with a Grade B idea to a Grade B team with a Grade A idea.[5] Therefore, although the decision to try to launch an NTBF may be triggered by the desire to exploit a specific invention, a detailed self-analysis of the strengths and weaknesses of the proposed venture team should be a first consideration.

Thus a new venture should coalesce around the technological entrepreneurial aspirations of one or a very small group of individuals, whom we may call the lead entrepreneur(s). For verbal convenience we will describe them as a group, which should collectively possess the technologically inventive and entrepreneurial attributes discussed in earlier chapters. Although the metaphor is a little unflattering, this group may be likened to the grain of sand or seed around which the "pearl" of a successful NTBF grows (figure 15.1). The seed must be its technological entrepreneurial and inventive capabilities. If the business idea is to be launched as a new venture it must, however, incorporate minimum functional management capabilities which constitute a second layer of the pearl.

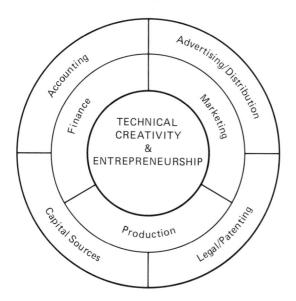

Figure 15.1 Venture Base.

One of the first tasks of this group, therefore, should be to complete a self-inventory of their collective management expertise to
identify any skills that may be lacking. This stock taking should,
of course, be conducted in the context of developing the venture idea
and business plan as discussed in the next section. Different industries and products have different specialist functional skill requirements, particularly in production and marketing. If the venture is
to succeed, the group should be "knowledgeable" of the industry and
possess the insight to identify the key determinants of business
success within it. If it lacks this knowledge and insight, it probably
lacks the maturity and experience to launch the venture. The stock
taking should identify what, if any, further management skills are
required. If further skills are required, they may be acquired either
by co-opting additional members with the required expertise into
the founding team or hiring such individuals on a full-time or part-
time basis, once the venture is launched. Outside financial support
may also be contingent upon the acquisition of its expertise. The
outer layer of the pearl represents the specialist expertise which,
though required, may be employed on a part-time consulting basis
during the early life of the venture. For example, in a study of Canadian small, high-technology business, this author found that most
did not hire a full-time accountant until several years after they
had been launched.[6]

The group's self-assessment process must not be confined to
only technological-entrepreneurial-managerial considerations, but
most importantly, must include behavioral aspects. The venture

team should be cohesive and once the venture is launched, practice a kind of "group marriage." The "moonlighting" period offers an opportunity for "courtship" during which the personal aspirations of individuals should be identified to determine their interpersonal compatibilities. They should be totally committed to the success of the venture, which should take priority over all other commitments, including (in a sense) their responsibilities to their families. At first sight this requirement appears morally outrageous and certainly requires some clarification.

The launching of a new venture invariably requires the team to work long hours, probably on a "seven-day week" basis, which leaves members little time to exercise their domestic responsibilities. Furthermore, it will require the investment of personal capital and the foregoing of a regular monthly income as a salaried employee. Typically, team members can expect to pay themselves low or no salaries during the earlier days of the venture, and their families may have to live at least partially on accrued savings, or the incomes of working spouses. Their families may experience a shortterm drop in their material standards of living. A commitment of this magnitude should only be made with the agreement and support of the team's spouses and families, *on both moral and psychological grounds.* Apart from the fact that both husband and wife must make such a commitment jointly, if they have a viable marriage, it is unlikely that an individual would be able to withstand the psychic strain of new venture creation, without familial support. This issue is discussed by Jolson and Gannon[7] and was confirmed by the Canadian study cited earlier. The NTBF founders interviewed in that study confirmed that, for the married workers, their successful development of the venture almost certainly required the indulgent support of their families. Moreover, many of the people interviewed said that their spouses worked during the initial phase of the venture development to provide the family income. Typical comments were as follows:

Had not the wives of my partner and myself been prepared to work in order to support our homes we would never have been able to get this company off the ground. We made **no** *profit, not even enough to pay ourselves salaries during the first two years of our operation.*

Our wives worked. There is no way we could have done without them.

The founders also need the support of others in their reference group (former work colleagues, other friends, and the community in general), if they are to succeed. Chapter 14 pointed out that both paternal and incubator organizational "demonstration effects" influence an individual's decision to become a technological entrepreneur, and community attitudes towards entrepreneurship are also strongly influential.

Timmons and his associates suggest that a cohesive venture team should be built around the concepts of organizational climate, interpersonal, and helping skills.[8] The climate should emphasize and reward responsibility and commitment to high performance standards through teamwork. It should therefore reflect some of the features of a creative climate, discussed in chapter 8, but focused upon the goals of the venture and teamwork, rather than individual R & D creativity.

The establishment and maintenance of a purposeful organizational climate is dependent upon the interpersonal and helping skills of team members. They should readily articulate problems, consciously develop their listening and participative skills, and be willing to help and accept help from others. This willingness to give and receive help is particularly important in a new venture situation because individual team members will typically lack the range of expertise and experience which the venture demands, and they will have to do a lot of "learning on the job." Leadership is also likely to be flexible because different situations may require the expertise of different team members. It is almost certain that the team will have to pay deliberate attention to these behavioral factors, especially as they are in conflict with the "individualistic" personality attributes of creative R & D personnel and entrepreneurs we identified in chapters 8 and 14. This potential conflict must be faced and accommodated if the team is to be cohesive.

This process of what could be called group introspection should enable the team members to clarify their personal goals in joining the venture. It is most desirable that individuals should identify and make explicit their personal goals to ensure that they are mutually congruent, given the requirement for managerial expertise in the venture team. The research evidence suggests that inventors and entrepreneurs are "achievement" oriented, in contrast to managers who are more "power and status" oriented. The former individuals seek self-actualization through R & D or business creativity. In contrast, the latter seek self-actualization through outward manifestations of managerial success. These latter manifestations are reflected in status symbols such as titles, office sizes and furnishings, company cars, and expense accounts. An NTBF offers little scope for "empire building" of this nature, and an individual who seeks to accrue such external status symbols will squander the scarce financial resources of the venture and disrupt team cohesiveness. Therefore, it is vital that all team members should recognize that their external status rewards can be expected to be modest during the first few years of the life of the venture.

This consideration brings us to the important issue of the reward structure for team members, and indeed, all employees. The design of the business plan, which is to be discussed in the next section, should identify specific business goals (expressed in terms of sales growth, return on capital employed, and so forth). Individuals will join the team, almost certainly making immediate material sacrifices, in the expectation of future psychological and material rewards. They will be co-opted into participation because they appear to be personally acceptable and they offer the required expertise, commitment, and financial investment. Obviously the split of the initial equity and other shareholdings must be agreed on between the members, together with some rules for future divisions and buy-back agreements. A member may later wish to leave the venture or may become ill or die. Clearly, stock "buy-back" rules should be laid down at the outset which, so far as possible, protect the interests of the team and the departing individual (or the estate). These rules should also allow for the contingency of inviting new members to join the team in the future, should the need arise. Some shares might be held as a "pool" available for purchase by new team members who may of course be employees. The team will also need to determine each of their individual salaries, based upon some group assessment of each individual's performance and contribution to the goals of the venture.

As the venture grows, it can be expected to take on a growing number of paid employees at all levels. Therefore, the team must plan a future reward structure which not only maintains its own cohesion, but also promotes a productive organizational climate. A pioneering spirit of a new venture can create a participative- and performance-oriented organizational climate for all personnel. The employee reward structure should therefore sustain its development and, as we suggested in the previous paragraph, it could allow for the longer-term contingency of offering the opportunity for share purchase to key and loyal employees.

15.4
The Venture Proposal and Business Plan

The previous section discussed the behavioral aspects of new venture formation. The processes of team selection and drawing up of both written legal and abstract psychological contracts must clearly interlock with the process of formulating a business plan. It is vital to formulate a business plan in as detailed and rigorous manner as the situation allows, for two quite practical reasons.

First, a person often desires to launch a NTBF out of frustration with working as a paid employee and out of a healthy desire to become his own boss. If this need for personal autonomy and independence is strong—and it must be if a person is willing to hazard the economic security of his family, in order to launch a business—there may well be an element of unconscious self-deception in an evaluation of the venture idea. The individual *wants* to become his own boss and he *wants* the business idea to look commercially viable, so that its pursuit can be viewed as an excellent opportunity, rather than an unacceptable risk to both family and career. The wish is indeed often father to the thought, so that the person runs the risk of unconsciously repressing critical aspects of the business idea in order to justify to himself the decision to attempt the venture.

Second, in this context, we are discussing the launching of an NTBF to innovate an invention or discovery that was probably made by one or more members of the venture team. The psychology of the act of creation often appears to dictate that the inventor has a strong "parental" attachment to the invention or "baby"—a parental instinct that can easily favorably bias the assessment of its intrinsic economic and social worth and make him resist the modification and/or degradation of the inventive idea, in the interests of commercial expediency. The untoward assertion of this instinct risks entrapping the new venture into a "technology push" mode of development leading to commercial fiascos of the kind quoted in chapter 9. This danger is reinforced by the prevalence of what Timmons, Smollen, and Dingee[9] call "mousetrap myopia." Emerson's claim that the world will beat a path to your door if you invent a better mousetrap is quite simply untrue. The world will only buy it after it has been promoted and marketed—so the world knows of its availability—and if it is proved to be superior by the Darwinian test of the marketplace. Because of his emotional attachment to his invention and probable commercial inexperience, the inventor is particularly prone to mousetrap myopia. Although this emotional attachment is understandable it is by no means forgivable, and the venture team should ensure that their assessment of their inventive idea is in no way shortsighted, but subject to critical commercial self-scrutiny.

Fortunately, even though the team may fail to recognize its vulnerability to the two forms of self-deception listed above, it will almost certainly be forced to develop a critically formulated business plan for a third imperative reason. Most NTBF's will require an initial financial investment over and above that which can be provided by venture team members, their families and friends, and must seek external sources of financing. Such financial support is typically provided by government agencies, banks, or venture capitalists. None of these sources has a philanthropic outlook, and they are only willing to invest money in a venture that appears to have a reasonable prospect of commercial success. This assessment of the prospect of commercial success of a venture will be based upon a critical evaluation of the business plan formulated by the venture team. Thus, in most circumstances, the venture team will be forced to formulate a business plan that will survive critical independent scrutiny if the proposed venture is to progress from the "moonlighting" to the "launch" stage.

The formulation and development of a detailed business plan will occupy a considerable proportion of the venture team's time during the "moonlighting" stage of the venture. Its format could correspond roughly with the procedure for project evaluation outlined in chapter 9, but incorporate more detail, since it constitutes a plan for developing a business operation in its entirety. Whatever the degree of novelty in the proposed innovation, it will virtually constitute a "revolutionary" innovation to the venture team, so they are likely to find White's conceptual framework useful when developing the plan. Prescriptions for compiling such plans are provided in numerous texts. Timmons and his associates[10] describe how to prepare one in some detail and provide a useful illustrative example. Other useful sources are Liles[11] and Scott and Blair.[12]

Funding agencies also usually provide guidelines for the preparation of new business proposals. Readers who are considering attempting to launch a new venture are advised to obtain a copy of one of the above texts or another of their own choosing. Therefore, I will only outline the main features and requirements of a well-prepared plan.

When preparing such a plan, two overriding points should be borne in mind. First, some funding agencies prefer to make their first contact with proposers "on paper," that is, through receiving a mailed or handed in copy of the business plan. They can then make a swift preliminary evaluation of the appeal of the proposal, before discussing it further with the venture team. Probably, over 50 percent of the proposals submitted to agencies fail to overcome this initial hurdle, so it is often cost-effective to complete this preliminary screening "on paper," rather than waste up to a whole day of everyone's time (plus traveling expenses) on face-to-face interviews. This means that the proposal should be well presented on paper and be lucidly and succinctly written omitting any irrelevant detail, so that the reader can readily identify its intrinsic merit. It implies that it be succinct but informative. The proposers will be asking for some tens or hundreds of thousands of dollars, and no responsible agency is going to part with this amount of money without a thorough and critical evaluation of the merit of the venture. It should not exceed 4,000 words in length (excluding appendices), and all material should be pertinent to the evaluation process.

Second, apart from the intrinsic merit of the proposed innovation and business, the appeal of the venture will be highly dependent upon the credibility of the venture team. They must not only demonstrate that they possess the experience and expertise to launch and manage the venture, but also give no reason for the reviewers to question their integrity. This means that any expected financial performance for the venture, or technical performance claims for products, should be as specific and accurate as realism allows, but err on the side of conservatism. Only quantitative estimates or data and opinions that can be cited or defended should be quoted. If the funding agency is seriously interested in funding the proposal, the venture team can be expected to be subjected to searching interviews before financial support is forthcoming. These interviews can be expected to expose any claims and opinions that cannot be defended and justified. Failure to justify statements made in the written proposal will undermine credibility in the integrity and competence of the venture team. In fact, writing a business plan may be compared with a research student writing a masters or doctoral dissertation. Ultimately, the dissertation will be passed, based upon its intrinsic quality and the writer's ability to defend it before an examining committee. A wise research student never makes any claim or expresses any opinion in a dissertation that he is not prepared to defend in an oral examination. To do otherwise would be to risk losing credibility to the examiner's eyes and having the dissertation rejected. A business plan should be seen in exactly the same light. If the venture team includes someone who has successfully completed a dissertation for a higher degree, the team should perhaps listen carefully to that person's opinions, when drawing up their business plan. Timmons and his associates suggest the following framework for designing a business plan.

Company Name and Address

A company, like a baby, needs a "given" name so that it can be discussed by all as a separate entity. Remember that like a baby, providing that it survives, it will grow into maturity with an identity of its own. Therefore give it a name and a "home" (that is, an address) as soon as possible. This suggests to others that the team has confidence in its offspring!

Executive Summary

This summary should be prepared last, but appear immediately after the company name, and so forth, summarizing what follows. It should identify the business idea and the industry in which it will operate, indicate any inventive novelty that is being developed and the market opportunity it will exploit, as well as the competence of the founders to exploit this opportunity. Financial performance projections for the first few years of the venture should be summarized, together with the financing required and the equity and security being offered.

Technology-Market Niche

This section should describe the industry (including its current economic climate) in which the company will compete and what the founders identify as key determinants of success in it. It should describe the products, their proprietary positions, and their competitive advantages.

Marketing

The market research and analysis should be as detailed and comprehensive as the current state of knowledge allows. The intended marketing strategy should be described including distribution and selling methods, servicing and warranty procedures, promotion plans, advertising, and pricing. The target market share and sales should be estimated over the first two or three years of operation. The procedures for "ongoing" market monitoring should be described.

Technical Development and Manufacturing

The evolution and current state of development of the innovative idea should be first described. The program and timetable of future technical development should be specified, identifying the anticipated problems and how they may be overcome. A timetable of development tasks, in the form of a Bar Chart or PERT network should be included, which preferably illustrates the tasks that have been completed to date, to convey a sense of accomplishment to the reader. The manufacturing strategy should be described including the production, location, facilities and layout, manufacturing methods, labor requirements, sources for raw materials and bought-in components, and the impact any sales seasonality will have on production planning, scheduling, and inventory policies.

Human Resources

This section of the plan should describe the backgrounds and past achievements of the venture team, the organization structure, and role each will play in the venture. It should also specify what other key management and professional personnel are required, and how these personnel will be recruited. It will specify salaries to be paid to founders and staff. The venture team should plan to pay themselves relatively modest salaries during the "start-up" phase of the venture as a demonstration of their commitment. Requirements for part-time professional services should be specified, together with the likely sources of these services. The use of reputable professional advisors also strengthens the appeal of the proposal.

Financial Plans

Financial forecasts for the first few years of intended operations should be prepared including profit and loss statements, pro forma cash flows, balance sheets, and a break-even chart. The cost and credit control system should be specified.

Contingency Planning

The anticipated risks and threats to the proposed venture should be identified and contingencies for dealing with them described. The ability to anticipate potential problems and set up contingency plans for mitigating their impact demonstrates management foresight.

Overall Timetable

The foregoing analysis should be incorporated into an overall timetable or schedule, probably expressed in the form of bar charts or PERT network. As was stated when discussing development and manufacturing, this schedule of tasks can reflect the work that has already been completed, to demonstrate a record of proven accomplishment.

Funding Requirements

The proposal can end (apart from appendixes) with an analysis of the funding that is being requested, broken down into individual years. If the proposal is being submitted to a venture capital institution, it will also specify the support that has been obtained from other sources—the venture team, families and friends, bank loans, and so forth. It will state the securities being offered and the proportion of equity the venture team are prepared to surrender in return for funding. The description of the organization will have specified the membership of the board of directors, including a nominee from the funding agency. The capital structure of the proposed venture should be specified. Last, but by no means least, the proposal should specify in some detail why the money is needed and how it would be spent!

All the above sections should be succinct, with "details" included as supplementary appendices.

15.5
Sources of Financial Support

Interlocking with the development of the venture team and the business plan will be the search for financial support, and this may be the most frustrating aspect of new venture creation. It is therefore useful to comment on sources of funding for the launching of new technological ventures. These can be roughly classified as follows: the venture team, plus their families and friends, independent private investors, banks, government business support programs, and venture capitalists. We will discuss each of these in turn.

The nucleus of financial support will most certainly have to be provided by the venture team, supplemented by whatever else is available from families and friends. Clearly if the team members are unwilling to put their own money into the venture, no one else will! In a few instances these immediate sources are sufficient to launch the venture, but these instances will be very much the exception rather than the rule. In this chapter we are discussing the creation of a new venture by technologically creative entrepreneurs who have typically previously worked as paid employees in incubator organizations. Not only will they have a limited business and management background but, in general, they will also have limited financial resources. The remuneration system for a professional scientist, engineer, or manager makes it very difficult for him to accrue large amounts of personal savings. He is probably required to undergo full-time education until his early or middle twenties, perhaps having to start paying back the money borrowed to finance that education, as soon as he finally starts working. Furthermore, by this time or shortly afterwards, he may be married with a family and a mortgage to support.

The researchers on the profiles of technological entrepreneurs reported in chapter 14 suggested that they were typically in their thirties, by which age they have gained a few years of industrial experience and acquired the knowledge and confidence to start a new venture. Given the relatively short working life to date and perhaps additional domestic obligations, it is likely that such an individual will have savings of, at most, some tens of thousands of dollars available for a new venture investment. Furthermore, once he resigns his current job to launch the venture (unless the spouse can earn the family income) the individual may well have to support his family for a year or more out of these savings, before the venture generates sufficient cash flow to enable him to draw an economic salary. These considerations suggest that the venture team will only have a modest amount of seed capital to invest in the proposed venture. This seed capital may be supplemented by unsecured borrowings from families and friends (probably at low or zero interest rates), but since these individuals are also likely to have modest savings, the venture team will be doing well if it can raise a sufficient sum to launch the venture, so much support will have to be sought elsewhere.

In the absence of sufficient philanthropic support from the immediate family and friendship circles of the team members, the next step should be to look for other private backers. Between them, the team members may interact in quite diffuse professional and social networks, which may include individuals with relatively substantial private wealth, which they have inherited or earned through professional or entrepreneurial activities themselves. The team should engage in judicious and tactful "word-of-mouth" advertising of their need for financial support, in the hope that an as yet unknown backer might be forthcoming. There may, for instance, be a retired successful businessperson in the community (particularly in the United States) who is willing to invest a portion of accumulated wealth in the sponsorship of a young, untried entrepreneurial team with a promising technical invention. Potentially successful teams, like Napoleon's successful marshals, do sometimes attract such "luck," and they would be foolish if they did not attempt to do so! If such a potential backer is forthcoming, obviously both he and the team should engage in detailed discussions to determine if their respective needs are complementary, in a "courtship" process comparable to that discussed earlier. The potential backer may decide that the proposed venture is too risky or the team may find him personally unacceptable or his requirements for financial support (in terms of loan interest charges or equity participation) are unacceptable. Alternatively, the team may be "lucky" and find an ideal financial sponsor whom they may also wish to invite to join them as a team member. It should be recognized that the probability of such a fortuitous outcome is quite low, but nevertheless it is a possibility that should be explored.

If, as is the most likely outcome, the venture team and private investors are fully explored without raising sufficient "start-up" finance, the team will be required to approach professional financing agencies that are most visibly represented by the clearing banks. The attitude of the clearing banks towards NTBF's varies between countries, but some overall comments can be made. All clearing banks are in the business of lending money to "attractive" ventures, but many have a prudent and conservative concept of "attractiveness." In Britain and Canada, the clearing bank business is largely performed by an oligopoly of nationally organized corporations who (rightly or wrongly) are viewed as highly conservative by technological entrepreneurs. In the United States, the local banking community is more sympathetic, because clearing banks are not national institutions, but federally regulated local business, usually in intense competition with each other to lend money. However, in any country, a proposal to launch an NTBF by a commercially inexperienced entrepreneurial team is unlikely immediately to appear as an attractive business loan proposition to the average "downtown" bank manager. Banks prefer to make short- or medium-term loans to established local businesspeople who can offer substantial security.

Alternatively, they welcome the opportunity to support the start-ups of entrepreneurs seeking to launch proven business ideas, such as fast-food outlets, and offering security in the form of a lien on the retail premises. Unproven technology, coupled with unproven entrepreneurs, looks to be a high-risk situation to a bank manager. In practice, particularly if government financial support is also forthcoming (see immediately below), clearing banks are likely to provide medium-term loans or lines of credit at market interest rates, to the extent of the realizable assets that can be offered by the entrepreneurial team, but the capital securable on these assets will probably be insufficient for its needs.

Governments of most developed nations recognize the important role played by the smaller high-technology businesses in stimulating growth and employment (recall chapter 14) and the problems of technological entrepreneurs when trying to launch new ventures. Some developed countries therefore offer a range of support programs at national, regional, and local government levels to help technological entrepreneurs. Simultaneously with exploring investors and banks, the team should identify and, if possible, exploit the government support programs for which they qualify. In the 1970s the U.S. National Science Foundation and Small Business Administration introduced the innovation center concept[13] and similar centers have been set up in other countries. Such a center (usually associated with a technological university) offers financial support and a range of management and technical advisory services to prospective or existing NTBF's. Some governments are also prepared to act as guarantors of bank loans to NTBF's provided certain conditions are fulfilled. For example, Canada's Federal Development Bank will provide "start-up" loans to prospective NTBF's with rather less stringent requirements than chartered banks.

A complementary avenue for seeking seed money to support the launching of a new venture is R & D or manufacturing contracts (particularly from government). Some NTBF's have begun as contract R & D service companies, before moving into manufacturing, and this avenue normally requires considerably less initial capital. Furthermore, if the company's competence enables it to win a government manufacturing contract early in its life, it means that it can establish its first manufacturing capability against guaranteed sales for its product. Historically, the United States has offered a ready market for the performance of contract R & D and manufacturing by competent technological entrepreneurs, although the scope for this activity suffered some decline from the cutbacks in R & D spending in the 1970s. The Canadian and British governments have also sought to stimulate industry-based R & D activities through deliberate policy initiatives.[14]

Probably most prospective venturers will be required to negotiate with banks and/or government institutions for financial support to launch NTBF's. In doing so, they will discover that the time and effort required for this process may cause extreme frustration. The frustration partially arises from the differences in personalities and attitudes between entrepreneurially oriented individuals on the one hand, and banking officers and civil servants on the other. Through their career choices, banking officers and civil servants are likely to have "managerial" rather than "entrepreneurial" attitudes because they have chosen to work in large bureaucratic organizations. Furthermore, they must adopt fairly conservative attitudes towards financial risk because they must exercise responsible stewardship in managing other people's money. Such individuals, through temperament and a legitimate sense of fiscal stewardship, want to invest in concerns that appear to offer commercial viability at minimum risk. In contrast, individuals seeking to establish and run NTBF's are temperamentally oriented towards technological opportunism, entrepreneurship, and moderately risky ventures. There is thus an implicit value conflict between the two parties in such negotiations. One president of a small high-technology business seeking financial support from government expressed in this conflict as follows.

> *The entrepreneur and government employee have conflicting goal objectives. The government employee wants always to "cover his ass" through having proforma financial statements, 110% collateral and legal interests. His effectiveness is measured on whether or not he can cut the entrepreneur's loan request. In other words, to survive, he goes the "route of low risk"....The entrepreneur is just the opposite. He is a high risk taker and will mortgage everything he owns, including his own family, to obtain money for his business. He is not interested in sophisticated accounting systems or even record keeping. Time is of the essence with him, because proper timing means the difference between success and failure.[15]*

Although this quotation expresses a rather extreme viewpoint, and reflects an immoderate attitude towards risk when compared with other evidence cited in this or the previous chapter, it illustrates quite graphically the frustrations that technological entrepreneurs *feel* when dealing with funding agencies!

Entrepreneurs normally find that they have to make multiple submissions to two (or more) agencies to obtain adequate funding, and this can present them with what Grasley and they themselves (to this author) have described as a "Catch 22" problem.[16] They are forced by circumstances to negotiate openly with perhaps two agencies simultaneously only to be told by each agency that it will provide funding, *once the other agency has already done so.* In some cases, the venturers have escaped from this logical entrapment by negotiating simultaneous funding, but such experiences can, to say the least, also induce intense frustration!

The final source of funding cited at the beginning of this section was venture capitalists. Venture capitalist or venture capital institution are terms that can of course be applied to any individual or institution providing financial capital to any new or existing business venture. However, they are more often used to describe financial institutions that aim to invest money in a particular type of venture. As stated earlier, banks prefer to invest their money by making short- or medium-term loans (up to a few years) to businesses that are traditional rather than innovative in character. They prefer to invest in a proven business idea with a readily measurable commercial risk (as reflected in the historic performances of similar businesses in identical industry markets), but, probably, a quite modest growth potential. Their loan may be offered on a fixed-term repayment schedule, against a solid security (such as the property or life insurance of the borrower), at an interest rate only a few percentage points above the bank or minimum lending rate. They thus aim to make a modest low-risk profit on the money which they hold "in trust" from their own depositor-customers. In contrast, the objectives of venture capital agencies are different. They wish to invest their money in ventures, including those that embrace novel business ideas, which may offer a *moderate* (but not high) commercial risk and *a high growth potential.* They want to invest equity or a mixture of equity and debt capital in a venture which, if it fulfills its promise, will enable them to sell off their investment after about five years for a capital gain of five times or more. By their nature, such agencies want to invest in promising technological innovations in industries, such as electronics and computers, in the fluid-growth stages of their life cycles. Venture capitalist institutions have proliferated in the United States to help fund the proliferation of NTBF's in such industries.

Although technological entrepreneurs will probably find that venture capitalists are the most sympathetic sources of potential seed capital, they can again expect to experience frustrations for several reasons. First, the availability of venture capital, like other institutional capital, fluctuates with economic and stock market conditions. Thus the chances of funding are markedly affected by the economic conditions prevailing when it is being requested. Second, although there are now numerous venture capital sources, even when economic conditions are favorable in a "bullish" market, technological entrepreneurs with promising innovative ideas find it difficult to attract finance *to launch* NTBF's. This follows from the objectives of venture capitalists outlined in the previous paragraph, namely, that they are looking for opportunities to inject capital into a company and then withdraw it (hopefully with a substantial capital gain), *after about five years.* They too face a "Catch 22" situation in funding the launching of brand-new technologically innovative ventures. If an NTBF launched with equity support from venture capitalists fails, then the venture capitalists lose most or all their investment. If, as hoped, the venture flourishes and the investment appreciates in the manner required, they still find it difficult to realize on this investment. The logical purchasers of their equity are the venture's founders or others who have joined it with its success and growth. However, it is unlikely that these people have the financial resources to do so. Their own assets must be tied up in the venture (or they would not have sought equity financing in the first place), and, since they have been initially successful, they will probably be seeking further capital to finance expansion. Thus, the venture capitalists' best hope for realizing their desired capital gain is to sell their equity to a third party, or to persuade the firm to allow a public stock issue. Both of the options may be unattractive to the firm's founders and difficult to implement, so that the venture capitalists find that the money is tied up in the firm for a much longer time than was originally intended. These considerations, plus the fact that they, like everyone else, prefer to invest money in a company with a visible "track record" means that venture capitalists invest a relatively small portion of their portfolios in launch or start-up situations. They prefer to invest in an existing firm with a few years of successful operations or "track record" completed, which needs capital to expand and consolidate its success.

Despite these reservations, the venturers will be advised to explore the venture capital market as a potential source of funding. As stated earlier, venture capital agencies have proliferated in the United States which provides the main source of venture capital. Sources of U.S. venture capital are described elsewhere.[17] Some U.S. venture capital agencies also operate "offshore" subsidiaries. Venture Founders Ltd., who offer short courses to prospective entrepreneurs based upon Timmons, Smollen, and Dingee[18] have offices in Canada, Sweden, and U.K. while McQuillan and Taylor indicate the locations of other Canadian venture capital sources.[19] Most Western European countries have similar agencies.

At the beginning of this section, it was stated that prospective technological entrepreneurs may find the acquisition of start-up capital to be the most frustrating aspect of launching an NTBF. By now, readers may be wondering how start-up capital for independent NTBF ever gets raised at all! The truth is that if the proposed venture looks to be technically and commercially viable, and the venture team has the entrepreneurial/project championing determination to launch and prove their innovation, they will raise the required money *somehow or other*. Raising capital, like inventive genius, also requires the infinite capacity for taking pains. A technically creative entrepreneurial venture team that has the infinitely painstaking determination to generate an invention and innovate it into an NTBF will have the determination to find some way of raising the money which it requires. Somehow or other it will put together a "package" of equity and debt capital from some combination of the sources listed above, and manage to launch its venture. Assuming that this goal has been achieved, we now discuss some of the more common problems and mistakes made by NTBF's during their first year or two of operations.

15.6
Common Pitfalls of Initial Operations

In the previous few sections, we have discussed the requirements for launching a potentially successful NTBF, namely, the incorporation of a technical invention into a commercially promising business idea which will be exploited by a balanced cohesive venture team, who has obtained financial funding based upon a well-documented sound business plan. Given the frustrations and problems the team will have had to face and overcome to launch the venture, it might be thought that, once it moves from the "moonlighting" to the "entrepreneurial" stages their problems will dwindle in difficulty. In reality nothing could be further from the truth, for the entrepreneurial stage represents a more or less permanent period of "crisis management" when the team will have to overcome an apparently unending succession of hurdles. Echoing the comment of Sir Winston Churchill on the Allied victory at the Battle of El Alemain in World War II, the launching of the venture does not mark the beginning of the end, but the end of the beginning of the "victorious" foundation of a new high-technology business. Technological and commercial uncertainties will have to be successfully resolved and sound business practices established during the first year or two of operations, if the NTBF is to be consolidated on a sound basis—ready, if required, to exploit opportunities for expansion.The venture team members are likely to have the technical resources and experience to resolve technical uncertainties. But their probable lack of management and business experience does expose them to the risk of making other mistakes, which have been quite frequently observed in small businesses run by an inexperienced management team.

First of all during the "moonlighting stage," all members of the venture team can have expected to work long hours, in order to develop the venture in whatever spare time they had available, after they had fulfilled their paid duties. Therefore, once the venture is launched, and they become full-time participants in the business, they might be forgiven for thinking that they will find more spare time to spend with their families and/or in the pursuit of leisure interests. Although one would not wish to dissuade individuals from accepting their domestic responsibilities or pursuing such interests, the venture team is likely to find running the new business to be a time-consuming process. They may find themselves forced to withstand a continued bombardment of "problems" demanding their time and attention.

If the business is to avoid failure through inadequate attention to critical problems, the team must learn to identify priorities for attention and to budget its time effectively. The subject of time budgeting and management has become quite a popular management concern during the past ten years or so and there are several paperback books extant on the subject. One of the pitfalls faced by novice technological entrepreneurs, because of their prior technical backgrounds, is the tendency to concentrate on technical or "vocational" problems in the exclusion of "management" problems. Obviously the resolution of technical uncertainty requires the solution of a succession of technical problems. Because many venturers find technical problem solving to be more "fun," they risk concentrating on these at the cost of neglecting the development of sound management procedures, standards, and controls for the embryonic business. McClurkin discusses the errors of technology managers "wasting" their times on vocational work.[20]

Earlier it was suggested that "management" is likely to be the weakest of the critical functions skills in the venture team, and I implied that it is particularly important to build up these skills during the early years of the venture, if it is to succeed. Clarke gives a useful, succinct overview of the managerial requirements of an NTBF.[21] As he points out, the majority of business failures can be attributed to poor management. The problems of the venture team's management inexperience are compounded by the need for each to accept a wider span of management responsibilities than would be necessary in a larger established firm. It is therefore imperative that the team ensures that, between the members, they accept responsibility for all aspects of the management of the firm and that each member recognized, and accepts, his individual managerial roles and responsibilities. Some of the problems NTBF's may experience through poor management are now discussed.

Technological Obsessions—
The "Toy-Train" Syndrome

In chapter 2 and frequently throughout this book we have discussed the dangers of technology-push as against market-pull. I also warned of the dangers of placing too great an emphasis on vocational as opposed to management tasks. Both of these pitfalls may be attributed to technological obession. Clearly the desire to innovate must be based upon pride in technological inventiveness and creativity. This pride, however, should not be allowed to cloud management and commercial judgment. An obsession with technological achievement can lead to the expenditure of time, money, and effort in developing a product that is overspecified for the market niche it is intended to fill, such as the development of a "Cadillac" for a "Model T" market. Venture teams that are too absorbed in their technology may, in reality, be regressing back to their childhoods, and trying to run their company as a technologically sophisticated toy train set. In a competitive market this could lead to financial disaster. Either the product will be too expensive to sell, or it will be priced competitively at a level that is too low to recover the total cost incurred. This pitfall will be avoided if the product-market niche(s) are specified as clearly as is practicable from the beginning, and internal and external technological and market developments continually monitored, to ensure that the product-market focus is being held.

Ill-Considered Diversification

During the first years of operations, the venture team will be particularly concerned for the financial viability of the venture; therefore, they will be particularly loath to ignore an opportunity to generate income. Earlier we suggested that the provision of R & D contract or other consultancy services provide a useful source of early income for a new venture. Sometimes the successful completion of a consulting contract or sale of a manufactured product will lead to the satisfied customer requesting a further service or product, which the venture team are no more than marginally technically qualified to provide, because it is outside their possibly limited range of experience to date. Given the pressure to generate income, there is a great temptation to accept the work. Just as it is commercially imprudent to be technologically obsessive and "overspecify" products, it is also clearly both imprudent *and unethical* to accept work that the firm is not qualified or experienced to undertake. An inadequate or (worst still) incompetent performance *outside* its prime field of expertise, will ruin the reputation of an NTBF *within* its field of expertise, and news of such malpractice travels fast! Therefore, in the interests of preserving the professional integrity and the good name of the venture, the team should clearly delineate their area of technological competence and not be tempted to undertake work outside it.

On several occasions earlier in the book, I have pointed out that technological innovations have frequently proved successful in product-market niches different from those originally intended for the invention. Similarly, new high-technology companies have exploited their area of technological competence in different markets than originally intended. For example, it may be recalled (chapter 9) that the Polaroid Corp. originally planned to exploit its technological expertise by manufacturing and marketing polarized light car headlights but then switched to cameras. If the venture team discovers, after a time, that its technological competence may be employed in a more satisfying and profitable manner in a further or different product-market niche than was originally intended, it may do well to try to occupy the new niche. However, such a radical shift in product-market strategy should only be made after conscious deliberation. Too often, in their understandable eagerness to generate income, new companies seek to diversify their efforts into different product-market niches in a willy-nilly ill-considered manner, thereby manufacturing a poorly coordinated product range based upon a nonexistent venture strategy.

Poor Pricing

The development of a technically overspecified product may dictate a price that is too high for the market to bear but, provided the technical specifications are congruent with its market niche, there is also great danger in pricing the product too low. First of all, the price should be set to ensure an adequate cash flow into the company, to ensure a satisfactory return on the capital employed after operating costs have been covered, plus (ideally) a surplus that may be invested in future innovative growth. Often, during the first year or two of operations, venture teams fail to measure and control costs adequately, so they are underestimated. Therefore, although the team thinks it can sell a product at lower price than the competition and still generate a sufficient cash flow, it may have simply underestimated costs. It may only discover this error after a year or more of operations when the firm is already virtually bankrupt. Secondly, persons who are naive about market behavior believe that pricing a product somewhat lower than the competition is bound to generate high sales. In fact, the converse may well occur. Some purchasers (perhaps erroneously!) view price as a measure of quality, and a NTBF with no track record and an as yet unproven product, will be treated cautiously in the market place. Prospective purchasers may view a lower price as indicative of an inferior product and so pay a higher price for a competitive product with which they are already familiar. If the team sets a low price and, in consequence, generates low sales, the firm will suffer a major shortfall from estimated revenue—again, with potentially disastrous consequences. Product pricing should be set in the context of the overall marketing strategy and, it goes without saying, should be sufficiently high to generate the estimated required cash flow.

Sales Equated with Order Predicted

Customers, when discussing their future anticipated (as opposed to current actual) product orders with a company's sales representative, tend to exaggerate these requirements—possibly as a form of "polite self-interest." First a customer may provide an optimistic estimate of future requirements out of a gracious desire to make the salesman feel that he is selling a good product, which people want to buy. Second, the customer wants to guarantee the longer-term supply of components for his own manufacturing requirements, so may (consciously or otherwise) exaggerate the longer-term demand to ensure that adequate supplies will be produced. Customers' longer-term predictions of product requirements, as opposed to shorter-term "firm" purchase orders, should be treated as "optimistic" and discounted accordingly (possibly by 50 percent or more). Failure to discount sales estimates obtained in this manner may lead to an overproduction that can have dire consequences for a financially vulnerable new venture.

Overreliance on Part-Time Staff

In section 15.3 it was pointed out that a new venture will probably have to employ some part-time people in its early years, to ensure that it can call upon the entire range of expertise required for effective management. Because of both time pressures and a realistic recognition that they lack expertise in all aspects of running a business, the venture team risk delegating too much responsibility to part-timers, who, by definition cannot be expected to give a total commitment to the new venture. Although part-time managerial and professional help can be immensely valuable, the venture team must ensure that it is retaining ultimate managerial accountability for running its "own" business, and must not become "dependent" (in the addictive sense) upon outside help. A venture capitalist or other investor may provide some expertise if only to protect the investment, but the team should always be learning on-the-job to ensure the long-term management autonomy of the company. This does not preclude the permanent retention of some part-time support, and the use of professional services and management consultants. Also, a judiciously selected nonexecutive part-time director, retained on a modest honorarium, can often enact a valuable "coaching" or alter-ego role. A (possibly retired) suitably experienced business or professional person can be a continued source of wisdom to the team, warning them of potential pitfalls and acting as a sounding board for their ideas.

After the pitfalls of inadequate product-market definition and naive marketing practices, nonexistent or inadequately designed monetary measurement and control systems are probably the next most common source of problems for a new venture.

During its first few years of operations, an NTBF is unlikely to require, or be able to afford, the full-time services of a professional accountant. Obviously accounting skills will be a mandatory requirement in preparing the business plan and, once the venture is launched, in fulfilling auditing requirements. The fulfillment will, however, be totally inadequate for measurement and control purposes. Moreover, auditors and other part-time accounting professionals are unlikely to be able to provide the services required to develop and implement an adequate measurement and control system. The definition and measurement of the costs of developing, manufacturing, and marketing new high-technology products requires significant sustained attention, and part-time professionals cannot be expected to undertake this task. In his study of Canadian technological entrepreneurs the author found that many of the interviewees had developed their own financial measurement and control systems (possibly with the advice of professionals and after reading a textbook on the subject), more or less from first principles.[22] It is the author's (albeit subjective) observation, that the scientific and engineering training of these individuals made them particularly adept at designing such systems, since it is essentially an exercise in empirical observation, measurement, analysis, and design. The performance of the exercise also gave each of them a useful insight into the pattern of monetary flows in his company.

Amazingly, some companies sustain several years of operations, without making serious attempts at cost measurement and control. If they are manufacturing several products or product lines, they cannot be in a position to know which are their most profitable products and which (if any) are being manufactured and marketed at a loss. Therefore, they cannot truly identify the business they *should* be in. Sloppy cost measurement and control can also go hand-in-hand with other management "sins of omission." Cost analysis should inevitably involve value analysis—the identification of the cost and value added for each bought-in or manufactured component which is incorporated in the final assembled product. It can also highlight faults in (or the absence of) quality-control procedures. Finally, it can draw attention to excessive inventory levels and accounts receivable, both of which can tie up significant amounts of the scarce working capital required by the company. Many companies (both large "old" ones and small "new" ones) are prone to accumulating too much "fat" in raw material, work-in-progress, and finished goods inventories. Many also allow their customers credit beyond the agreed time duration, through failure to chase down "bad pay-ers." Some customers are adept at delaying payment for as long as possible, to provide themselves with the maximum interest-free cred-it that their suppliers will unwittingly allow them! Failure to pay due attention to all these costs measuring and control aspects can make the difference between bankruptcy or survival for a small company early in its life.

15.7
Consolidation and Growth to Maturity

Many NTBF's may remain, through choice and/or necessity, small and hopefully "beautiful." Others may seek to grow and become mature and stable firms. Cooper suggests that small companies grow through three stages.[23]

The first is the start-up stage when the firm is founded and its ini-tial technology-market niche is identified. The second, early-growth stage is when the technology-market niche strategy is tested and the president (or venture team) maintains direct contact with major activities. As was stated above, some firms stabilize at these stages, while others continue to the third stage. This later-growth stage is characterized by manufacturing diversification and the institution of one or more levels of middle management and some delegation of decision making. Some problems of such growth indicated in section 15.3 are discussed further here.

Growth requires the continued judicious exploitation of techno-logical entrepreneurial opportunities and runs the continual risk of following success by failure. Assuming that, by now, the venture team has acquired the commercial maturity and judgment to iden-tify suitable growth opportunities and to avoid willy-nilly diversifi-cation, its financing often presents real problems. The financing of the growth of entrepreneurial organizations in service industries, such as fast-food chains, may frequently be achieved by the geo-graphic replication of the product-market concept and franchising although these too have to cross a "Bermuda Triangle" of growth. This path is less readily available to the technologically innovative organization, although "standardized" manufacturing plants may be replicated in different locations to satisfy a growing demand. Therefore, it may have to cross its own "Bermuda Triangle," bal-ancing the risks of being underfinanced during a rapid growth phase against being overfinanced if, and when, sales level out. Moreover, until it has achieved a sufficient size and fiscal perfor-mance to attract the attention of the financial community, it is un-likely to be able to attract the required capital by "going public" and selling shares on the stock market. Thus growth must be sustained either from the profits put back from the initial operations, or from secondary funding from the original or related financial sources.

In section 15.5 it was pointed out that venture capitalists prefer to provide a secondary capital funding after the venture has a proven track record and the lender can expect to obtain his desired capital gain after a few more years when the venture may indeed go public. Other lending agencies also prefer to provide secondary funding to "proven" ventures so, given that the venture has achieved a good performance so far, it is generally easier to obtain "expansion," as opposed to "launch" capital. It does, of course, require the preparation of a detailed venture plan and business proposal, as before, but this is a valuable discipline anyway.

If the venture team pursues a continued growth policy, sooner or later, it may face the issue of going public or merging with another company. Providing the desired growth can be maintained, it is probably best to delay this decision for as long as possible, since more "mature" companies are viewed more favorably, and hence valued more highly, on the open market. Going public is a time-consuming and hazardous process and as Howard points out, a significant proportion of companies that do so, regret the decision afterwards.[24]

Similar comments apply to a merger deal, since an injudiciously designed merger agreement can deprive the team of much of the wealth generated by their "sweat equity." In the last analysis, the team will have to decide whether the benefits of a public offering or merger outweigh any loss of equity and control incurred. The more the equity and control of the company is fragmented, the more valuable it becomes to "take over" later.

Given acceptable expansion capital is forthcoming, expansion will typically imply hiring more staff, possibly with more management and bureaucratic orientation, to monitor management standards and control procedures. The venture founders and the original cadre of paid employees will have to accept new members into their "club," while maintaining a purposeful organizational climate and zeal for further innovation. The reward structure of salaries, status, and (possible) stock options will have to reflect the need to repay "old retainers" for loyal service and to attract new key personnel. Achieving such a balance may be far from easy, and unless tactfully managed, the interpersonal disruptions caused may lead to the disintegration of the venture.

15.8
Illustrative Examples

At the end of the previous chapter, I summarized Litvak and Maule's study of the backgrounds of Canadian technological entrepreneurs.

Those researchers also studied the fortunes of forty-seven small technology-based Canadian companies. They first studied these companies in 1970/71 and then performed a follow-up study in 1980.[25] To conclude this chapter I will summarize those authors' analyses of the experiences of the companies over the ten-year period.

Litvak and Maule categorized these experiences as failures, marginal survivors, and successful survivors, and we examine each of the categories in turn. Eighteen companies could be categorized as failures because they were declared bankrupt, in a state of receivership, or had ceased to do business for several years. At the time of their demise, the original founders were at least marginally participating in the management of sixteen of these firms, and twelve had attracted venture capital investment with (in some cases) the input of new management expertise. When ownership and control by the original founders were significantly reduced, the managerial role of the technological entrepreneur was often the first to be constrained or eliminated. The majority were single-product companies and, while some of these products were technically successful, most lacked significant technical novelty and failed to achieve substantive market penetration. User needs, training, and after-sales servicing, together with associated operating and capital costs, were often underestimated. In at least one-half of these failures, the owner managers were insufficiently prepared managerially, psychologically, and financially to cope with the protracted and mounting problems of launching and consolidating a venture. A few experienced alcoholism, breakdowns, and marriage breakups.

The authors defined nine companies as marginal survivors because, over ten years, they had yet to show any or (at best) spasmodic profits. A key difference between these firms and the failures was the sheer determination and endurance of their managers. Thus, their survival may have been dependent on their maintaining a "Dunkirk spirit" organization climate, but it should also be noted that new ventures in larger corporations can take over ten years to become profitable (see chapter 16). Five of these companies are family owned with the original technological entrepreneurs still active. Three of the remainder are privately owned by a few people, while the fourth which went public is controlled by individual venture capitalists and its chief executive officer. The founding entrepreneurs have left two of these four companies. All nine companies have narrow product lines and have achieved only modest annual sales, six under $1 million, only one over $5 million.

All the twenty successful firms had achieved annual sales in excess of $1 million, five in excess of $10 million. Higher sales could be attributed to several factors, when compared with the failures and marginal survivors. These successful companies had diverse product lines which incorporated a higher degree of technological novelty. They concentrated their technical and market efforts on specialized, expanding industrial markets where they enjoyed comparative advantages and avoided competition with larger companies. They sought out export markets and set up foreign subsidiaries, mainly in the United States. They were less obsessed with their product technologies and less prone to "mousetrap myopia." Eight of these successful companies had publicly traded shareholdings, and two had lost their original founders and become subsidiaries of larger Canadian-owned companies. Of the other six of these, three went public at the instigation of their founders and three under pressure from venture capitalists holding a substantial equity share in them. Most of these six firms were dissatisfied with the experiences of "going public." In the remaining twelve companies, ownership was retained within a family or small group with some infusion of outside capital. In thirteen of these twenty firms which had successfully reached Cooper's later-growth stage of diversified manufacturing, some of the original founders retained a significant participation.

15.9
Final Comments

At the beginning of this chapter I suggested that many readers had unfulfilled dreams to set up their own high-technology companies. Reading this chapter from beginning to end (itself a daunting task!) may have shattered some of these dreams! We hope not. Launching and nurturing a high-technology venture to commercial success is undoubtedly a challenging, protracted, and hazardous enterprise, but history demonstrates that *it can be done*. Many men and women have demonstrated this fact. Most readers will hold (or be obtaining) academic or professional qualifications at the bachelor's or higher degree levels. Most readers will agree that the acquisition of such qualifications is also a challenging, protracted, and (at times!) hazardous enterprise. If an individual has the ability and tenacity of purpose to obtain a good grade or good class of degree from a good university, he or she has the latent capacity to succeed in life. If this capacity is matched with a personal aptitude for technological entrepreneurship, he or she should be able to participate in a successful venture. If, as is to be expected, the venture grows successfully to reach maturity and stability, it must then find ways of institutionalizing innovation and entrepreneurship, if it is to avoid "bureausis" and stagnation. Some of these ways are discussed in the next chapter.

References

1. Norman Fast, "A Visit to the New Venture Graveyard," *Research Management,* 22, no. 2 (March 1977): 18-22.

2. Jeffrey A. Timmons, Leonard E. Smollen, and Alexander L. M. Dingee, *New Venture Creation: A Guide to Small Business Development* (Homewood, Ill.: Richard D. Irwin, 1977).

3. Jeffrey A. Timmons, "New Venture Creation: Models and Methodologies"; and Herbert Keirulf, "Additional Thoughts on Modelling New Venture Creation," chap. 7 in *Encyclopedia of Entrepreneurship,* ed. Calvin A. Kent, Donald L. Sexton, and Karl H. Vesper (Englewoood Cliffs, N.J.: Prentice-Hall, 1982), pp. 126-39.

4. Timmons, Smollen, and Dingee, *New Venture Creation,* p. 12.

5. Gene Bylinsky, *The Innovation Millionaires: How They Succeed* (New York: Charles Scriber's Sons, 1976), chap. 1.

6. M.J.C. Martin, J.H. Scheibethut, and R.C. Clements, *Technology Transfer from Government Laboratories to Industry,* Technological Innovation Studies Program Report no. 53 (Ottawa: Department of Industry, Trade and Commerce, November 1978).

7. M.A. Jolson and M.J. Gannon, "Wives—A Critical Element in Career Decisions," *Business Horizons* 15, no. 1 (February 1972): 83-88.

8. Timmons, Smollen, and Dingee, *New Venture Creation,* p. 368.

9. *Ibid,* p. 171.

10. *Ibid,* chap. 14.

11. Patrick R. Liles, *New Business Ventures and the Entrepreneur* (Homewood, Ill.: Richard D. Irwin, 1974).

12. Donald S. Scott and Ronald M. Blair, eds., *The Technical Entrepreneur* (Ontario: Press Porcepic, 1979), 203-15.

13. R.M. Colton and G.G. Udell, "The National Science Foundation Innovation Centres—An Experiment in Training Potential Entrepreneurs and Innovators," *Journal of Small Business Management* 14, no. 2 (April 1976): 11-20.

14. *The Make or Buy Policy* (Ottawa: Ministry of State for Science and Technology, 1975); *A Framework for Government Research and Development* (London: HMSO Cmnd. 4814, 1971).

15. Martin, Scheibethut, and Clements, *Technology Transfer.*

16. R.H. Grasley, "Financing the New Enterprise," in Scott and Blair, eds., *The Technical Entrepreneur,* pp. 185-201.

17. Stanley M. Rubel, ed., *Guide to Venture Capital Sources,* 3rd. ed. (Chicago; Capital Publishing Corp., 1974). U. Dothan, *Survey of Venture Capital Industry* (Chicago: Venture Capital Assoc., May 1977).

18. Timmons, Smollen, and Dingee, *New Venture Creation.*

19. P. McQuillan and H. Taylor, *Sources of Venture Capital in Canada,* 2nd. ed. (Ottawa: Department of Industry, Trade and Commerce, 1977).

20. D.T. McClurkin, "Managing Your Management Time," in Scott and Blair, eds., *The Technical Entrepreneur,* pp. 249-54.

21. Thomas E. Clarke, "Managing Your Technology Based Company," in Scott and Blair, eds., *The Technical Entrepreneur,* pp. 239-48.

22. Martin, Scheibethut and Clements, *Technology Transfer.*

23. A.C. Cooper, "Strategic Management: New Ventures and Small Business," *Long-Range Planning,* 14, no. 5 (October 1981): 39-45.

24. Graemar K. Howard, "Going Public When It Makes Sense," in Rubel, ed., *Guide to Venture Capital Sources.*

25. I.A. Litvak and C.J. Maule, *Canadian Entrepreneurship; A Study of Small Newly Established Firms,* Technological Innovation Studies Program, Report no. 1 (Ottawa: Department of Industry, Trade and Commerce, October, 1971). I.A. Litvak and C.J. Maule, *Entrepreneurial Success or Failure—Ten Years Later,* Technological Innovation Studies Program, Report no. 80 (Ottawa: Department of Industry, Trade and Commerce, October 1980).

16

Small in Large Is Also Beautiful: Stimulating Intracorporate Entrepreneurship

To encourage innovation, industry must now re-discover the fact that any innovation starts with a dream, and only people have dreams. It must accommodate to the individual in its organizational planning, and allow some of those dreams to materialize.

Joseph W. Selden, "Organizing for Innovation" in
**Innovation and U.S. Research: Problems and
Recommendations**

16.1
Introduction

In chapter 14 the spin-off phenomenon was discussed as well as the personality characteristics of technical entrepreneurs. We saw that, in many instances, a key factor stimulating independent spinoffs from larger companies was a person's frustration with general management and the environment for entrepreneurship in the incubator organization. In the previous chapter, we examined the problems of launching and operating an independent spin-off or NTBF, and concluded by suggesting that when it grows to maturity it runs the risk of losing its innovative entrepreneurial drive. Because "entrepreneurship" is a vital constituent in the innovation chain reaction, well-established, high-technology firms that fail to stimulate and nurture entrepreneurship also run the risk of becoming technologically moribund, losing their capacity for innovation. At the end of chapter 13 I briefly referred to the need to set up independent or quasi-independent ventures to exploit innovations, and forge the entrepreneurial link in the innovation chain equation. Indeed Macrae has proposed the notion of the organization as a confederation of entrepreneurs.[1] At that point we stated that we would return to this important topic in chapter 16, having reviewed the literature on "spin-offs" and NTBF entrepreneurship.

16.2

305

Dynamic
Conservatism
Versus
Entrepreneurship
in Large
Organizations

Dynamic Conservatism Versus Entrepreneurship in Large Organizations

Donald Schon is a leading thinker and writer who has explored many facets of technological and social change.[2] He argues that social systems, including high-technology corporations, have a built-in tendency to maintain a "stable state" and resist change, and this tendency is much stronger than an inertial resistance to change.

> *The system as a whole has the property of resistance to change. I would not call this property "inertia" a metaphor drawn from physics—the tendency of objects to move steadily along their present courses unless a contrary force is exerted on them. The resistance to change exhibited by social systems is much more nearly a form of "dynamic conservatism"—that is to say, a tendency to fight to remain the same.[3]*

We have already commented on some of the causes of dynamic conservatism in previous chapters, but it is useful to review them here. Collier[4] provides the following summary:

1. Financial resources are usually allocated by discounted cash flow (DCF) criteria which favor short-term incremental investments in the company's present business. Other innovations require a longer development lead time (recall chapter 2), before generating net positive cash flows and a significant ROI, which must be discounted over a longer time period. This observation is born out by Biggadike in an analysis of new ventures undertaken by large corporations.[5] He found that, on average, they took eight years to become profitable and ten to twelve years before their ROI's equaled those of the corporations' mature businesses.

2. Product engineering is required to produce designs which work reliably and safely in the user environment (recall chapter 13). Product engineers are therefore typically conservative because they prefer to support new product (or process) designs which incorporate incremental changes in proven designs, rather than risk their professional reputations in unproven significantly new product or process design concepts.

3. Manufacturing achieves highest efficiency in routine operations on an assembly line basis. Quoting Collier, "This the exact antithesis of what happens when a new product is introduced. To manufacturing, the new product is a monkey wrench tossed into the gears of a highly efficient operation."

306

Small in
Large Is Also
Beautiful:
Stimulating
Intracorporate
Entrepreneurship

4. Trade unions resist any process improvements that reduce labor requirements, and when new products are introduced they seek to loosen work content standards.

5. Salespeople, who are usually paid partially by commissions on sales revenue generated, know that time is money. New products require them to invest extra time in educating themselves and their customers, so they thereby receive less commission per hour of sales effort invested.

6. The reward system for general managers is typically based upon annual profits or ROI on corporate resources managed. They are therefore rewarded for achieving short-term rather than long-term profit. Moreover, apart from the greater inherent risks involved, the rewards associated with the profits from any longer-term, more radical, innovations are unlikely to accrue to the manager making the original investment because he is likely to have moved on to other responsibilities before they are achieved.

These factors largely explain the dynamic conservatism of many high-technology organizations which has to be overcome if they are to be technologically innovative. As we discussed in chapter 4, such organizations must stimulate "innovation" as well as "operations" management if they are to remain commercially healthy. Fortunately (from the viewpoint of large corporations) not all entrepreneurially motivated individuals wish to set up their own independent ventures. Some are attracted by the relatively high community status enjoyed by individuals holding middle management or senior management positions in larger recognized corporations. Moreover, such individuals' ambitions may motivate them to seek advancement in larger organizations that offer rewards in terms of increased formal power and internal status. Thus larger high-technology organizations can retain a reservoir of entrepreneurial talent, and we will now consider alternative organizational mechanisms for stimulating innovative and entrepreneurship within them. We begin by examining the "Texas Instruments" approach, which encourages managers and technologists to enact both "innovation" and "operations" roles, in quite considerable detail. We will then refer to a less detailed overview of the approaches adopted by other corporations. The chapter will be concluded by reviewing the problems inherent in intracorporate entrepreneurship.

16.3
Texas Instruments—Objectives, Strategies, and Tactics—OST

It almost goes without saying that Texas Instruments (TI) has been one of the most successfully innovative high-technology corporations since World War II. Founded in 1930 as a geophysical services company, it established laboratory and manufacturing facilities in 1946 and entered the then-infant semiconductor industry in 1952. Since then it appears to have followed a consistent offensive strategy, pioneering the development of silicon semiconductor technology, first with discrete semiconductor devices, and then integrated circuits and a range of products embracing this technology. Innovation is described as a way of life at TI, and the following quote by one of the company's senior managers illustrates this philosophy:

307

Texas
Instruments—
Objectives,
Strategies, and
Tactics—OST

> *We are convinced that useful products and services as well as long-term profitability are the result of innovation. Further, we feel that profitability above the bare compensation for use of assets can come only from a superior rate of innovation and can no longer exist when innovation is routine. This is why our long-range planning system is fundamentally a system for managing innovation.* [6]

TI's long-range planning system designed to institutionalize innovation is called Objectives, Strategies, and Tactics, or OST. As Jelinek states, it is difficult to capture on paper a true feeling for the organizational climate created by OST.[7] Indeed, parodying the title of a popular BBC television program of the early sixties, OST appears to be "not so much a system more a way of life" at TI. Reading his own account of its conception and inception, it appears that the system owes much to the impressive personality and managerial insight and judgment of Patrick Haggerty, the sometime president and chairman of TI.[8] OST was conceived and instituted in TI in the 1960s, during what might be loosely labeled its adolescent formative years and now provides a framework for managing the corporation at virtually all levels. It is relevant to note that OST was instituted at a formative stage of organizational development, and the installation of a similar system in a mature larger company might be very difficult. Despite this reservation, a brief examination of its features is in order here, since OST offers salutary ideas for managers in other high-technology firms.

When TI entered the semiconductor industry, virtually as a small novice company, it achieved considerable technological and marketing success through developing silicon transistor technology. By 1959-60, top management recognized that as the organization grew out of its success, it risked losing its innovative drive and "edge" through the onset of "bureausis," and its success had been based upon well-conceived strategies and well-executed tactics in support of these strategies.

Top management thinking echoed the view of Gardner who argues that societies and institutions tend to lose their vitality as they mature, and that to avoid this decay, they must build in a capacity for self-renewal, based upon individual self-renewal.[9]

308

Small in
Large Is Also
Beautiful:
Stimulating
Intracorporate
Entrepreneurship

It is necessary to discuss not only the vitality of societies but the vitality of institutions and individuals. They are the same subject. A society decays when its institutions and individuals lose their vitality.

Does this mean that there is no alternative to eventual stagnation? It does not. Every individual, organization or society must mature, but much depends on how this maturing takes place.... In the ever-renewing society that matures is a system or framework within which continuous innovation, renewal and rebirth can occur.

As Haggerty states, "The objectives, strategies and tactics system at Texas Instruments is just that—an attempt to create for one organization a system or framework within which continuous innovation, renewal and rebirth can occur."[10] Historically, TI's operations have been managed through a group divisional structure and Product Customer Centers (PCC's). In 1978 there were four groups, thirty-two divisions and more than eighty PCC's with annual sales between $10 million and $100 million. The management of each center holds responsibility for marketing, manufacturing, and new product development and is judged annually on its profit and return on assets performance. Although the PCC concept should ensure that products are developed that satisfy market needs, it is vulnerable to stagnation and "bureausis," for the reasons discussed earlier. If a manager is judged annually by his operating performance in terms of profits and returns on assets, he is less than likely to wish to promote innovations that require a longer than one year lead time, incur significant development costs, and (if they are really innovative) carry a definite risk of commercial failure anyway. Furthermore, the administration of the innovation process becomes more cumbersome as an organization grows, and the industry maturing process may make profitable innovation more difficult. Therefore, a greater premium is placed on administrative rather than innovative skills, so that managers seek to demonstrate the former rather than the latter skills in order to obtain promotion. These views were expressed by one of TI's senior managers as follows:

OST constitutes a formalization and institutionalization of the informal approach to long range planning used by senior corporate management in the nineteen fifties, but extended throughout the organization. Implemented in conjunction with PCC's it creates a form of matrix organization in which all staff can wear "two hats" as both innovation and operations managers.[11]

309

Texas
Instruments—
Objectives,
Strategies, and
Tactics—OST

This matrix structure is illustrated in figure 16.1. Individual PCC's are aggregated into divisions, and then groups, based upon generic technologies or markets, which report to corporate management. It is the OST structure, constituting a hierarchy extending downwards from corporate objectives to tactics, which requires some explanation.

The objectives are broadly stated goals at the corporate level and each business area. The strategies define long-term general courses of action in pursuit of these goals. The tactics represent the relatively short-term projects in support of the strategies, to which are allocated individuals with specific responsibilities and resources. It is the hierarchy of individuals who are responsible for the objectives, strategies, and tactics that collectively constitute the innovation management function. The purpose of this hierarchy is to create a climate which stimulates the generation of innovative ideas or proposals throughout the organization and to evaluate these proposals in terms of corporate and group objectives and strategies.

Given the fecundity of inventive and innovative ideas within an organization, the problem becomes that of rank ordering proposals in terms of their impact on strategies and objectives and (as in R & D project selection) funding the highest ranking of these—which become known as tactical action plans (TAP's)—until the

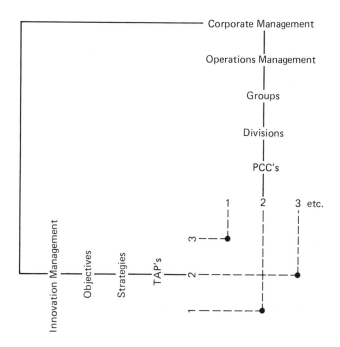

Figure 16.1 Texas Instruments Matrix.

310

Small in
Large Is Also
Beautiful:
Stimulating
Intracorporate
Entrepreneurship

current funds allocated for TAP's has become exhausted. TI's long-term economic and innovative performance is clearly dependent upon the effectiveness of this screening and selection process.

Starting from the bottom and working upwards, individual proposals are screened and ranked at the tactical, strategic, and objective levels. At any point in time, higher-level, innovation management specifies the total funds available at various hierarchical levels, so that a funding cutoff can be established. Proposals which look promising, but cannot be supported in competition with others because of insufficient funds, are placed in a "creative backlog" so that they may be supported should more funds become available. Furthermore, the selecting/screening process is repeated at the higher hierarchical levels. "Strategy" managers review the selections made at the tactical levels and "Objective" managers review the selection made at the strategic levels. In some cases these higher-level managers may change the lower-level rankings, possibly in light of wider strategic and objective considerations. This repeated review of rankings is important because it ensures that a potential radical, as opposed to incremental innovation which may have a potentially broader impact upon strategies and objectives, receives adequate consideration at the appropriate hierarchical level. It also gives higher-level managers an opportunity to identify potential synergies, between proposals coming from separated lower-level sources. Some earlier funded proposals (or TAP's), having duration times that extend into the current cycle, are reviewed as part of the selection process to see if they have fulfilled their earlier promise, or are now less congruent with revised strategies and tactics. They may consequently be terminated and the unspent funding reallocated to other proposals. TAP's thus face a continued competitive review process.

OST is a procedure for institutionalizing an organizational culture conducive to maintaining innovation and entrepreneurship as the organization grows. However, because it institutes an innovation management hierarchy which overlays the operations management hierarchy, it inevitably duplicates bureaucracy, since OST itself is a bureaucratic process. Although OST has succeeded in institutionalizing an innovative spirit in TI, its bureaucracy inhibits some creative entrepreneurial individuals from championing their ideas competitively over the successive hurdles of review process. This will apply particularly to the younger person who may be inexperienced and impatient with "company politics" and who may produce an idea for a radical innovation which, though offering commercial promise, is incongruent with the priorities of the current OST program. The company therefore also has an *Ideas Program* designed to accommodate such exceptions. It is introduced to its staff as follows:

*The IDEA Program provides an opportunity for initial feasi-
bility demonstration of concepts that do not fit within the
immediate OST thrusts. This program will appeal to those
of you who want to be entrepreneurs and innovators because
it will provide an environment in which you and your ideas
can flourish.*[12]

311

Texas
Instruments—
Objectives,
Strategies, and
Tactics—OST

Quoting the words of Patrick Haggerty,

*Every effort is made to keep the IDEA System simple. There
are no approval cycles, no delays, no reviews, and no reports.
TI had named about 50 key persons around the company
who have a proven record and have shown that they know
what a good idea is when they see one. Each of them is available to listen to new ideas and on their own, they can fund,
without any further discussion or approvals, up to $25,000.*

Two very successful products which were initially funded from
the Ideas Program were the $19.95 digital watch launched in 1976,
and the "Speak and Spell" learning aids based on TI engineer
Gene A. Frantz's idea for electronically synthesizing the human
voice using large-scale IC technology.

One impressive feature of the semiconductor industry, which
is still little more than a generation in age, has been the rapidity
with which unit prices (first with discrete devices and the IC's) have
fallen as output yields have risen. In this situation, a company that
can maintain continued increases in labor and resource productivities is more likely to maintain a competitive cost advantage.
Furthermore, for a given market price level, manufacturing operations which are more efficient will incur lower costs, thus creating
more surplus "profit" to be reinvested in innovative ventures. Therefore once OST had become established, TI expanded it to include
a People and Asset Effectiveness Program (P & AE) which the then
president of TI, Mark Shepherd, Jr., described to stockholders in
these words in 1973:

*People and Assets Effectiveness is an improved ability to
solve customer's problems through increased productivity
of people and assets. It means improved planning and control systems. It means the application of automation to
design, manufacturing and to management problems. And,
most importantly, it means the fullest use of all the talents
of their minds as well as their hands—to be involved in the
planning and control of their work, not simply the doing....*

312

Small in
Large Is Also
Beautiful:
Stimulating
Intracorporate
Entrepreneurship

The P & AE Program is, in effect, incorporated in the OST Program with goals set for increased productivity and growth. Proposals for improving productivity are evaluated through a similar selection process and, as the final sentence in the above quotation implies, it is designed to elicit contributions from "blue collar" as well as "white collar" workers. TI was one of the early companies to join the "job enrichment" movement, encouraging both manufacturing and clerical work teams to devise their own workplace organizations and procedures. The participative climate for stimulating innovation thus permeates all levels; this is an important factor because successful technological innovation often requires innovative thinking at the technician/craftsman level as well as the scientific/engineering level.

OST generates an organization culture that has two important features relevant to our present discussion. First, since individuals know that their proposals will be evaluated in the context of corporate objectives, strategies, and tactics, they will be forced to conduct their own evaluations of their ideas, in this context, thus impelling them to seek a reconciliation of their personal goals with corporate goals. OST thus provides a framework for matching individual and organizational goals and ensures that top management's perception of corporate goals permeates the organization, thus increasing the likelihood of the effective implementation of a corporate technological strategy. Second, it forces individuals to wear "two hats" as both operations and innovation managers. This second feature will be discussed in a little more detail.

As was stated earlier, operations at TI are managed through PCC's. This management structure is decentralized, and local PCC managers have the autonomy to run their own "businesses"—being held accountable for their own performances against criteria agreed between corporate management and themselves. Operating performance is measured using conventional reporting criteria of profit-and-loss statements and annual operating budgets. That is, when wearing their operations manager's hats, individuals report in a conventional manner. Successful proposals or TAP's are funded from a separate strategic budget and evaluated independently, against agreed criteria so that operations and innovation programs are separated, then consolidated in an overall profit-and-loss statement (figure 16.2). Typically, at any point in time 75 percent of the managers have innovation as well as operations management responsibilities and report simultaneously through innovation management and operations management hierarchies that are largely overlapping. That is, an individual may report to a divisional manager wearing his operations management "hat" and a strategy manager wearing his innovations "hat." In some cases, this individual may be a "two-boss" person (as in the conventional matrix

313

Texas
Instruments—
Objectives,
Strategies, and
Tactics—OST

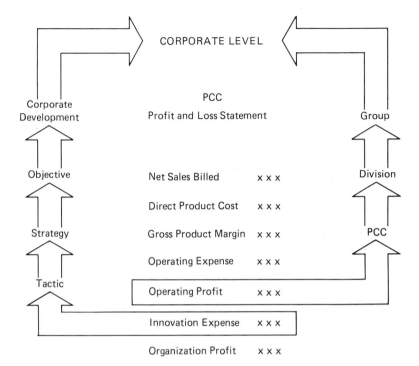

Figure 16.2 Operations and Innovation P & L Statement.

structure) since divisional and strategy managers are different individuals. However, in many cases, an individual may report to only one boss, since his superior will also be wearing "two hats" and enacting both the divisional and strategy manager's roles. Whichever situation obtains, the dual-operations and innovation reporting structure permeates the entire management hierarchy. However, the enactment of an innovation as opposed to operations management roles imposed differing behavorial requirements.

Conventionally, it is generally recognized that a manager's spans of responsibility and authority should coincide, if he is to perform effectively. However, virtually by definition, an innovative project or TAP will require the innovation manager to exercise a wider span of responsibility than his authority allows. The separation of responsibility and authority requires him to exercise entrepreneurial or project championing skills to secure through negotiation and cooperation rather than authorization the organizational resources that the project requires. The OST system creates a demanding but satisfying climate for technological entrepreneurship, and an individual may develop a successful management career based upon proven performance in innovation as well as

314

Small in
Large Is Also
Beautiful:
Stimulating
Intracorporate
Entrepreneurship

operations management. This is summed up in the works of George H. Heilmeier, a vice-president of TI who formerly headed the Pentagon's Advanced Research Projects Agency.[13]

TI seems to have discovered the style that allows a multi-billion-dollar corporation to grow at 15% a year. To match that, companies are going to have to share information with more people. They're going to have to maintain entrepreneurial spirit, and management is going to have to have good visibility in all the company.

16.4
Separate Venture Operations

As stated earlier, OST was instituted in TI in a relatively early stage of that organization's development, and might be difficult to replicate elsewhere, particularly in a mature corporation. We therefore now examine some of the ways other mature high-technology organizations have stimulated innovation and entrepreneurship. These are essentially based upon "separate venture" approaches that seek to replicate some of the positive features of the climates of spin-off companies discussed in the previous two chapters. Such approaches became popular in the late sixties and early seventies. Hanan cites Dow, Westinghouse, Monsanto, Celanese, Union Carbide, 3M, Du Pont, and General Mills as corporations committed to the venture philosophy.[14] At the time of that writing (1969), Hanan reported that 3M had launched over two dozen ventures (with six current operating divisions having grown out of the venture system), while Du Pont was committed to ventures in thirty to fifty new developmental areas.

There are a number of quite compelling reasons for exploiting an innovation by establishing a separate venture operation. We have already cited most of these reasons already, but others follow.

1. The needs to exploit applications engineering opportunities and changes in societal attitudes towards R & D. These are discussed immediately below and are reflected in the venture operations of British Oxygen.

2. The need to exploit a radical/revolutionary innovation of a central R & D facility, that is, one that does not fit into the technology-matrices of existing product market lines, so that the company lacks an established manufacturing expertise to do so.

3. The fecundity of R & D invention that may overload the capacities of the operations divisions to exploit innovations.

4. The enhancement of a project team into a venture team which may bypass the technical and bureaucratic barriers to technology transfer and provide a satisfying outlet for the autonomy needs and entrepreneurial drives of staff. It thus provides a useful mode for stimulating organizational entrepreneurship and compensates for the bureaucratic arteriosclerosis which often develops as the organization matures.

5. The attempt to prevent key members of the project team from resigning from the company to establish an independent venture to exploit the innovation.

6. Attitudinal factors in operations management that are partially reflected in the above and were discussed earlier. That is, qualities that make individuals successful operations managers—the administrative and personal abilities to coordinate workers, material, and money into a continually profitable manufacturing operation—are contrary to those required by the successful entrepreneur. Therefore, operations management may take a jaundiced view of a radical innovative project which promises to dislocate current manufacturing operations, consume scarce resources, and erode perhaps the next several years profit-and-loss records, all with (to them, at least) an unacceptably high risk of failure.

A separate venture organization may be established in one of several ways, the preferred choice being dependent upon the nature of the innovation in relation to the company's other product lines, patenting, and financing considerations and the strengths and weakness of the planned venture team, among other situational factors. Alternative models for establishing a venture are now discussed.

16.5
R & D Venture Organizations

When discussing technological strategies in chapter 4, it was pointed out that many companies follow mixed strategies (that is, two or more technological strategies simultaneously) to achieve corporate goals. We demonstrated in that chapter that the enactment of an offensive/defensive strategy required all-round excellence throughout the technological base of the organization. On the other hand, to take another example, an "applications engineering" strategy can be enacted from a truncated technological base, since it requires a D-intensive organization capable of swiftly exploiting specialist applications opportunities or new "market needs" perceived by field sales and servicing staff. Furthermore, changes in corporate, government, and social attitudes to science over the

316

Small in
Large Is Also
Beautiful:
Stimulating
Intracorporate
Entrepreneurship

last decade or so have encouraged companies to perform "market needs"-oriented "development" rather than new knowledge-oriented "research," to achieve corporate growth goals. This attitudinal change was also reflected in the "consumer-contractor" principle for research prescribed by Lord Rothschild in Britain.[15]

The R-intensive central R & D facility of an offensive/defensive innovator is unlikely to employ staff with the orientation and speed of responsiveness to achieve the timely development results required to implement an applications engineering strategy. For the same reasons, the facility may display a slow inertial response towards the changes in corporate and public research attitudes cited above. These considerations have led some companies to establish semi-independent satellites to perform contract R & D (with an emphasis on the "D" rather than the "R") on a profitable venture basis.[16] Such a venture can operate as a semiautonomous entrepreneurial enterprise, performing commissioned development work for an operations facility in pursuit of an applications engineering opportunity, contract R & D on the open market (that is, for any other "customer" who is willing to pay for their services), and possibly the development of new products separate from the company's current product/market lines. The British Oxygen Company went even further in 1971 when they replaced their central R & D facility with a "new venture secretariat" which selected R & D projects, which were then performed on operations divisional sites.[17] These projects were established as separate semiautonomous venture units with responsibilities for marketing their own products, the operation divisions merely acting as landlords.

16.6
Internal Individually or Jointly Sponsored Ventures

An internal venture, perhaps sponsored by one, two, or more operating divisions, is one approach which has proved to be useful in several large corporations. This format may be particularly appropriate when the innovation contains features which impinge upon the technologies, product lines, markets, or uses of two or more operating divisions.

Hillier cites Radio Corporation of America's (RCA's) development of magnetic video tape which was considered commercially unattractive by a division.[18] Their broadcasting affiliate of National Broadcasting Company (NBC), however, anticipated that an RCA competitor would introduce video tape equipment which NBC would then have to purchase to maintain their technological competitiveness with the Columbia Broadcasting System (CBS) and American

Broadcasting Company (ABC), the other major U.S. commercial TV broadcasting networks. A sensible approach, in this situation, was for the equipment division and NBC jointly to sponsor a "video tape equipment" venture. It was developed internally in the equipment division, partially financed by NBC.

In section 16.4 it was pointed out that 3M had developed numerous ventures, including some which have developed into autonomous divisions. 3M sees itself as a "people-oriented" corporation and appears to offer a climate that informally encourages entrepreneurship. Individuals are encouraged to spend about 15 percent of their time on developing their own new product ideas. If they believe that these ideas offer attractive venture opportunities to the company, they can seek out sponsorship within their own divisions.[19] Thus if they work in R & D, they may seek an R & D sponsored venture (rather as in section 16.5 above) or, if they work in an operating division, seek sponsorship there. If their parent division refuses to support them, they may seek sponsorship from another division or from a New Business Ventures Division which is associated with Central Research. Consequently, if they possess a good idea plus the required personal entrepreneurial qualities, they are likely to find sponsorship somewhere within the corporation. As Park expresses it:

> You get the sense that it's de rigeur to jostle a bit for the limelight—especially if you want to become a project manager. Sometimes, at division level, the spot-light wavers—there are so many things to push along the managerial scrutiny must, perforce, skip lightly over some of the newer ideas. But even then, the man with an idea—especially if he's done his homework and can answer some hard-nosed questions—can bring his dream to an arena where the candlepower is a bit more intense. This is 3M's New Business Ventures Division....

In deciding whether to sponsor a new venture, 3M recognizes the critical importance of the entrepreneurial qualities of the lead entrepreneur or "product champion" in the 3M vocabulary. Quoting one former 3M technical manager,

> The typical product champion at 3M turns out to be one-quarter technical man and three-quarters entrepreneur. He pushes a product idea by what amounts to conning the management—somewhat exaggerating its chances for success, hoping to get enough money to make the product fill the bill.[20]

318

Small in
Large Is Also
Beautiful:
Stimulating
Intracorporate
Entrepreneurship

Clearly experienced top management knows how judiciously to discount an individual's exaggeration and when to support him. As one 3M President put it,

> *We're in the business of gambling on individuals. If a man has been a pretty good judge in the past of what he said he was going to do and has done it, then if he comes in and says I want to start importing moon dust, I guess I'm likely to let him try. In like manner, if a fellow who's been around here a great many years comes forward and says this year we've got a tremendous breakthrough that's going to cause all sorts of things to happen, and he's had four or five years in a row of not delivering on what he said, well, I'm likely to give him an argument.*

Du Pont uses a similar approach to stimulate entrepreneurship.[21] Du Pont's R & D activities are broadly categorized as exploratory research, improvement of established business, and new venture development. "Exploratory research" is long term and "improvement of established business" can be roughly equated with incremental innovations in existing markets—innovations that may be readily accommodated in the extant management structure of current operating divisions. When looking at more radical innovations in possibly new markets, other considerations obtain. "In many cases, however, promising new products or systems are visualized in R & D programmes which do not fit well into the marketing and manufacturing capabilities of the department. It is then that a New Venture Development is considered."[22] At Du Pont, new venture groups are set within an extant department of the firm, but there is no attempt to standardize the approach. Again, in deciding to sponsor a new venture, the emphasis is placed on the venture's CEO, called the "New Venture Manager," who enacts the role of the lead entrepreneur in the Timmons and his associates terminology, or the project/product champion in the critical functions/3M terminology. The emphasis on the entrepreneurial qualities of the venture leader is a recurrent theme in most of the articles on this subject. Hanan quotes one company president:

> *I have an unfailing test for identifying entrepreneurial types. I throw every candidate right in with the alligators. The establishment man complains he can't farm alligators in a swamp. The entrepreneur farms 60% of the alligators, markets another 30% for everything but their squeal, drains their part of the swamp, and leases the land for an amusement park overlooking "Alligatorland". The other 10% of the alligators? That's his delayed compensation.[23]*

319

External
Supported
Ventures or
Severed
Ventures

16.7
External Joint Ventures

Another approach is an external joint venture sponsored by the parent company and an independent second company. This format may be appropriate when the innovation contains features that infringe upon the other distinct markets or industries. Alternatively, our innovative company may buy a major or controlling share in the second company and use the latter's industry/market knowledge and image to spearhead its thrust into new markets. This is one typical strategy for interindustry invasions, such as chemicals into the textile industry or electronics into the watch industry. EMI used it to introduce their CAT scanner technology into the medical electronics market, as was described in chapter 1.

16.8
External Supported Ventures or
Severed Ventures

A further approach, which has been pioneered by GE in the United States, is to establish a virtually independent company outside the walls of its parent.[24] It is a "spin-off" venture that provides a compromise between either the internal or external joint ventures discussed above and the formation of a totally independent entity or "spin-off" company discussed in chapters 14 and 15. The severed venture approach adopted by GE has the following features:

1. The severed venture team contributes its skills and some personal capital to the development of the innovation and holds an equity share in the venture.

2. The parent company holds a minority equity share based upon its contribution of equipment, inventories, and patents to the venture. GE does not contribute any capital to the venture, but (in principle anyway) a parent company could make a minority venture capital contribution, without attaining a majority equity position and effectively creating an internal venture.

3. Further equity capital is obtained from the venture capital market or other sources.

The problems faced by GE are broadly identical with those faced by other high-technology corporations and discussed earlier. Their rationale for setting up ventures are noted below.

An innovation may appear to be commercially promising, but offer too small a market and therefore an unattractive direct oppor-

320

Small in
Large Is Also
Beautiful:
Stimulating
Intracorporate
Entrepreneurship

tunity to the corporation, for two reasons. First, a small market does not offer the commercial returns to justify the "opportunity cost" of investing further scarce corporate management and physical resources in the innovation. Larger returns can be obtained by investing these resources in other investment opportunities. Second, if such a large corporation were to enter such a small market it would probably dominate it with the possible consequent invocation of antitrust legislation. On the other hand, if a promising innovation cannot be sold or licensed to a third party, its abandonment represents a potential loss to society at large (which does not enjoy the benefits of the innovation) and an actual loss to the corporation (since it has to "write off" the investment to date).

By formally "spinning off" the innovation into a severed venture launched by its former employees, GE still maintains the potential for social benefit of the innovation, and the recovery of corporate sunk investment from its minority equity share in the venture.

The approach was pioneered by David J. BenDaniel, who was foundation head of GE's Technical Ventures Operation which launches such ventures. To qualify for incorporation as a severed venture, an innovation must possess an identifiable market niche in which it offers a competitive advantage, which can be exploited as an opening wedge for expansion into a larger market—with the consequential potential for equity growth for the corporation. It must offer an entry into a market which the corporation wishes to "monitor" without further investment or formal entry. It must also be spun off by a severed venture team of former GE employees, which includes committed project champions or a viable entrepreneurial team—"men who believe in the business enough to invest their savings and risk their futures in it."[25] BenDaniel set up the Technical Ventures Operation in 1970 and by 1973 he estimated the GE had enjoyed a threefold growth in its equity shares in its severed ventures, from $3 to $4 million.

16.9
Problems of Venture Development and Intracorporate Entrepreneurship

In the foregoing sections we have reviewed a number of mechanisms used by the larger high-technology firms to stimulate innovation and entrepreneurship, and it would be appealing to be able to report that these mechanisms have been uniformly successful. Unfortunately, this is not so, and we must now examine some of the problems of new venture development in high-technology companies.

321

Problems of
Venture
Development and
Intracorporate
Entrepreneurship

Vesper and Holmdahl studied the 100 companies in the Fortune 500 list with the largest sales which had also won awards for introducing the "most successful new technical products."[26] They found that 65 percent of their respondents used venture management approaches compared with 36 percent for a sample drawn from top 100 and 25 percent for a sample drawn from all the Fortune 500 firms. Thus venture management approaches do appear to be positively associated with successful technological innovation. However, despite these sanguine observations, other studies have highlighted difficulties with such approaches.

Hlavacek studied twenty-one product innovation failures which were nurtured using internal venture management approaches.[27] These failures were selected from twelve corporations in the Fortune 500 list. He conducted in-depth interviews with the venture managers concerned and the corresponding top management (either at divisional or corporate level) to whom they reported. He found that the responses from the two groups were consistent. The causes of failure are summarized below.

Top management most commonly cited sunk costs as growing too large, whereas venture managers cited that top management was too impatient for "results."

Top management next cited poor market evaluations and too small a market. This citation suggested that minimum ROI (the 3M criterion) rather than market size should be the criterion. Alternatively (although that author does not directly suggest it), the GE severed venture approach may be appropriate. Note also that both groups cited distribution difficulties which again suggests an inadequate attention to marketing problems.

Top management often cited too narrow venture management experience as a cause of failure, which suggests injudicious entrepreneurial team selection procedures. Interestingly, venture managers cited that top managers (particularly if they were divisional as opposed to corporate managers) adopted insular viewpoints and saw new ventures as potential "threats" to existing operations. Again this latter problem can be ameliorated by the 3M alternative sponsorship approaches.

Note that all the above problems might well have been avoided with a rigorous new venture planning process as discussed in chapter 15, especially if top management defined and maintained a clear mission for corporate venture operations. Hlavacek makes seven suggestions for more successful venture management which are comparable with the recommendations of other writers.

1. Both corporate and divisional managements should be made aware of the long-term growth benefits of venture operations.

322

Small in
Large Is Also
Beautiful:
Stimulating
Intracorporate
Entrepreneurship

2. Top management should develop a venture charter which specifies the functions, procedures, and boundaries of venture management.

3. Uniform formats for composing and reviewing venture business plans, within agreed time frames should be instituted.

4. A limited number of ventures, with independent budgets, at varying stages of development and maturity, should be sponsored.

5. Multiple sources of internal sponsorship for ventures should be maintained.

6. Top management should frequently review the organizational status of each venture in the light of its performance and changing circumstances and change this status if warranted. Thus a venture might be upgraded to new divisional status, absorbed into an existing division, restructured as a joint or severed venture, sold or liquidated.

7. Product champions and high-caliber key team members should always be selected to lead and manage such ventures.

Even with a well-conceived and delineated business plan and a strong venture team, internal ventures are vulnerable to failure from more subtle causes. Fast analyzed the life span of eighteen new venture groups established and/or operating in the early seventies, with a 50 percent survival rate in 1976.[28] He described the two main causes of venture failure as the *strategic reversal* and the *emergence trap.* We will deal with each cause in turn.

The first of these reflects a conflict between the typical lead time requirements for developing new ventures and dynamics of organizational economic and political climates. Fast postulates that four main driving forces dictate the launching of new internal ventures: a corporate strategy emphasizing diversification; a corporate money glut or positive cash flow which exceeds promising investment opportunities; an unfavorable outlook for the firm's main line of business; and entrepreneurially oriented or risk-taking top management. When these four forces coincide, a venture group is probably established. When they disappear, that is, when the firm experiences a strategic reversal, the venture is probably disbanded.

A new venture will typically be created in the fourth or fifth phases of Bright's innovation process—the development/pilot production stages. Thus, given the time scale of the process, it may well be three to ten years before the venture becomes commercially viable. *It is very unlikely that either the internal or external environment of the firm remains unchanged over this time period.* Economic environments and business conditions typically experience cyclical changes over such time periods and corporate strategies often change also, possibly associated with changes in top management. Then the "benign" corporate climate may change for the worse, and venture activities may experience a strategic reversal in their fortunes. An organizational climate of this kind may be severe enough to terminate a venture.

323

Problems of
Venture
Development and
Intracorporate
Entrepreneurship

Fast cites the example of a major chemical company which established a new venture division in 1968, when the first four driving forces pertained. By the early seventies, the climate changed. The demand for two of the company's major products increased dramatically, so it decided to concentrate on these. The company was also suffering from a shortage of cash so that capital had to be tightly rationed. The entrepreneurially oriented president had been succeeded by a more conservative individual. Consequently, top management now decided to spin off or wind down its separate ventures and disband the new ventures operation. EMI's sad experience with the CAT scanner (chapter 1) provides another example.

Thus, even if they are showing commercially promising results, internal ventures run the risk of premature termination from quite normal fluctuations in corporate climates. Fast found that there was a strong correlation between the percentage changes in the corporate profits over the period 1965-76 and changes in the number of venture groups extant in the sample studied.

The second cause of venture failure is associated with what can be called "envious and threatening success." If a venture group does not experience or survives a strategic reversal and launches several successful ventures it faces other hazards. Its success may lead to top management favoritism. The venture group may be viewed as top management's "blue-eyed boys" or managerial corps d'elite, and the leading contenders for top management succession. The expanding requirements of the new ventures may lead to territorial infringements. A growing venture may establish its own engineering, production, and marketing functions, both domestically and "offshore," which impinge upon the territories of established divisions.

Thus, by being successful it will be perceived as a threat by established power centers within the corporation. They will be resentful and jealous and perceive it as a dangerous new contender for scarce corporate resources and top management patronage. At best these power centers will seek to subvert and undermine its status. They may well seek to destroy it and be successful.

Fast suggests guidelines congruent with Hlavacek's, to steer a new venture group between the Scylla and Charybdis of the strategic reversal and the emergent trap, but even if the group successfully navigates its metaphorical Straits of Messina, there are other, more personal psychological factors that inhibit intracorporate entrepreneurship.

324

Small in
Large Is Also
Beautiful:
Stimulating
Intracorporate
Entrepreneurship

Smollen suggests that the "risk-rewards" structure in many corporations discourages successful venture development.[29] In chapter 14 it was shown that entrepreneurs are moderate risk takers and they often feel intracorporate, as opposed to independent entrepreneurship is a riskier and less rewarding situation. This is because the entrepreneur has less control over his environment in the corporate situation. Despite his readier access to corporate financial or other resources, if a strategic reversal occurs, the intracorporate entrepreneur lacks maneuverability in seeking alternative financial funding sources. The entrepreneur is also likely to receive less financial reward if he is successful. A successful independent entrepreneur may enjoy significant to outstanding capital gains from his share in the equity of the successful company. Although an intracorporate entrepreneur may hold stock options in the parent corporation, even an outstanding performance of a venture which he launched may have only a modest impact on corporate stock values, so the entrepreneur's reward is correspondingly small.

It is theoretically possible to offer the intracorporate entrepreneur a special salary/bonus package linked to the performance on his own venture (particularly it is a spin-off severed venture), but this is often difficult to implement without offending other managers in the parent corporation. Cook reports that some corporations have successfully employed such compensation schemes,[30] and Smollen argues that the financial rewards offered for success should be large compared with losses incurred by failure. It was also pointed out earlier that an intracorporate entrepreneur is likely to be "hierarchically" ambitious compared with his independent counterpart. Although successful venture management may offer a vehicle for advancement to top management, it does face the perils of the "emergent trap." If a venture fails or is only moderately successful, the intracorporate entrepreneur may find his corporate ambitions permanently constrained.

These considerations suggest that there are no simple ways of stimulating intracorporate entrepreneurship, but that appealing risk-reward structures should be devised and that venture managers should be allowed "one or two failures" (provided they are not too disastrous financially), without penalty to their long-term career prospects. This tolerance of failure is one of the features of excellent companies which we discuss in the final chapter.

References

1. Norman Macrae, "The Coming Entrepreneurial Revolution: A Survey," *The Economist*, 25 December, 1976.

2. Donald A. Schon, *Technology and Change: The New Heraclitus* (New York: Delacorte Press, 1976); Donald A. Schon, *Beyond the Stable State* (New York: W. W. Norton, 1973).

3. Schon, *Beyond the Stable State*, p. 320.

4. Donald M. Collier, "Research Based Venture Companies—The Link between Market and Technology," *Research Management* 17, no.3 (May 1974): 16-30.

5. Ralph Biggadike, "The Risky Business of Diversification," *Harvard Business Review* 57, no.3. (May-June, 1979): 103-111; also reprinted in *Readings in the Management of Innovation*, Michael L. Tushman and William L. Moore, eds. (Mashfield, Mass.: Pitman Publishing, 1982), pp. 537-48.

6. Richard P. Olsen and Mitchell P. Diamond, "Innovation at Texas Instruments," ICCH 9-672-036 (1971).

7. Mariann Jelinek, *Institutionalizing Innovation: A Study of Organizational Learning Systems* (New York: Praeger Publishers Inc., 1979), chap. 5.

8. Patrick E. Haggerty, "The Corporation and the Individual," address in a series of lectures on *The American University: Community and Individual* (Dallas, Tex.: University of Texas, 1979).

9. John Gardner, *Self Renewal; The Individual and the Innovative Society* (New York: Harper & Row Ltd., 1971).

10. Haggerty, "The Corporation and the Individual."

11. Olsen and Diamond, "Innovation at Texas Instruments."

12. Jelinek, *Institutionalizing Innovation*.

13. "Texas Instruments Show U.S. Business How to Survive in the 1980's," *Business Week,* 18 September, 1978, pp. 66-92.

14. Mack Hanan, "Corporate Growth through Venture Management," *Harvard Business Review* 47, no.1 (January-February 1969): 43-61.

15. *A Framework for Government Research and Development* (London: HMSO Cmnd. 4814, 1971).

16. Richard M. Hill and James D. Hlavacek, "The Venture Team, A New Concept in Marketing Organization," *Journal of Marketing* 36, no.3 (July 1972): 44.

17. Brian C. Twiss, *Managing Technological Innovation,* 2d ed. (London: Longman Group, 1980), pp. 197-78.

18. James Hillier, "Venture Activities in the Large Corporations," *IEEE Transactions on Engineering Management* 15, no.2 (June 1968); 65-70.

19. Ford Park, "Start More Little Businesses and More Little Businessmen," *Innovation*, no.5 (1969).

20. Ibid.

21. A. B. Cohen, "New Venture Development at Du Pont," *Long-Range Planning* 2, no.4 (June 1970); 7-10.

22. Ibid.

23. Mack Hanan, "Venturing Corporations—Think Small to Stay Strong," *Harvard Business Review* 54, no.3 (May-June 1976): 139-48.

24. Sharon Sabin, "A Nuclepore, They Don't Work for GE Anymore," *Fortune* (December 1973): 145.

25. Ibid.

26. Karl A. Vesper and Thomas G. Holmdahl, "How Venture Management Fares in Innovative Companies," *Research Management* 16, no.3 (May 1973): 30-32.

326

Small in
Large Is Also
Beautiful:
Stimulating
Intracorporate
Entrepreneurship

27. James D. Hlavacek, "Towards More Successful Venture Management," *Journal of Marketing* 38, no.4 (October 1974): 56-60.

28. Norman Fast, "A Visit to the New Venture Graveyard," *Research Management* 22, no.2 (March 1979): 18-22.

29. Leonard E. Smollen, "Entrepreneurship in the Existing Enterprise," (unpublished paper, Institute for New Enterprise Development, Belmont, Mass., 1975).

30. F. W. Cook, *A Piece of the Action—Venture Management Compensation* (New York: Frederick W. Cook & Co.).

17

Epilogue: Institutionalizing Technological Innovation and Entrepreneurship

Innovative firms tend to have a style of management that is open to new ideas, ways of handling staff that encourage innovation, systems that are customer focussed and which reward innovation, skills at translating ideas into action and so forth. ... One has to really believe an organization cares in order to invest the energy needed to help it change. Such commitment derives from superordinate goals ... this is probably the most underpublicized "secret weapon" of great companies.

Richard Tanner Pascale and Anthony G. Athos,
The Art of Japanese Management

17.1
In Search of the "Secret Weapon" of Great Companies

It would be utopian to expect to end this text by suggesting that there is some ubiquitous guaranteed system for managing the corporate technological innovation and entrepreneurship process. Like the proverbial "free lunch," such a system cannot be expected to exist this side of paradise. Nevertheless, after acknowledging this reservation, it is possible to suggest some general guidelines for enhancing the process.

In chapter 2 (section 2.8 and figure 2.5) we viewed innovation in the context of an open systems cybernetic framework that identifies potential synergies between evolving technological knowledge and market needs. In chapters 14 and 15 we saw that new entrepreneurial companies are typically set up to exploit such synergies. In chapter 16 we argued that in larger mature corporations a form of Gresham's Law applies whereby concern for present operations tends to drive out concern for future operations or innovations; we also considered the approach used by some corporations to mitigate the impact of this "law." The Texas Instruments' approach, which encourages individuals to enact both operations and innovation management roles, is one such approach. Jelinek pro-

328

Epilogue:
Institutionalizing
Technological
Innovation and
Entrepreneurship

vides a detailed analysis of the evolution of OST (and its extensions) at TI in the context of the overall evolution of management systems in large-scale organizations during the past century.[1] She suggests that OST and its extensions—the institutionalizing of the insights of Patrick Haggerty and others when TI was a relatively small company—enabled it to grow large without losing its commitment to innovation. OST constitutes an organizational learning system that enables the organization to innovate and grow continuously, exploiting technology-market synergies under corporate management guidance even after the founding generation of corporate managers have retired from day-to-day involvement in the firm. However, TI's unhappy incursion into the consumer electronics industry suggests that the presence of such an institutionalized organizational learning system does not guarantee invariable success.

Other corporations have also displayed impressive track-records for innovations over many years, so it is useful to look for some common features in these successes. Frohman studied nine companies that have sought, successfully or otherwise, to exploit technology.[2] In the successful companies he identified three conditions that reflect considerations cited throughout this book:

1. The majority of top managers have technical backgrounds and experience and were involved in the development of technological strategies. Frohman argues that technological development must be placed in the hands of a manager who is comfortable wielding it.

2. A potential pitfall for technically trained managers is the toy-train syndrome cited in chapter 15. That is, they emphasize technological novelty and innovation without proper consideration for business goals and market needs. The successful companies selected projects that supported business goals, reinforced technological leadership, and solved customers' problems.

3. The successful companies have planning/decision systems and organizational structures that reinforce technological strategies. Senior managers assume responsibility for technological planning and decision making that is integrated with business planning. The chief technical officers of the companies participate in these planning activities and, in the two most successful companies, report directly to their presidents. Also, the companies' corporate, promotion, and hiring practices reinforce these strategies. They operate dual promotion ladder systems discussed in chapter 12 and recruit new technical staff from the best talent available in the top schools in the country. They create a corporate culture that supports the priority attached to technological innovation.

Pascale and Athos draw broadly similar conclusions from their study of the management practices of a sample of Japanese and

U.S. firms.[3] Their conclusions are expressed in their 7-S framework of management. They suggest that successful Japanese and U.S. firms (including Texas Instruments and 3M) have the style, staff, skills, systems, structure, and strategy to stimulate and nurture innovation. These six S's are embedded in the seventh—the superordinate goals or values enshrined in the organizational culture that elicit a commitment to innovation from the firm as a whole and each of its individual members. There are alternative approaches to enshrining this culture, as reflected in the contrasting approaches of TI and 3M, but Pascale and Athos suggest that this is the "secret weapon" of the successfully innovative Japanese and Western companies.

Probably the best analysis of the corporate practices for stimulating innovation and entrepreneurship, also based upon the 7-S framework, is the best-seller *In Search of Excellence,* by Peters and Waterman.[4] The only reason their book has not been cited earlier is that this author did not have the opportunity to read it before the first sixteen chapters of this text had been written. However, the overall thrust of these chapters is broadly congruent with Peters and Waterman's admirable, detailed analysis and conclusions.

The inherent potential for conflict between the needs of present and future operations or innovations implies a dualistic approach to management. On the one hand, present operations require sustained implementation of the established systems and procedures that define the essential identity of the firm as it is expressed in the products it makes and the market it serves, or as Peters and Waterman put it, "that [provide] the essential touchstone which everybody understands, and from which the complexities of day-to-day life can be approached."[5] This sustained implementation is typically represented by a product division. On the other hand, innovation requires the flexibility and fluidity to respond to opportunities created by new inventions and new technology-market synergies. It typically requires the (possibly temporary) formation of small project teams or task forces to pursue innovation-entrepreneurship in a "small business" atmosphere.

Peters and Waterman studied 62 highly regarded U.S. companies (including Amdahl, Digital Equipment, Hewlett-Packard, IBM, TI, Eastman Kodak, Johnson & Johnson, Proctor & Gamble, 3M, and Boeing) to try to delineate some common characteristics of corporate excellence. They observed that 43 of these companies (including those cited above) share common characteristics of excellence including the following:

1. A core of shared values that define the corporate culture. These core values are not mere "motherhoods," but are religiously adhered to. In chapter 16 we saw that TI emphasizes an innovative culture, while 3M's culture emphasizes product championing and entrepreneurship. IBM's culture emphasizes customer service, Digi-

330

Epilogue:
Institutionalizing
Technological
Innovation and
Entrepreneurship

tal's stresses quality and reliability and P & G's emphasizes quality in consumer markets and competing brand managers. Peters and Waterman also identified simultaneous tight-loose properties similar to the dual requirements cited earlier. That is, such organizations are rigidly controlled and directed but *at the same time* encourage autonomy, entrepreneurship, and innovation from everyone. This culture is not achieved through bureaucratic rules and negative controls, but through a common faith in the shared value systems of the organization. It could possibly be described as almost a non-theistic, non-sectarian religious faith—a comparison made by Pascale and Athos in describing good Japanese companies. The shared values are usually established early in the life of the organization, often by its founders, who (consciously or otherwise) generate an organizational mythology. For example, at Hewlett-Packard there are reputed to be numerous "Bill and Dave" (Hewlett and Packard) stories. Peters and Waterman suggest that such stories (mythological or otherwise) play an important role in defining and conveying the organization's shared values or culture. Thus they help to define and maintain the productive organizational climate as the company grows.

2. Although Peters and Waterman emphasize that excellent companies are well-led, they claim that such leadership does not imply a large central corporate bureaucracy. Rather such companies emphasize decentralization or lean staff and simple form. That is, they employ relatively few corporate staff and maintain a fairly stable, simple form usually represented by individual product divisions. For example, Johnson & Johnson (a $5 billion company) is broken up into 150 independent divisions or subsidiary companies, with average turnovers of just over $30 million, that are aggregated into eight groups of up to twenty companies by geography or product. Flexibility and fluidity are maintained by frequent reorganizations of temporary project teams, task forces, and small businesses to pursue innovative ventures, reflecting the team functions we have discussed earlier (particularly in chapter 12). Given the loose properties cited earlier, this approach fosters entrepreneurship and multiple attempts at innovation. Excellent companies recognize that both organizational and technological evolutions occur through the "technological mutations" and the trial and elimination-of-error learning discussed in chapter 2 of this book, so tolerance of failure is a key feature of their cultures. Based upon successful "technological mutations," innovative ventures that grow large enough can become new product divisions. Thereby both organizational and technological learning and evolution are encouraged.

3. At the beginning of chapter 8 it was stressed that creativity is required throughout the innovation process. Excellent companies are "people-oriented" and encourage creative efforts from all employees. Peters and Waterman state that a most pervasive theme in these companies is respect for the individual and the truly unusual ability to achieve extraordinary results through ordinary people. Often this people orientation has its roots in the early days of the company. The founders of H-P decided it should not be a hire-and-fire company, and they have maintained full employment through bad days as well as good. Needless to say, this attitude should encourage creative problem-solving from all involved in the innovation process and inspire the corporate loyalty supposed to be a unique feature of Japanese companies. This attitude also helps maintain a productive organizational climate as it grows. As the authors express it:[6]

331

In Search
of the
"Secret Weapon"
of Great
Companies

> *The top performers create a broad, uplifting, shared culture, a coherent framework within which charged-up people search for appropriate adaptions. Their ability to extract extraordinary contributions from very large numbers of people turns on the ability to create a sense of highly valued purpose. Such purpose invariably emanates from love of product, providing top-quality services, and honoring innovation and contributions from all.*

4. Excellent companies are also especially good at the market monitoring required in the open systems view of innovation. They are customer-oriented and obsessively concerned with the quality, reliability, and service offered by their products. They employ salesmen as market gatekeepers, monitoring their customers' changing requirements (especially the lead-users) to identify unmet market needs or new technology-market niches. 3M, Digital, and H-P among others are cited as companies that excel at "nicheman-ship"—the tailoring of products to specific market niches.

Interestingly, Peters and Waterman imply that many of their excellent companies do not follow the offensive strategy described in chapter 4. They cite H-P, Digital, IBM, and P & G as companies that are proud to follow a defensive strategy and to be second to reach the market. By such delays these companies seek to produce products that are more reliable and better satisfy user needs than those initially provided by the offensive innovators. This does not imply that they are niggardly in their R & D expenditures. H-P, Digital, IBM, and P & G are among the industry leaders in R & D spending. What it may imply is that they simultaneously engage

332

Epilogue:
Institutionalizing
Technological
Innovation and
Entrepreneurship

in technology and market monitoring to ensure that new technology really does satisfy customer needs. This observation is also supported by the experiences of GE and Gillette's Paper Mate Division in the second and third cases described in chapter 1. Both of these companies successfully challenged the efforts of offensive innovators from the United Kingdom and Japan respectively, by developing their own competitive products.

5. Finally, excellent companies "stick to the knitting" most of the time. That is, they stay in the technology-market segments within which they have achieved excellence. They recognize their limitations. When they abandon this precept they fail. We have already referred to TI's unsuccessful sally into the consumer electronics market, notably with digital watches and personal computers. Earlier, other major electronics corporations experienced similar failures in trying to enter the computer market. Recalling the second case in chapter 1, EMI's failure to consolidate upon its leadership in introducing CAT-scanner technology can also be partially attributed to a failure to observe this maxim.

17.2
Final Comments

The preface of this book began by commenting that public and business concern for technological innovation had been aroused by the current problems of declining economic growth coupled with high inflation and increased unemployment. Many writers now take the view that the stimulation and implementation of technological change offers the best means of at least ameliorating, if not eradicating, these problems (see, for example, Freeman, Clarke, and Soete, *Unemployment and Technical Innovation*[7]).

In chapter 2 (figure 2.1) it was suggested that the innovation chain equation or process provides a fruitful analogy for viewing the innovation-entrepreneurship process, which provides the driving force of technological evolution. In using this analogy, it is important to recognize that the chain process is not the product of inanimate chemical kinematics. Rather, it is initiated and enacted through the wisdom, insight, and efforts of a team of talented human beings. If the above writers are correct, this chain equation may also be viewed as the driving force of continued economic (and hopefully cultural) improvements in the human situation. The historical enactment of this chain equation was illustrated in chapter 1 through Marconi's successful development of the revolutionary innovation of wireless telegraphy. His technological entrepreneurial efforts (coupled with those of others) created the radio industry, which has subsequently evolved through many further innovations (some of which have also been revolutionary) into today's microelectronics and computer industries. It goes without saying that the economic and cultural impact of this evolutionary trajectory have been considerable.

Revolutionary innovations are frequently pioneered by small companies (led by entrepreneurs such as Marconi) that enact this chain process. Revolutionary innovations are, however, comparatively rare events. The overall path of technological evolution is sustained by a myriad of normal innovations (using the Kuhnian terminology introduced in chapter 2). Often such normal innovations are also introduced by small companies launched by individuals specifically to exploit them—individuals who also seek to satisfy their creative aspirations by enacting this chain process. In chapter 15 we reviewed the problems faced by such new technology ventures.

Such ventures help sustain the evolution of technology and, when successful, provide meaningful employment for others. Almost invariably, successful ventures grow and growth increases organizational complexity and bureaucracy. Given the fecundity of inventiveness of good scientists and engineers, it also increases potential innovative opportunities. The prime innovation management task facing larger companies is therefore to select and sustain through the chain process those inventions which offer the most promising innovative opportunities congruent with corporate goals and resources. In the main body of this book we have examined some of the problems and issues pertinent to the effective management of this task. In the final two chapters we have seen that, despite differing organizational arrangements, successful innovative companies have certain features in common. They foster a corporate culture or climate in which individual creativity, inventiveness, and entrepreneurship can thrive, that is, a culture that nurtures the chain process. However, their senior managements provide frameworks to ensure that innovative efforts are congruent with corporate goals and satisfy customers' needs with high quality and reliable products. That is, they exhibit the simultaneous loose-tight properties cited by Peters and Waterman.

The opening sentence of the preface commented that concern for technological innovation had also been aroused by the *Defi Japonais* and, over the past few years, there have been numerous books published that sought to reveal the secrets of the Japanese challenge to the West's management skills. When this book was begun, it was planned to include a chapter on the Japanese art of innovation management. However, by the time it approached completion, it was recognized that such a chapter would be redundant. One reason for Japan's deserved success has been through the ability of its managers to apply traditional skills to the management of current

334

Epilogue:
Institutionalizing
Technological
Innovation and
Entrepreneurship

operations (especially in the areas of quality control and inventory management). Although Japan has been outstandingly successful in introducing innovative products in emergent technology-market niches (notably in the consumer electronics and robotics markets), I cannot accept that that country possesses some unique attributes or "secret weapon" hidden from the West. As was stated earlier, such a "secret weapon" is common to both Japanese and Western companies. It is my personal observation (shared, I am sure, by many readers) that the healthy "animal spirits" (to use Keynes' words) of technological innovation and entrepreneurship are alive and well in North American and Western European companies. Long may they continue to thrive.

At the end of the preface, I expressed the sentiment that this book would have been worthwhile if it helped just one individual become a successful technological entrepreneur. To any such readers, whether they are planning or have already launched their own ventures, or work in larger companies, go my good wishes for success in their endeavors.

17.3
References

1. Mariann Jelinek, *Institutionalizing Innovation: A Study of Organizational Learning Systems* (New York: Praegar, 1979).
2. Alan L. Frohman, "Technology as a Competitive Weapon," *Harvard Business Review,* 60, no.1 (January-February 1982): 97-104.
3. Richard Tanner Pascale and Anthony G. Athos, *The Art of Japanese Management* (New York: Simon & Schuster, 1981).
4. Thomas J. Peters and Robert H. Waterman Jr., *In Search of Excellence* (New York, Harper & Row, 1982).
5. Ibid, p. 308.
6. Ibid, p. 51.
7. Christopher Freeman, John Clarke, and Luc Soete, *Unemployment and Technical Innovation* (Westport, Conn.: Greenwood Press, 1982).

293-32

INDEX

ABCO Ltd., 170
Abernathy, William J., 25-27, 247, 253
Absorbent strategies, 64
The Achieving Society, 262
The Act of Creation, 135, 142, 146
Advisory information services, 45-46
After-sales services, 45, 57
Allen, D. H., 201, 202
Allen, T. J., 224
Allenstein, B., 83
Allison, G. T., 109
Ambrose, James, 10, 12
American Cyanamid Division, 188
Anglo-American Telegraph Co., 6
Ansoff, 52
Apollo space program, 255
Applied Imagination, 151
The Art of Conjecture, 72
Athos, Anthony G., 328
Atkinson Morley Hospital, 10

Baker, N. P., 182
Baker, W. J., 3
Baruch, J. J., 32
Basley, J. J., 263
Beastall, H., 91
Becker, H. S., 110
Bell, Alexander Graham, 4, 6
Bell Canada, 110, 112
Bell, D. C., 190
BenDaniel, David J., 320
Biggadike, Ralph, 305
Binary relevance trees, 91
Bisociation, 140-141, 153
Blair, Ronald M., 282

Blake, R. R., 214
Blowatt, K. R., 167, 169-170
Bobis, A. H., 188
Borland, Candace, 263
Bozeman, Barry, 109
Brainstorming, 151-153
Bright, James R., 75
  innovation process, concept of, 17-20
  innovation related activities, 41-42
  project profile, 184
British Oxygen Company, 316
Brockhouse, Robert H., 262
Burns, Thomas, 147, 204
Business plans, 280-285

Carr, Edward Hallet, 73
Carson, Rachel, 106
Cash flow profile, 21, 48
CAT scanner, 9-12, 36
Cetron, Marvin, 184
Challis, E. J., 114
Chambers, J. A., 145
Chambers, John C., 190
Chesley, G. R., 201
Cincinnati Milacron, 101-102
Clarke, Thomas E., 121, 293
The Closing Circle, 106
Club of Rome study, 72, 83, 100
Collier, Donald M., 305
Commoner, Barry, 106
Composite forecasts, *see* Scenarios
Computer simulations for TF, *see* Systems dynamics models
Computer, solid state, 9

Computerized axial tomography, *see* CAT scanner
Contract research, 42-43
Control Data Corporation, 63
Convergent thinking, 150
Cook, F. W., 324
Cooper, Arnold C., 258-260
Cooper, Robert G., 172-178
Cosmetic innovations, 66
Cost
  and envelope curves, 77-79
  measurement and control for NTBFs, 297-298
  R & D budgets, 121-125
  vs. time for innovation process, 20-22
Creative symbiosis, 29
Creativity
  brainstorming, 151-153
  creative climate, 147-149
  creative people, finding, 144-146
  stimulating techniques, 87-93, 149-157
  thinking, process for, 134-144
  see also, Innovation, technological; Technological entrepreneurs
Crippen, Albert, 8
Crookes, Sir William, 4, 19
Cross-impact matrices, 91-93

Danilov, V. J., 256
Davies, 221
Davis, Stanley M., 244
Day, 110
de Bono, Edward, 155

de Bresson, C., 28
de Forest, Lee, 7
De Jouvenal, Bertrand, 72
Defensive strategy, 59-61, 65
Delphi method, 84-86, 110, 191
Dependent strategies, 63-65
Development strategy, 44-45
Dickinson, P. J., 258
Dill, D. D., 222
Dingee, Alexander L. M., 274
Divergent thinking, 150-155
Dominant design, 26
Doriot, George, 276
Dow, Herbert, 4
Downstream coupling, 57-59
Dowty Group, 67
Drahem, Kirk P., 256
Drucker, Peter, 74, 94, 239
Dual-ladder structure, 216-221
DuPont Corporation, 314, 318

Ebert, R. J., 201
Edison, Thomas, 4
Education services, 45-46
EMI Ltd, 9-13, 36
Empiricism, 137
Entrepreneurs, see Inventor-entrepreneurs; Techno-logical entrepreneurs
Entrepreneurship vs. dy-namic conservatism, 305-306
Envelope curves, 77-79
Environment
    impact statements, 109-112
    impacts of technology, 72, 106-108
    and normal innovation evaluation, 171
Etzioni, Amatie, 109

Extrapolative techniques, see Technological forecasting

Fairchild Semiconductors, 256-257
Faraday, Michael, 3-4, 19
Feldman, Philip, 110
Fiber-tip porous pen, 13-16
Fischer, David W., 110
Fisher, J. C., 80
Fisher-Pry substitution model, 80-81
Flair pen development, 13-15, 37
Fleming, J. A., 5
Fluid state of development, 26
Forrester, Jay, 83
Frantz, Gene A., 311
Free choice period for entre-preneurs, 268
Freeman, Christopher, 52, 70, 201, 254
Frohman, Alan L., 328
FS, see Future studies
Fundamental research, 41
Fusfield, Alan R., 101, 183, 223
Future studies (FS), 71-75

Gabor, Denis, 72, 98
Gannon, M. J., 278
Gap analysis, 129-132, 166-167
Garber, J. D., 147
Gardner, John, 307
Gasse, Yvon, 262
Gear, A. E., 190
Gee, Edwin A., 253
Gender
    of creative people, 145-146
    of technological entre-preneurs, 268n
General Electric Co., 12, 101-102, 319-320

Gerstenfeld, Arthur, 147, 168-170, 240-241
GERT network, 207-209
Getzel, J. W., 145
Gibbons, Michael, 109, 254
Gilbert, Humphrey, 3
Gillette Company
    Flair pen development, 14-15, 37
    stainless steel blade de-velopment, 67
Gladstone, William Ewart, 5
Gompertz function, 76-77
Goodyear Co., 101-102
Gordon, J. J., 110
Gordon, Theodor J., 72, 91
Gordon, William, 153
Governmental support for innovation, 35
Graphical evaluation and review technique, see GERT network
Grasley, R. H., 289-290
Graubar, S. R., 32-33
Green, E. I., 137
Greenhouse effect, 106-107

Hagan, Everett E., 262
Haggerty, Patrick, 307, 308, 311
Hall, D. L., 263
Halo effect, 183
Hanan, Mack, 314
Hansen, E., 145
Haywood, H., 91
Heilmeier, George H., 314
Helmer, Olaf, 72, 84
Hertz, David B., 187
Hertz, Heinrich, 4, 19
Hewlett, William, 255
Hewlett-Packard, 255, 329, 331
High-technology industries
    conservatism vs. entrepre-neurship, 305-306
    evolution, theory of, 25-27

innovation planning, 48-50

research choices, 40-43

spin-off phenomenon, 58-59, 255-260

Texas Instruments OST system, 205, 306-314, 328

Hillier, James, 316

Hirschmann, Winfred B., 246

Hlavacek, James D., 321

Holmdahl, Thomas G., 321

Honeywell, Inc., 90

Houndsfield, Godfrey, 9

Howard, Graemar K., 299

Hudson, L., 145

Hudson, Ronald G., 190

IBM, 57, 329

ICI Ltd., 217-218

Imitative strategy, 26, 61-62, 65

In Search of Excellence, 329-332

Incubator organizations, 255, 274

Innovation chain equation, 20

Innovation, nontechnological, 66-69

Innovation, technological absorbent, 64

Bright's concept, 17-20

Cooper's evaluation process, 172-178

cosmetic, 66

defensive strategy, 59-61, 65

dependent, 63-65

described, 2-3

evaluation criteria, 34-36

evaluation process, 33-34

imitative strategy, 61-62, 65

interstitial strategy, 62-63, 65

management function, 48-50, 173-178

market analysis, 169-170

normal, 30-31, 166-171

offensive strategy, 52-59, 65

open system, 33-34, 48-49

opportunistic offensive, 58-59

project championing, 225

research, *see* Research and Development

revolutionary, 30, 159-166

technical development phase, 170-172

and technological base, 41

time and cost characteristics, 20-22

time sequence for cash flow, 48

traditional strategies, 65-66

see also Creativity; Technological entrepreneurs

Interstitial strategy, 26, 62-63, 65

Inventing the Future, 72, 98

Invention, scientific creative process, 134-144

described, 2-3

Inventor-entrepreneurs, 4

see also, Technological entrepreneurs

Isle of Wight, 5-6

Jackson, P. W., 145

James Bay Development Project, 110

Jantsch, Erich, 72, 101

Jelinek, Mariann, 307, 327

Jermakowicz, Wladyslaw, 245

Johnson and Johnson, 190, 330

Johnston, Robert G., 190

Jolson, M. A., 278

Jones, Harry, 75, 101

Kahn, Herman, 72, 93

Kamien, Morton I., 253

Kaplan, Norman, 147

Keirulf, Herbert, 274

Kepner, C. H., 150

Klein, Harold E., 94, 97

Koestler, Arthur, 134, 135, 139-141

Kohler, Wolfgang, 135

Kolodny, Harvey F., 243-244

Kovac, F. J., 80

Kuhn, Thomas S., 145

paradigm concept, 22-24

Lateral thinking, 155-157

Lawrence, Paul R., 244

Layton, Christopher, 3

Leopold, L. B., 110

Li, Y. T., 33

Licensing

for defensive innovators, 60

for offensive innovators, 56-57, 65

services, 45

Liles, Patrick R., 282

The Limits to Growth, 72, 83, 100

Lin, 198

Linneman, Robert E., 94, 97

Linstone, Harold A., 80

Litvak, I. A., 265-267, 299-301

Lockett, A. G., 190

Lodge, Sir Oliver, 4, 19

Lorenz, Konrad, 33-34, 142

MacKinnon, Donald, 145

MacNulty, C. Ralph, 94, 95-97

Macrae, Norman, 304

Magic Marker, 14
Management
  for innovation, 48-50
  for technology transfer, 240-247
The Management of Innovative Technological Corporations, 119
Mansfield, Edwin, 202, 253
Marconi, Guglielmo, 4-8, 19, 36
Market pull, 31-34
Marketing strategies
  absorbent, 64
  cosmetic, 66
  defensive, 59-61, 65
  dependent, 63-65
  development, 44-45
  imitative, 26, 61-62, 65
  for innovation evaluation, 173
  interstitial, 26, 62-63, 65
  and normal innovations, 167-170
  for NTBFs, 284
  offensive, 52-59, 65
  pricing, 68
  product-market evaluation process, 176-178
  and revolutionary innovation, 163-166
  traditional, 65-66
Marquis, D. C., 244
Marshall, A. W., 202
Martin, Michael J. C., 170
Martino, comments of, 77-78, 92
Mason, R. M., 197
Matrix structures for transfer, 241-245
Mature state of development, 27
Maule, C. J., 265-267, 299-301
Maxwell, James Clerk, 4, 19, 154
McClelland, David C., 262

McClurkin, D. T., 293
McQuillan, P., 291
Meadows, D. L., 199, 202
Meckling, W. H., 202
Meisel, 221
The Mentality of Apes, 135
MGee, John F., 190
MITRE Corporation, 110-111
Monsanto, 101-102
Moore, 221
Mouton, J. S., 214
Mumford, findings of, 80

n-Ach in entrepreneurs, 262-263
Nadar, Ralph, 108
New technology business functions (NTBFs)
  business plan for, 280-285
  creation stages, 274-276
  financial support for, 285-292
  growth phases, 298-299
  initial operating problems, 292-298
  spin-off phenomenon, 255-260
  venture team formation, 276-280
Nobel, Alfred, 4
Normal innovations, 30-31, 166-171
Norris, K. P., 202-204
"Not invented here" syndrome, 236
November, P. J., 202
NTBFs, *see* New Technology Business Functions

Objectives, Strategies and Tactics (OST) system, 205, 306-314, 328
Offensive strategy, 52-59, 65
Ohio Nuclear, 12
Omphaloskepsis, 137

Open-system innovation, 33-34, 48-49
Operations research approach, 183
Opportunistic-offensive innovation, 58-59
Osborn, Alex, 151
Osborne, Adam, 255
OST, *see* Objectives, Strategies and Tactics system

Packard, David, 255
Paper Mate division, 14, 37
Paradigm, Kuhn's, 22-24
Parmenter, S. M., 147
Pascale, Richard Tanner, 328
Patents, 45
  for defensive innovators, 60
  for offensive innovators, 56-57, 65
Pearl function, 75-77
Pearson, A. W., 222
Peck, M. J., 202
Peper, Henry, 13, 15
Perkin, W. H., 4
PERT/CPM network, 91, 206-209
Peters, Thomas J., 329-332
Philp, J. McL., 113
Pilkington, L. A. B., 3
Planning, corporate
  for innovation management, 49-50
  R & D budget setting, 126-132
Polaroid Corporation, 168
Popper, Sir Karl R., 24-25, 172
Pound, W. H., 182
Powell, J. A., 9
Precursor trends, 79-80
Pricing strategies, 68, 295-296
Probert, D. E., 83
Proctor and Gamble, 66, 329-330

Program Evaluation and Review Technique/Critical Path Method, *see* PERT/CPM
Prototype production, 43, 55-56
Pry, R. H., 80
Pugh-Roberts Assoc., Inc., 223
Puzzle-solving science, 23

Q Sort method, 191-194

R & D, *see* Research and Development
Radio Corporation of America (RCA), 7, 316
Ramo, Simon, 66, 119
Rand Corporation, 72, 84-86
"The Rationality of Scientific Revolutions," 24
RCA, *see* Radio Corporation of America
Read, A. W., 190
Research and development
  budgeting considerations, 121-125
  contract, 42-43, 122
  creative climate, creating, 147-149
  creative people, finding, 144-146
  defensive strategy, 59-61, 122
  dual-ladder structure, 216-221
  external joint ventures, 319
  fundamental, 41
  gap analysis, 129-132
  individual and organizational conflicts, 212-214
  individual and organizational needs, matching, 226-231

locations, 119-121
offensive strategy, 53-56
planning, 126-132
production and marketing efforts, 43-46
project selection, 181-190
Q Sort method, 191-194
staff development alternatives, 214-216
staffing needs, 222-226
technological base concept, 40-41
uncertainty reduction process, 198-204
venture organizations, 315-316
see also, Innovation, technological
Revolutionary innovations, 30, 159-166
Revolutionary science, 23
Rhoades, R. G., 223
Risk analysis models, 187-189
Risk-return profiles, 188-189
Roberts, E. B., 223, 263
Roe, Anne, 145
Rosenthal, Sidney, 14
Rossen, Philip J., 170
Rossini, F. A., 109
Rotter, J. B., 263
Rubinstein, A. H., 201

Sahal, Devendra, 28-29, 80
Scenarios
  described, 93-95
  generating, 95-97
  and inventing the future, 98-99
Scherer, F. M., 202
Schmidt, A. W., 80
Schmidt-Tiedemann, K. J., 172, 178
Schmookler, J., 33
Schon, Donald A., 305
Schroder, H. H., 201

Schwartz, Nancy L., 253
Schwartz, S. L., 183
Scientific invention, *see* Invention, scientific
Scott, Donald S., 198, 282
Seiler, Robert E., 184
Shapero, Albert, 253-254, 265
Shepherd, Mark, 311
The Silent Spring, 106
The Sleepwalkers, 135
Smith, D. F., 80
Smith, J. J., 220-221
Smith, Kline and French, 190
Smollen, Leonard E., 274, 324
Sokol, Lisa, 265
Souder, William E., 183, 190-192, 197
Specific state of development, 27
Spital, Francis C., 101
Squeaky wheel selection method, 184
S-shaped curve, 75-77
Stalker, C. M., 147, 204
Stanford University, 255
Statistical decision analysis models, 189-190
Steele, Lowell W., 125, 235-236, 238
Step-wise growth, 28-29
Stewart, 52
"Stranger" hypothesis, 266
Strasser, G., 109
Synectics, 153-155
Systems dynamics
  Delphi method, 84-86, 110, 191
  models, 82-86
  morphological analysis, 87-89
  relevance trees, 89-91
Szabo, T. T., 220-221
Szakonyi, Robert, 205
Szonyi, A. J., 198

339

Taylor, F. W., 222

Taylor, H., 291

Technological entre-
preneurs
characteristics of, 261-
265, 265-267
free choice period, 268-
269
and spin-off phenomenon,
255-260
see also, Creativity; In-
novation, technological

Technological evolution
Abernathy-Utterback
treatment, 25-27
Kuhn's paradigm, 22-24
and market synergy, 33
Popper's methodology,
24-25
Sahal's treatment, 28-29
technosphere system, 33
timing, importance of, 29

*Technological Forecasting
in Perspective*, 72

Technological forecasting
(TF)
composite forecasts, *see*
Scenarios
concept of, 71-75
envelope curves, 77-79
Fisher-Pry substitution
model, 80-81
morphological analysis,
87-89
organizational location,
101-103
precursor trends, 79-80
relevance trees, 89-91
S-shaped curve, 75-77
systems dynamics mod-
els, 82-86
validity of, 99-101

Technological guideposts,
28

Technological innovation,
see Innovation,
technological
Technology assessment
and environmental im-
pact statements, 108-112
organizational frame-
work, 113-116
Technology monitoring,
74-75
Technology push, 31-34
Technology transfer
management framework
for, 240-247
problems of, 235-240
Texas Instruments, 99, 205,
306-314, 328, 329
TF, *see* Technological fore-
casting
Thomas, Howard, 201, 202
Thompson, P. N., 190
Thorn Group Ltd., 13
3M Corporation, 314, 317-
318
Time vs. cost for innovation
process, 20-22
Timmons, Jeffrey A., 274,
279
Torrance, E. P., 145
Townsend, J., 28
Traditional strategies, 65-66
Training services, 45-46
Trans-Atlantic wireless
communication, 6-7
Transition state, 26
Tregoe, B. B., 150
Triode tube, 7
Twiss, Brian C., 75, 101,
128, 201
Tyler, Chaplin, 253

Udell, G. G., 263
Uncertainty reduction process, 198-204
Unilever testing procedures,
113-114

Unsafe at Any Speed, 108
Utterback, J. M., 25-27, 253

Varian, Russel, 255
Vasarhetyi, 198
Venture capital for NTBFs,
290-292
Venture evaluation and
review technique, *see*
VERT
Ventures, new, *see* New
technology business
functions
VERT network, 208-209
Vertinsky, Ilan, 183
Vesper, Karl A., 321
Victoria, Queen of Eng-
land, 5-6
Vincent, David, 190

Wainer, H. A., 263
Waterman, Robert H., Jr.,
329-332
Watkins, D. S., 254, 258
Wayne, Kenneth, 247
Wells, H. A., 185-186
Wells, H. G., 71
Westinghouse Research
Laboratories, 217-218
Whirlpool, 101-102
White charger selection
technique, 184
White, George R., 160-165
Whitfield, 156
Wiener, Anthony J., 72, 93
Wilkinson Sword Edge, 67
Williamson, Byron, 263-264
Wills, G. S., 49, 87, 93
Winkofsky, E. P., 197
Wireless telegraphy, 4-8, 19
Wisemma, Johan G., 171

Xerox Corporation, 68, 188

The Year 2000, 72, 93

Zoppoth, R. C., 188
Zwickey, F., 89